COMPUTER INTEGRATED MANUFACTURING

COMPUTER INTEGRATED MANUFACTURING

A. Alavudeen
Assistant Professor
Department of Mechanical Engineering
Kalasalingam University, Krishnankoil
Tamil Nadu

N. Venkateshwaran
Assistant Professor
Department of Mechanical Engineering
Rajalakshmi Engineering College
Chennai

PHI Learning Private Limited
Delhi-110092
2015

₹ 350.00

COMPUTER INTEGRATED MANUFACTURING
A. Alavudeen and N. Venkateshwaran

ISBN-978-81-203-3345-1

The export rights of this book are vested solely with the publisher.

Fifth Printing ... ... **August, 2015**

Published by Asoke K. Ghosh, PHI Learning Private Limited, Rimjhim House, 111, Patparganj Industrial Estate, Delhi-110092 and Printed by Mudrak, 30-A, Patparganj, Delhi-110091.

CONTENTS

PREFACE

Automation plays an important role in the global economy and in day-to-day life. Engineers strive to combine automated devices with mathematical and organizational tools to create complex systems for a rapidly expanding range of applications and human activities. Specialised hardened computers, referred to as programmable logic controllers (PLCs), are frequently used to synchronize the flow of inputs from (physical) sensors and events with the flow of outputs to actuators and events. This leads to precisely controlled actions that permit a tight control of almost any industrial process. Currently, for manufacturing companies, the purpose of automation has shifted from increasing productivity and reducing costs to broader issues such as increasing quality and flexibility in the manufacturing process. In the last two decades, the developments in the field of computer graphics, electronics and computing technology have left a significant impact on manufacturing. This book has been written keeping these developments in mind.

In this book, we have attempted to present a balanced mix of theories and practices of automation. Computers are now being used extensively in the industry for both design and manufacture (CAD/CAM). The aim of this book is to introduce the subject of computing as an aid to design and manufacture, and to take the reader from the basics of computers to application in real engineering design and manufacturing. The book provides a tutorial and a reference source for students, and practising engineers and professionals who have an interest in the automation of manufacturing systems.

The book is intended for the undergraduate and postgraduate students of mechanical, production and industrial engineering. Researches and practising engineers will also find this book quite useful.

Chapter 1 describes the importance of computer aided manufacturing and design principles with the interfacing of CAD and CAM. Chapter 2 deals with group technology, its design, workflow in the group technology layout, and balancing of workflow in the layout. The main aim of group technology is to reduce work-in-progress and improve delivery performance by reducing lead times.

In an environment of fierce competition, industry cannot afford large amounts of waste – in time, material or production capacity. Cost, quality and responsiveness are the three requirements the industry must be aware of if it hopes to survive. This requires a new type of manufacturing system—the flexible or reconfigurable manufacturing system. So, in Chapter 3, adetailed explanation of various activities in flexible manufacturing systems is given.

Chapter 4 concentrates on shop floor control. In a typical job shop, knowing the status of a job is the key to efficient shop floor control. You need to know how new orders and change orders affect your shop load, and which jobs to run on which machine. Chapter 5 covers production planning and control. It provides the basic understanding and appreciation of the fundamental principles and methodologies relevant to planning, design, operation, and control of world-class productive systems.

Chapter 6 focusses on computer-aided process planning. Process planning translates design information into the process steps and instructions to efficiently and effectively manufacture products. As many computer-aided tools support the design process, computer-aided process planning (CAPP) has evolved to simplify and improve process planning and to achieve more effective use of manufacturing resources. Chapter 7 is on Just-in-Time (JIT) and lean manufacturing. The essential elements of lean manufacturing are described in such a way that one can easily understand and practice the lean manufacturing concept to achieve the lean manufacturing principles like zero waiting time, zero inventory,and so on.

Chapter 8 delineates database systems as database is required for storing and transferring the required information to and fro and across the factory. Therefore, the basics of how data are stored and transferred have to be known so that further speed of accessing or transferring can be done in future. Chapter 9 is concerned with the fundamentals of CNC, CNC programming, including step-by-step coverage of machining processes. Each topic is considered in detail, with the aim of showing the reader how machines are to be programmed and used, together with the advantage that they bring to the manufacturing industry.

Chapter 10 discusses the basic design of different components of CNC machines, along with characteristics and application of various drive systems. In a flexible manufacturing system, a driverless computer-controlled vehicle is equipped with guidance and collision-avoidance systems and is used to transport work pieces and tools between work stations. The basic function of automatic guided vehicle, its types and its principle of operation are discussed in Chapter 11.

Chapter 12 covers the basic configuration of robot, robot kinematics, operating methods of robot control units, and the various sensors used. Chapter 13 gives an introduction to modules of ERP. ERP facilitates integration of company-wide information systems with the potential to go across companies. Chapter 14 is on Product Data Management (PDM) or Product Information Management (PIM) systems. It discusses PDM systems and tools in great detail.

Chapter 15 deals with artificial intelligence. As the recent trend in automation is towards creating the factory of the future, the role of AI becomes important. Hence, we discuss AI basics here in sufficient detail. Chapter 16 explains Programmable Logic Control (PLC). Today, PLC is the leader in the field of automation. Therefore, we give an in-depth analysis of PLC in this chapter.

Chapter 17 dwells on concurrent engineering, which has emerged as a means to bring in rapid solutions to product design and development process. This strategy focusses on the optimization and distribution of a firm's resources in the design and development process to ensure an effective and efficient product development process. It mandates major changes within the organizations and firms that use it, due to the people and process integration requirements.

Chapter 18 deals with the LAN (Local Area Network). As the computer replaces the human in transferring and storing data, and the same information flows across the various departments of a firm, the networking of systems becomes essential. Thus, the basic concepts of networking are explained for the benefit of the reader.

I have tried to make the book as student friendly as possible.

Any constructive comments for improving the contents will be highly appreciated.

A. Alavudeen
N. Venkateshwaran

ACKNOWLEDGEMENTS

The first author, A. Alavudeen, is deeply indebted to Dr. C. Thangaraj, Vice Chancellor, Kalasalingam University, who was the first person to suggest that his thoughts were worth putting into book form. Dr. S. Saravana Sankar, Director Academics of the University, has long encouraged the author to write. He kindly read an early draft of this book, provided feedback, and offered his positive but candid views.

Professor M. Thiruchitrambalam, the first author's mentor, has always been a source of inspiration to him. Dr. J.T. Winowlin Jappes has been his severest critic but invariably tried to instill creative thoughts in the author and has been a true friend to him.

In particular, Alavudeen would like to thank Mr. Ganapathy Sankar for his help and support while writing this book.

Special thanks are due to the families and colleagues of both the authors for the patience and encouragement they have shown during the preparation of this book.

A. Alavudeen
N. Venkateshwaran

Chapter 1

INTRODUCTION

Computer Integrated Manufacturing (CIM) is manufacturing supported by computers. It is the total integration of Computer Aided Design (CAD)/Computer Aided Manufacturing (CAM) and also other business operations and databases. CIM has generally been replaced by Manufacturing Process Management in the wider field of **PLM (Product Lifecycle Management)**.

Besides CAD/CAM, the other components of CIM are CAP (Computer Aided Planning) and CAQ (Computer Aided Quality Assurance).

1.1 DEFINITIONS

CIM is the integration of total manufacturing enterprise by using integrated systems and data communication coupled with new managerial philosophies that improve organizational and personnel efficiency.

The Computer and Automated Systems Association of the Society of Manufacturing Engineers

> One needs to think of CIM as a computer system in which the peripherals, instead of being printers, plotters, terminals, and memory disks, are robots, machine tools, and other processing equipment. It is a little noisier and a little messier, but it's basically a computer system.
>
> —JOEL GOLDHAR, Dean, Illinois Institute of Technology

> CIM is a management philosophy, not a turnkey computer product. It is a philosophy crucial to the survival of most manufacturers because it provides the levels of product design and production control and shopfloor flexibility to compete in future domestic and international markets.
>
> —DAN APPLETON, President, DACOM, Inc.

> CIM is an opportunity for realigning your two most fundamental resources: people and technology. CIM is a lot more than the integration of mechanical, electrical, and even informational systems. It's an understanding of the new way to manage.
>
> —CHARLES SAVAGE, President, Savage Associates

CIM is nothing but a data management and networking problem.

—JACK CONAWAY, CIM Marketing manager, DEC

Since about 1970, there has been a growing trend in manufacturing firms toward the use of computers to perform many of the functions related to design and production. The technology associated with this trend is called CAD/CAM, or Computer Aided Design and Computer Aided Manufacturing. Today, it is widely recognized that the scope of computer applications must extend beyond design and production to include the business functions of the firm. The name given to this more comprehensive use of computers is Computer Integrated Manufacturing (CIM).

The preceding comments on CIM have different emphases. For example, Goldhar considers CIM a computer system, whereas both Appleton and Savage regard it as a management objective. In Conway's view, CIM is data management and communications. Although these individuals view CIM differently, the underlying message is the same: orchestrated use of the various resources improves productivity and quality.

An attempt to define CIM reminds us about the story of a group of blind persons trying to describe an elephant by touching it. Each has a different description depending upon which part of the elephant's body was touched. In the case of CIM, several definitions have been attempted. The one put forward by Shrensker (1990) for the Computer and Automated Systems Association of the Society of Manufacturing Engineers (CASA/SME) is perhaps the most appropriate. According to him,

CIM is the integration of the total manufacturing enterprise through the use of integrated systems and data communications coupled with new managerial philosophies that improve organizational and personnel efficiency.

CAD/CAM is based on the capability of a computer system to process, store, and display large amounts of data representing part and product specifications. For mechanical products, the data represent graphic models of the components; for electrical products, they represent circuit information, and so forth. CAD/CAM technology has been applied in many industries, including machined components, electronic products, equipment design, and fabrication for chemical processing. CAD/CAM involves not only the automation of the manufacturing operations, but also the automation of elements in the entire design-and-manufacturing procedure.

Computer-aided design makes use of computer systems to assist in the creation, modification, analysis, and optimization of a design. The designer, working with the CAD system rather than the traditional drafting board, creates the lines and surfaces that form the object (product, part, structure, etc.) and stores this model in the computer database. By invoking the appropriate CAD software, the designer can perform various analyses on the object, such as heat transfer calculations. The final object design is developed as adjustments are made on the basis of these analyses. Once the design procedure has been completed, the computer-aided design system can generate the detailed drawings required to make the object.

Computer-aided manufacturing involves the use of computer systems to assist in the planning, control, and management of production operations. This is accomplished by either direct or indirect connections between the computer and production operations. In the case of direct connection, the computer is used to monitor or control the processes in the factory. Computer process monitoring involves the collection of data from the factory, the analysis of the data, and the communication of process-performance results to plant management. These measures increase the efficiency of plant operations. Computer process control entails the use of the computer system to execute control actions to operate the plant automatically, as described above. Indirect connections between the computer system and the process involve applications in which the computer supports the production operations without actually monitoring or controlling them. These applications include planning and management functions that can be performed by the computer (or by humans working with the computer) more efficiently than by humans alone. Examples of these functions are planning the step-by-step processes for the product, part programming in numerical control, and scheduling the production operations in the factory.

1.2 BASIC CONCEPTS

Computer integrated manufacturing includes all the engineering functions of CAD/CAM and the business functions of the firm as well. These business functions include order entry, cost accounting, employee time records and payroll, and customer billing. In an ideal CIM system, computer technology is applied to all the operational and information-processing functions of the company, from customer orders through design and production (CAD/CAM) to product shipment and customer service. The scope of the computer system includes all activities that are concerned with manufacturing. In many ways, CIM represents the highest level of automation in manufacturing.

The conceptual goal of modern factories is computer integrated manufacturing. CIM denotes data-driven automation that affects all components of the manufacturing enterprise: design and development engineering, manufacturing, marketing and sales, and field support and service. Computer-aided design systems were first applied in the electronics industry. Today, they feature three-dimensional modelling techniques for drafting and manipulating solid objects on the screen and for deriving specifications for programs to drive numerical-control machines. Once a product is designed, its production process can be outlined using computer-aided process planning systems that help select sequences of operations and machining conditions. Models of the manufacturing system can be simulated by computers before they are built. The basic manufacturing functions—machining, forming, joining, assembly, and inspection—are supported by computer-aided manufacturing systems and automated materials-handling systems. Inventory control systems seek to maintain an optimal stock of parts and materials by tracking inventory movement, forecasting requirements, and initiating procurement orders.

The technological sophistication of manufacturing information systems is impressive, and it increasingly includes applications of robotics, computer vision, and expert systems—a part of Artificial Intelligence (AI)—which comes under computer science. The core of the CIM concept is an integrated database that supports the manufacturing enterprise and is linked with other administrative databases.

In a sense, CAD/CAM allows the mass production system to manufacture customized, handmade articles. The machinery can be adapted to a particular product through computer programming, enabling work on small batches to achieve many of the economies, previously available only through mass production of identical objects. Computer aided design itself makes possible the testing of production methods and the design of the product by running tests (of such factors as ability to withstand stress, for example) through the computer. If necessary, the product design or the process can be modified without going to the expense and time required for building actual prototype models.

CIM is recognized as the *islands of automation*; these 'islands' have been created in many enterprises and can be categorized into four groups. These groups are listed below along with typical islands of automation found in these groups.

1. **CAD/CAM**
   - Group technology
   - Computer aided engineering
   - Computer aided design
   - Computer aided manufacturing
2. **Manufacturing planning and control**
   - Inventory control
   - Shop loading
   - Capacity planning
   - Master scheduling
   - Purchasing
3. **Factory automation**
   - Computer-aided manufacturing
   - Robotics
   - NC/DNC/CNC
   - Flexible manufacturing systems
   - Automated materials handling systems
   - Automated test equipment
   - Process controllers
4. **General business management**
   - General and cost accounting
   - Marketing
   - Order entry
   - Decision support
   - Labour collection
   - Payroll

Issues

One of the key issues regarding CIM is equipment incompatibility and difficulty of integration of protocols. Integrating different brand equipment controllers with robots, conveyors and supervisory controllers is a time-consuming job which has a lot of shortening. Quite often, the large investment and time required for software, hardware, communications, and integration cannot be financially justified easily.

Another key issue is *data integrity*. Machines react clumsily to bad data, and the costs of data upkeep as well as general information systems departmental costs are higher than those involved in a non-CIM facility.

Yet another issue is the attempt to program extensive logic to produce schedules and optimize part sequence. There is no substitute for the human mind in reacting to a dynamic day-to-day manufacturing schedule and changing priorities.

Computer integrated is no panacea, nor should it be done with religious fervour. It is an operational tool that, if implemented properly, for all manufacturing problems, will provide a new dimension to competing: quickly introducing new customerized high quality products and delivering them with unprecedented lead times, swift decisions, and manufacturing products with high velocity.

1.2.1 Nature and Role of the Elements of CIM System

Nine major elements of a CIM system are shown in Fig. 1.1. These are:

1. Marketing
2. Product design
3. Planning
4. Purchase
5. Manufacturing engineering
6. Factory automation hardware
7. Warehousing
8. Finance
9. Information management

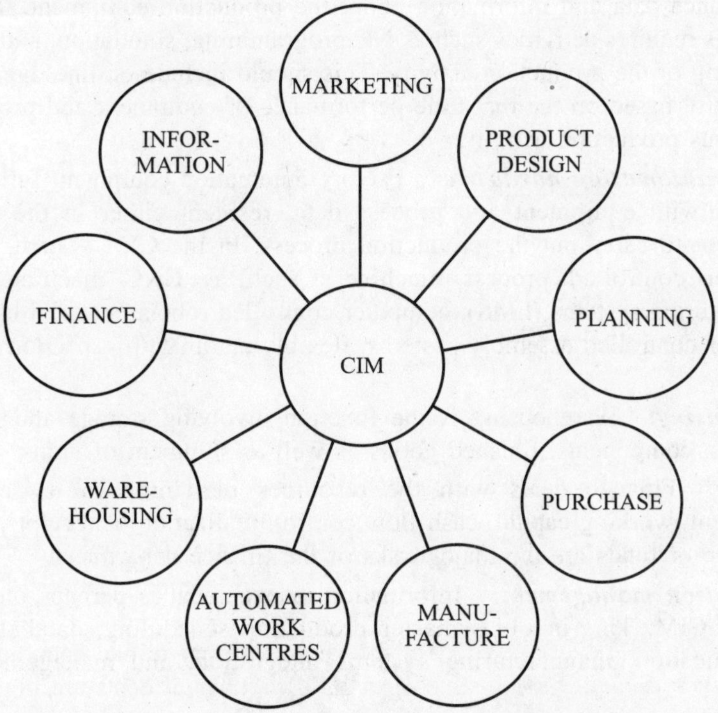

Figure 1.1 Major elements of a CIM system.

(i) ***Marketing:*** The need for a product is identified by the marketing division. The specifications of the product, the projection of manufacturing quantities, and the strategy for marketing the product are also decided by the marketing department. Marketing also works out the manufacturing costs to assess the economic viability of the product.

(ii) ***Product design:*** The design department of the company establishes initial database for production of a proposed product. In the CIM system, this is accomplished through activities such as geometric modelling and computer aided design by taking into account the product requirements and concepts generated by the creativity of the design engineer. The design process should be constrained by the costs involved in the actual production and by the capabilities of the existing production equipment and processes.

(iii) ***Planning:*** The planning department takes the database established by the product design and enriches it with production data and information to produce a plan for the production of the product. In a CIM system, this planning process should be constrained by the production costs and by the production equipment and process, in order to generate an optimized plan.

(iv) ***Purchase:*** The purchase department is responsible for the purchase orders and follow-ups, receiving the items, arranging for inspection, and supplying the items to the stores for eventual supply to manufacture and assembly.

(v) ***Manufacturing engineering:*** Manufacturing engineering is the activity of carrying out the production of the product, involving further enrichment of the database with performance data and information about the production equipment and processes. In CIM, this requires activities such as NC programming, simulation, and computer aided scheduling of the production activity. This should include on-line dynamic scheduling and control based on the real-time performance of equipment and processes to assure continuous production activity.

(vi) ***Factory automation hardware:*** Factory automation equipment further enriches the database with equipment and process data, resident either in the operator or the equipment to carry out the production process. In the CIM system, this consists of computer controlled process machinery such as CNC machine tools, flexible manufacturing systems (FMS), computer controlled robots, material handling systems, computer controlled assembly systems, flexibly automated inspection systems, and so on.

(vii) ***Warehousing:*** Warehousing is the function involving storage and retrieval of raw materials, components, finished goods as well as shipment of items.

(viii) ***Finance:*** Finance deals with the resources pertaining to money. Planning of investment, working capital, cash flow control, realization of receipts, accounting and allocation of funds are the major tasks of the finance department.

(ix) ***Information management:*** Information management is perhaps one of the crucial tasks in CIM. This involves master production scheduling, database management, communication, manufacturing systems integration, and management information systems.

It can be seen from Fig. 1.2 that CIM technology ties together all the manufacturing and related functions in a company. Thus, implementation of CIM technology basically involves integration of all the activities of the enterprise.

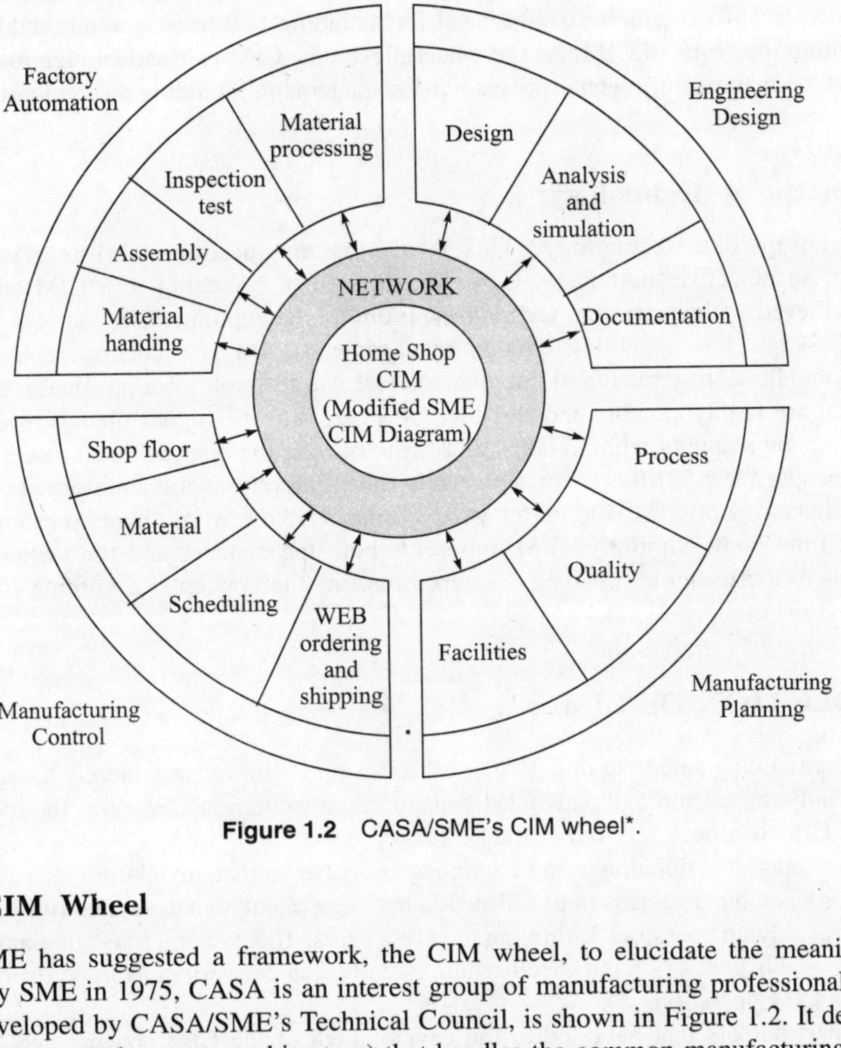

Figure 1.2 CASA/SME's CIM wheel*.

1.2.2 CIM Wheel

CASA/SME has suggested a framework, the CIM wheel, to elucidate the meaning of CIM. Formed by SME in 1975, CASA is an interest group of manufacturing professionals. The CIM wheel, developed by CASA/SME's Technical Council, is shown in Figure 1.2. It depicts a core on centre (integrated systems architecture) that handles the common manufacturing data and is concerned with information resource management and communications. The radial sectors surrounding the core (wheel hub) represent the various activities of manufacturing, such as design, material process and inspection. These activities are grouped under three categories — manufacturing planning and control, product/process, and factory automation—as depicted in

**Source:* Courtesy of the Society of Manufacturing Engineers. The CIM "Wheel" developed by the Technical Society of Manufacturing Engineers' Computer and Automated Systems Association (CASA/SME). Copyright * 1985 SME, second edition, revised November 5, 1985. Reproduction of this CIM Wheel is from the Society of Manufacturing Engineers (SME).

the wheel's inner rim. The outer rim represents the upper management functions, grouped into four categories: strategic planning, marketing, manufacturing, human resource management, and finance.

The CIM wheel depicted in Figure 1.1 is the expanded version of an earlier model. The outer rim was added in 1985 to emphasize the need for including both management and technology functions within the scope of CIM. As the wheel illustrates, CIM is broad enough to encompass all aspects of the manufacturing enterprise and its management, including those of personnel and finance.

1.2.3 Concept or Technology?

Is CIM a concept or a technology? This is a common question. CIM is a concept, an environment, an objective, a strategy. However, the aim of operating in a CIM environment cannot be achieved without modern technology. From this viewpoint, CIM can also be thought of as a technology. For upper level/higher management, CIM is a concept, a blueprint for success; for middle management and line managers, it is a technology, a physical realization of resources that are highly capable and flexible. CIM can help the higher management improve long-term strategic planning while it helps those who operate the plant to do so more efficiently.

"Some people view CIM as a concept, while others merely as a technology. It is actually both. A good analogy of CIM is *man*, for what we mean by the word "man" presupposes both the mind and the body. Similarly, CIM represents both the concept and the technology. The concept leads to the technology which, in turn, broadens the concept," according to Vajpayee and Reiden (1991).

1.3 EVOLUTION OF CIM

CIM has been evolving since the mid-1970s; however, until 1980 it was merely a concept. The 1980s, especially the second half, saw CIM expand into a technology. By now, the industry has realized that CIM is a necessity rather than a luxury.

Computer integrated manufacturing continues to evolve so that any claim that a "true" CIM plant exists is debatable. Progress in this direction has been phenomenal, several full-blown CIM plants will probably be operating within the next few years. Today, numerous companies market an array of products that, when put together intelligently, can convert an average manufacturing facility into a CIM operation.

The primary factors that have led to the development of the CIM concept and associated technologies include the following:

1. Development of numerical control (NC)
2. The spread and cost-effectiveness of computers
3. Manufacturing challenges (e.g. global competition, high labour cost, regulations, product liability, and demand for quality products)
4. The capability-to-cost attractiveness of microcomputers

1.3.1 Development of Numerical Control

Numerical control is a technique for controlling machine tools and other equipment using symbols that include numbers. Its origin and development are interesting. In 1947, the US Air Force faced a machining problem concerning turbine blades. Parsons Corp. took the challenge of developing a new method for moving the tool, using numbers, to achieve the desired accuracy. Parsons worked with the Massachusetts Institute of Technology (MIT) to produce the first NC milling machine in 1954. Since then, NC has developed, along with microelectronics and computers, into a sophisticated technology that drives modern machining centres, flexible machining cells (FMCs), and flexible manufacturing systems (FMSs).

1.3.2 Emergence of Computers

While NC was evolving on its own, computers were being developed independently. Among the various factors behind the evolution of CIM, the advent of computers undoubtedly was the most important one. A computer is an effective tool for carrying out repetitive tasks. Moreover, it can store large amounts of data and handle various types of mathematical analyses. Manufacturing involves these three tasks.

Initially, computers were introduced in manufacturing to support business functions such as payroll and accounting. These were then common applications of the computer in service industries. Next, computers were extended to inventory control. Further inroads into manufacturing had to wait until computer graphics matured as a technology. Today, with enhanced capabilities and lower costs, computers represent an attractive technology.

Early applications of computers in administrative and financial functions of manufacturing were feasible due to their cost-effectiveness. These functions involve well-defined, repetitive data-handling tasks that are suitable for computers. In the 1950s and 1960s, however, no one could foresee that computers would be integrated with production functions.

As expected, business computers accelerated the 'mushrooming' of data processing departments reporting to various management functions, especially finance. Over the years, these departments became so large and specialized that the idea of computers at the shop floor seemed unlikely. However, today, plant computers are a reality. In fact, in high-tech companies, the plant's computing needs are affecting the choice of business computers. Chapter 4 is devoted entirely to computers since they represent the brain of CIM.

1.3.3 Computer-Aided Design

A product must be designed before it can be manufactured. Many manufacturing companies may seem to be engaged in production only, with no visible design department. A closer look, however, reveals that an outside designer is involved. Design involves both creative and repetitive tasks. The repetitive tasks within a design are appropriate for computerization. In the early 1970s, using computers for tasks such as those for drawings and documentation proved cost-effective. Since then, there has been an explosive growth in computer-aided design (CAD), especially with advances in computer graphics. The activity of design and drafting is commonly expressed by the term CADD (computer-aided design and drafting) to emphasize the drafting task. CAD is extremely important; with products conceived at the design stage, CAD is, in fact, the cornerstone of CIM. In recognition of this, a separate chapter on CAD has been included in this text.

1.3.4 Computer-Aided Manufacturing (CAM)

By the time the use of in design computer began, NC technology had matured to become cost-effective for applications in machining. An important task in NC is *part programming*. A part program is simply a set of statements comprehensible to the control unit (MCU) that oversees slide and tool movements and other functions. In the case of components with complex geometries, part programmers had to carry out lengthy calculations, for which it was logical to use compilers. This gave rise to MCUs with built-in microprocessors—the building blocks of computers. The use of computers in extending the applications of NC technology, especially to part programming, was earlier termed computer-aided machining (CAM), and the associated technology was called computer numerical control (CNC). Later, CAM became an acronym for computer-aided manufacturing (note that manufacturing is a broader term than machining). CAM was coined to distinguish computer use in tasks other than design. With increasing use of computers in nondesign functions other than NC, the meaning of CAM has broadened over the years. Earlier, it used to denote computer use in part programming only. Today, it means any nondesign function of manufacturing that is computer aided. In fact, in the literal sense at least, CAM is as broad in scope as CIM, except that the latter emphasizes integration. Since lack of integration is the main stumbling block to further use of computers in manufacturing, CIM is currently more talked about than CAM.

1.3.5 From CAD and CAM to CAD/CAM

As the use of computers in design and machining broadened under CAD and CAM, it became evident that certain tasks were common to both. For example, both design and manufacturing require data on tolerances. Part geometries created during CAD can readily be saved in a database for later use. Why reenter such information again at the MCU? Though manufacturers knew that design and manufacturing represent a continuum, in the beginning CAD and CAM developed independently of each other. Later, when the drawbacks of CAD and CAM compartmentalization were realized, the logical move from CAD and CAM was to CAD/CAM. The slash between CAD and CAM was meant to reinforce the shared functions of design and manufacturing. By now, the two have come close enough not to require the slash; perhaps the term CADCAM should replace CAD/CAM.

1.3.6 CASA/SME CIM Wheel (Background)

1. **1985**–The Computer and Automated Systems Association of the Society of Manufacturing Engineers (CASA/SME) published its vision of enterprise-wide teaming—the CIM Enterprise Wheel.
2. **1992**–New Manufacturing Enterprise Wheel. This updated version preserves the understanding gained from the previous CIM Enterprise Wheel. The old Wheel looked primarily at automation and integration inside the enterprise. The new Wheel looks outside the organization as well. It adds understanding in these six areas: The central role of a customer-oriented mission and vision is to strive for continuous improvement.

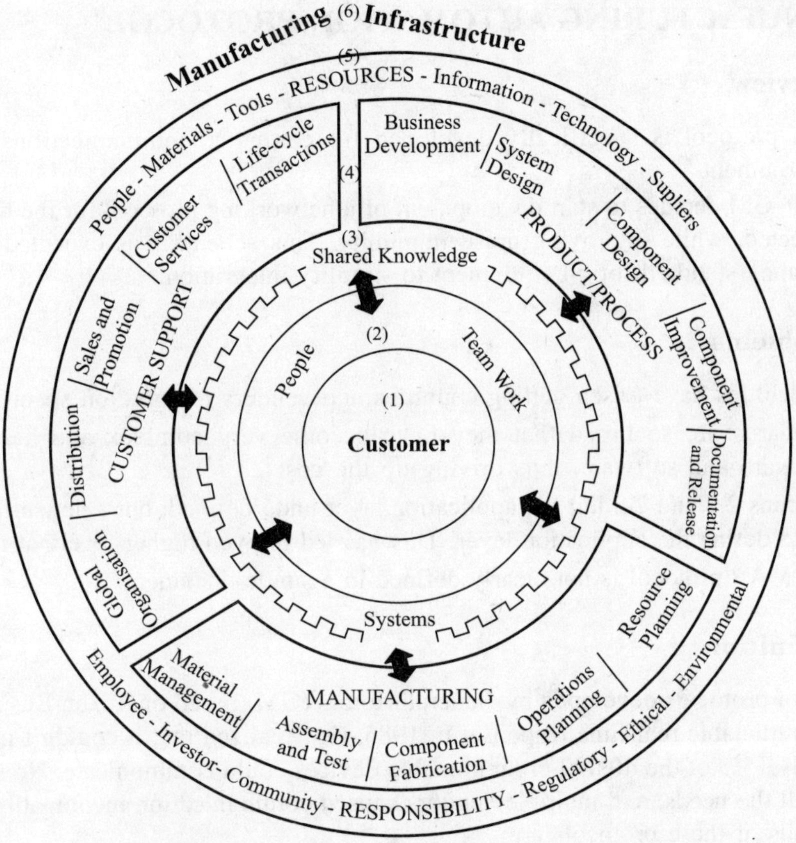

Figure 1.3 CASA/SME CIM Wheel (From CASA/SME).

3. The importance of teams and human networking in the new manufacturing environment.
4. The continuing importance of computer tools now increasingly distributed and networked. This includes tools to support networking and concurrent engineering.
5. A focus on key processes and best practices throughout the enterprise, from marketing through design, manufacturing, and customer support.

CIM wheel or five elements

1. The central role of the customer and evolving customer needs.
2. The role of people and teamwork in the organization.
3. The revolutionary impact of shared knowledge and systems to support people and processes.
4. Key processes from product definition through manufacturing and customer support.
5. Enterprise resources (inputs) and responsibilities (outputs).

1.4 MANUFACTURING AUTOMATION PROTOCOL

1.4.1 Overview

- This protocol is specifically developed for computer communications in a factory environment.
- **1980**–GM decides to start development of a networking protocol for the high data rates expected, while improving noise immunity. This scheme was expected to provide a common standard for all equipment to simplify integration.

1.4.2 Problems

- Difficulties have arisen getting countries and vendors to agree on specific standards.
- Standards are so broad that they have become very complex and hard to develop hardware and software, thus driving up the costs.
- Versions 2.1 and 2.2 left the application layer underdefined, but a new effort in version 3.0 to define the application layer. This has led to even higher levels of disagreement.
- The MAP protocol is not clearly defined in a single document.

1.4.3 Definition

MAP: A set of protocols developed by General Motors (GM) based on Token Bus (IEEE 802.4) and giving predictable real-time response. In 1985, GM realized that it couldn't meet its needs for CIM. Only 15% of the 40,000 programmable devices could communicate. No single vendor could meet all the needs, and multiple vendors caused communication incompatibilities.

The details of these protocols are:

- Physical Alternatives (Layer 1 OSI Model)
- Data Link Layer (Layer 2 OSI Model)
- Application Layer

1.4.4 Physical Alternatives (Layer 1 OSI Model)

1. **Networks: Broadband**
   - Many simultaneous communications on the same wire at the same time, using varied frequencies for transmit channels and receive channels.
   - Uses a head end demodulator to retransmit the signal
   - Each station needs MAP hardware
   - For very large factories
   - Uses a 1 inch (2.5 cm) thick cable
   - Maximum 10 Megabits/second per channel
2. **Carrierband**
   - Less expensive
   - Uses smaller cables
   - Rate of 5 to 10 Megabits/second (single channel).

3. **Other Equipment**

   Bridge: Allows connections between different network types (such as a connection between Broadband and Carrierband)

1.4.5 Data Link Layer (Layer 2 OSI Model)

Real-time response was required, so token passing is used.

1. Token passing allows a 'token' to be passed between computers; whichever computer has the token is given control of the network.
2. The token is continually passed to each machine on a network.
3. If a token passing ring is broken, the computers will wait, and eventually each half of the broken ring will have tokens being passed.
4. Extensive algorithms have been developed for control of token-passing, and resolution of conflicts.

Preferred standards

The preferred LAN is IEEE 802.4 token-passing bus uses standard ISO network, transport, session, and presentation layers.

1.4.6 Application Layer

1. FTAM (File Transfer Access Method) for file transfer.
2. MMS (Manufacturing Message System) for content of messages bound for robotic equipment.
3. ACSE (Association Control for Service Elements) to program communication.
4. Set of protocols should make communication between heterogeneous devices, not only in moving data around, but also in communicating the meaning of that data.
5. MAP has been delayed because of changes in the standard between 2.1 (1985) and 3.0 (1988).
6. A version of MAP, EPA (Enhanced Performance Architecture) has been developed, and is supported by the process control industry, and CNMA (Communication Networks for Manufacturing Applications).
7. CNMA has lobbied for the inclusion of Ethernet (IEE 802.3) as an alternative for the standard.

1.5 DETAILS FOR TOP

TOP (Technical and Office Protocols) is designed for the office environment.

1. Preferred LAN is IEEE 802.3 Ethernet, but allows for Token Bus (IEEE 802.4) and Token Ring (IEEE 802.5). Fibre optical media will be added when they are defined.
2. Uses same ISO standards for network, transport, session and presentation layers that MAP uses.

Comparison of MAP Layers with ISO/OSI Reference Model

ISO/OSI Reference Model	MAP Version 3.0
Application	Directory Services Network Management MMS FTAM ACSE
Presentation	Presentation
Session	Session
Transport	Transport Class 4
Network	CLNS ES-IS
Data Link	IEEE 802.2
Physical	IEEE 802.410 M-byte Broadband or 5M-byte Carrier Band

Abbreviations used

ACSE	Association Control Service Element
CLNS	Connectionless Network Service
ES	End System
FTAM	File Transfer and Access Method
IS	Intermediate System
MMS	Manufacturing Message Specification

Figure 1.4 ACSE/MAP

1.6 APPLICATION LAYER

1. ACSE and FTAM are the same as MAP; MMS is not included.
2. (MHS) CCITT X.400 message handling service provides electronic mail transfers of all types of documents.
3. ISO Virtual terminal protocol to allow a terminal on any machine to log onto any other machine.
4. ODA/ODIF (ISO Office Document Architecture and Office Document Interchange Format) to permit formatted documents to be transmitted independently of the package used to create them. (Text and Raster or Line graphics)
5. CGM (ISO Computer Graphics Metafile) to permit the exchange of geometrical graphics.

6. IGES (ANSI Initial Graphics Exchange Standard) to exchange product design information among CAD/CAM workstations.

1.7 MAP AND TOP

1. It was suggested by some that the two protocols MAP for CAM, and TOP for CAD be separated at the database level.

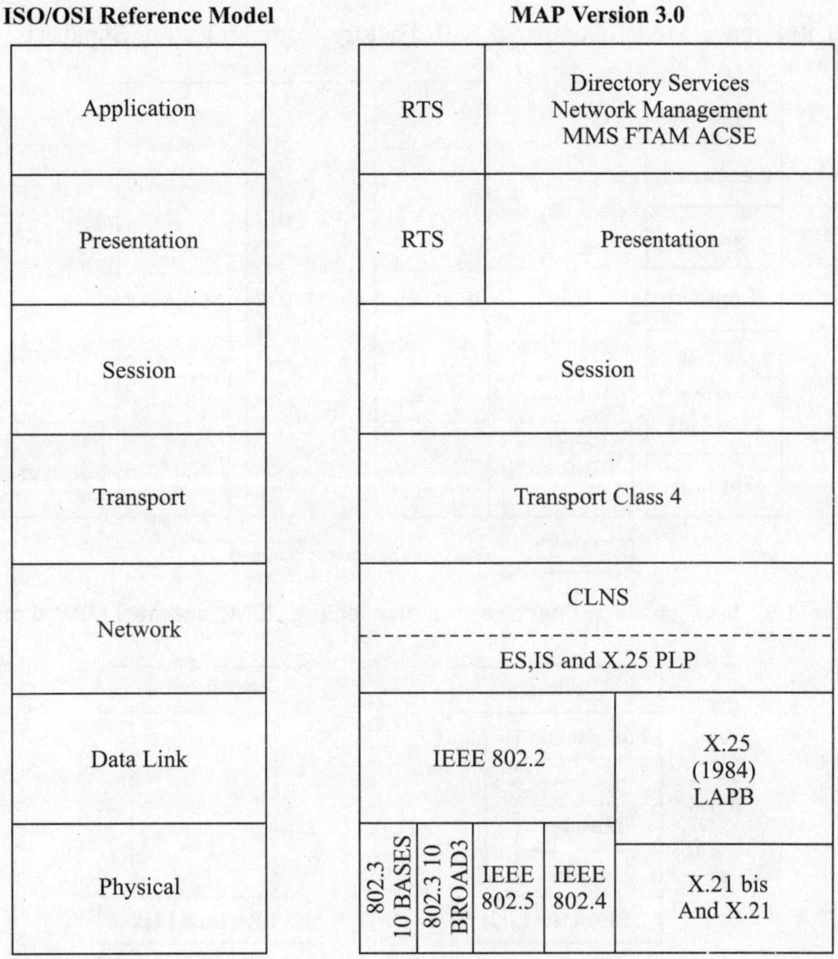

Abbreviations used

ACSE	Association Control Service Element	LAPB	Link Access Protocol-Balanced
CLNS	Connectionless Network Service	MHS	Message-Handling System
ES	End System	PLP	Physical Layer Protocol
FTAM	File Transfer and Access Method	RTS	Reliable Transfer System
IS	Intermediate System	VT	Virtual Terminal

Figure 1.5 Comparison of TOP protocol layers with the ISO/OSI reference model.

2. STEP (Standard for the Exchange of Product Data), based on PDES, is an international standard to facilitate the transfer of data in CAD/CAM.
3. GOSIP (Government Open Systems Interconnection Profile) has been developed by the US Government (NIST) to provide an open systems standard for their computers. GOSIP is almost a subset of TOP 3.0.

1.8 MAP AND OTHER STANDARDS

ISO/OSI Reference Model Compared with Factory Communication Standards

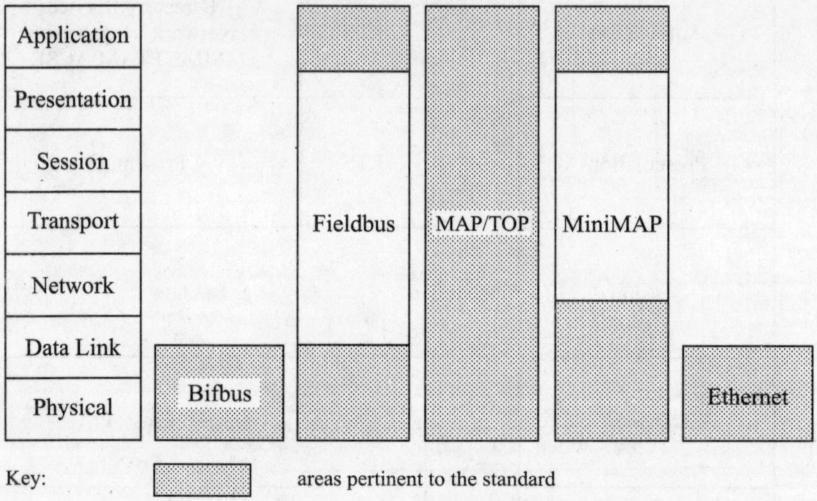

Figure 1.6 MAP enhanced performance architecture (EPA) and mini MAP protocol stocks.

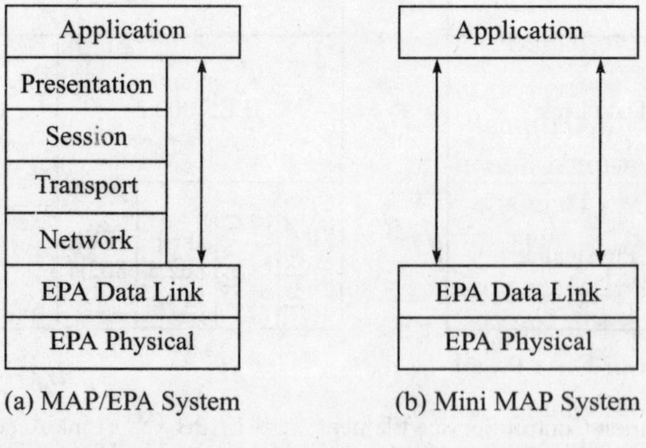

Example of a MAP implementation

1. One proposed MAP-compliant approach to computer integration is the introduction of an FMS into the plant-wide data communications network.

2. All seven MAP layers reside and run on each board set of this board-level MPA interface.

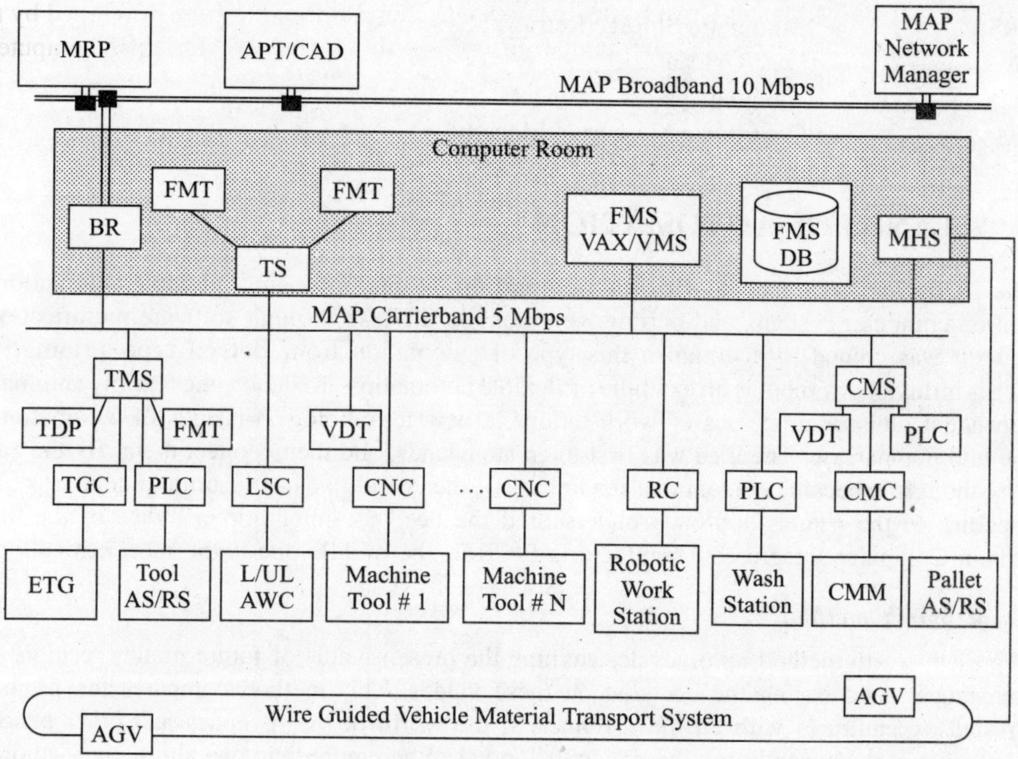

Figure 1.7 MAP implementation.

MRP	Material Requirements Planning
APT/CAD	Automatic Programmed Tool/Computer Aided Design
BR	Bridge
FMT	FMS Terminal
TS	Terminal Server
FMS DB	FMS Database
MHS	Material Handling System
TMS	Tool Management System
CMS	Coordinate Measuring System
TDP	Tool Data Panel
TGC	Tool Gauge Control
PLC	Programmable Logic Control
LSC	Load Station Controller
VDT	Video Display Terminal
CNC	Computer Numerical Controller
RC	Robot Controller

CMC	Coordinate Measuring Control
ETG	Electronic Tool Gauge
AS/RS	Automatic Storage/Retrieval System
L/UL	Load/Unload
AWC	Automatic Work Changer
CMM	Coordinate Measuring Machine

1.9 ISLANDS OF AUTOMATION

In the 1970s, another concept—**flexible automation**—came into being. Flexible automation is based on computer systems and permits variations in the task through software modifications. The term was coined to distinguish this type of automation from **Detroit-type automation**, which is inflexible. A robot is an example of flexible automation. To denote the flexibly automated independent units of processes or workstations, a new terminology—islands of automation—came into popular use. The idea was first to create islands, and then connect them. By the early 1980s, though, it became evident that the linking of the islands was an insurmountable task. The difficulties of the islands approach underscored the need for integration. Today, of the three words in Computer Integrated Manufacturing (CIM), the middle one is the most critical.

Present worth method

The present worth method involves determining the present value of future money receipts and disbursements and basing the decision on these values. This method compares the project's estimated expenditures with all the revenues. If the worth of the revenues and other benefits exceeds that of the expenditure, the project is funded. When more than two alternatives compete with one another, the largest present worth is selected. A variation of the present worth method is the *future worth method*, which compares a project's estimated expenditures and benefits at a future time.

Uniform annual cost method

This is a variation of the present worth method. In this method, all the cash flows are converted to J equivalent uniform series of annual payments similar to mortgage payments on j estate properties. The project is approved if the annual worth of the revenues exceeds that of the costs.

1.10 CHALLENGES AND TRENDS

The developments that led to CIM represent the response to the challenges faced by manufacturing in the 1970s and 1980s. This section examines these developments and the resulting trends.

1.10.1 Manufacturing in Transition

Manufacturing is currently undergoing a transition (Naisbitt, 1984) which, though evolutionary from a technology viewpoint, is revolutionary in its impact on the economy and employment.

In which countries will these facilities be located and who will own them? Which countries will belong to the group of industrialized nations? Will there be fewer members in this group? How will CIM affect the standard of living and quality of life? These are all important questions.

Assuming that manufacturing will follow in the footsteps of farming and that CIM will make this happen, jobs lost in the manufacturing sector are expected to be replaced by new positions in information technology. Information technology industries trade in data, information, and knowledge, and fall in the service sector.

1.10.2 Profit and Performance

The objective of manufacturing is profit for the owner or shareholders of the company. Profit margins must be competitive with alternative investment opportunities. As private manufacturers face lot of competition, their goal is to offer quality products at competitive prices. They also aim at lower unit cost which will ultimately determine the price. The ultimate goal of the private enterprise is to maximize profits. In government-supported operations, where profit is not the primary motive, low unit cost is usually the objective.

Companies assign managers and directors the responsibility to make a profit. Managers make decisions that are aimed at benefiting the company. But, since decision-making often depends on the future, which is uncertain, managers do make mistakes. JIM helps managers make better decisions by providing timely, accurate, and reliable data and information.

1.10.3 Productivity and Automation

A major advantage of treating manufacturing as an input-output system is that it directs company efforts toward increasing output, as much as possible, for a given level of input. The term commonly used to assess these efforts is **productivity**.

Thus productivity is the ratio of the output value to the cost of all the inputs. It is thus an index with no dimension or unit. Productivity is measured over a specific time and compared with its value (baseline) from a past period.

Besides overall productivity, we sometimes consider partial productivity, such as labour productivity or capital productivity. If an existing process is modified or redesigned requiring fewer person-hours (low input) to produce the same number of units (unchanged output), then obviously labour productivity has improved. But what about the cost (capital) of the redesign and modification? Could it be that capital productivity has deteriorated? An improvement in one kind of productivity, such as labour productivity may not necessarily increase the overall productivity, which really is the ultimate objective. Low labour productivity may be one of the several reasons behind poor company performance; however, careful managers do not ignore the other related issues, which may at times be difficult to assess.

Sometimes, comparisons are made about worker productivity in different countries. For example, in 1987, US labour productivity was \$37,800, which is the GNP divided by the number of people employed. The corresponding figures for Germany and Japan were \$36,600 and \$35,700, respectively. Such a comparison is valid, however, only if the countries being compared are on equal ground technologically (as they were in this case). The comparison loses its value if the countries' production facilities are vastly different in terms of technology and sophistication, for example, those in Mexico and the United States—the former, unlike USA, does not possess sophisticated technology in many areas.

REVIEW QUESTIONS

1. Describe the need for CIM and the issues addressed by CIM.
2. What are the different types of manufacturing? Make an assessment of the extent of computer control in specific cases of each type of manufacturing.
3. What are the various activities of a manufacturing plant which can be carried out through computer control?
4. Discuss the main elements of CIM systems.
5. Differentiate among physical integration, application integration and business integration. Give specific examples of each.
6. Define CIM.
7. What is the importance of CAD/CAM?
8. List out the nature and role of the elements of the CIM system.
9. Discuss the importance of the CIM wheel.
10. What do you understand by the term MAP?
11. Why is a TOP necessary?
12. What do you mean by islands of automation?

Chapter 2

GROUP TECHNOLOGY

2.1 INTRODUCTION

A technique used to group parts into families based on similar shapes, materials, or processing requirements must be used to categorize parts made in the manufacturing facility. By grouping similar parts, the manufacturing process for a family of parts is simplified. Group technology is a philosophy that implies the notion of recognizing and exploiting similarities in three different ways:

1. By performing like activities together
2. By standardizing similar task
3. By efficiently storing and retrieving information about recurring problems.

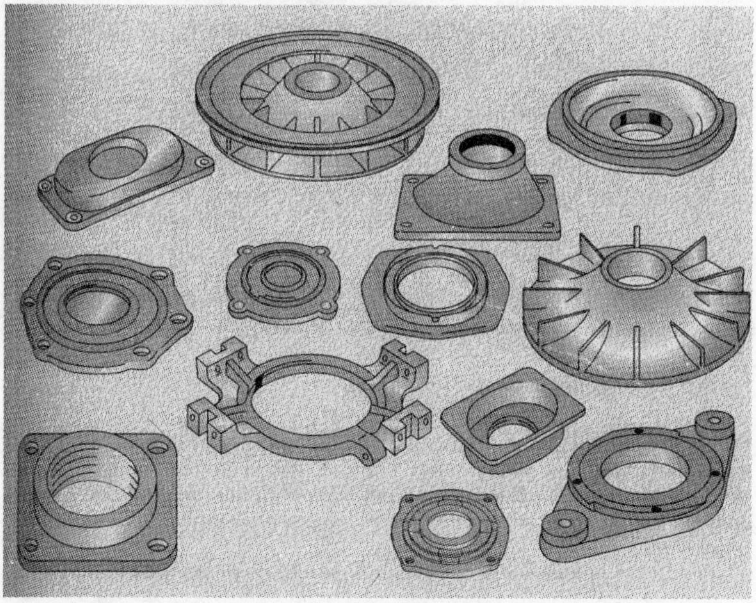

Figure 2.1 A family of parts.

Machine cell:

(a) Dissimilar machines are laid out in a sequential path so that each family of parts flows in a predictable, not jumbled, pattern.

(b) The machines are laid out in a U-shape or straight line. The advantage of a U-shape is that a single multifunctional operator can stand in the centre of the U and service many different machines.

(c) Several cells may exist in the factory. Each different family of parts may have its own cell.

(d) Cells should be arranged to minimize material movement.

(e) Machines that are shared among several cells should be located close to the cells that share them.

(f) The layout *within* cells corresponds to a product layout (assembly line) while the flow of products *between* cells follows a process layout. Cells solve a major problem by converting the jumbled flow pattern of many parts into an assembly line flow pattern with a few jumbled movements between cells.

Figures 2.2 and 2.3 show a "before" and "after" layout using cells.

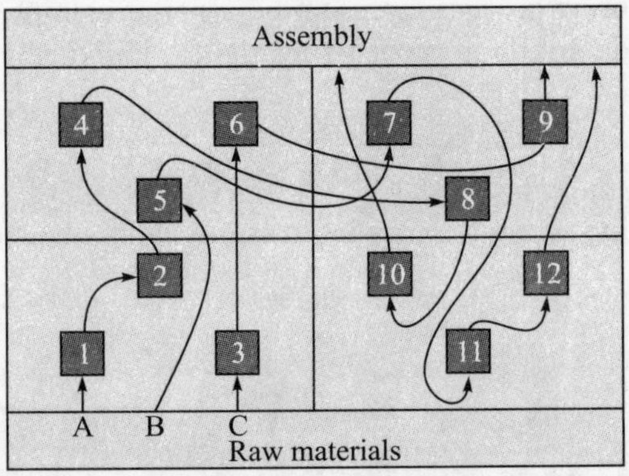

Figure 2.2 Process layout.

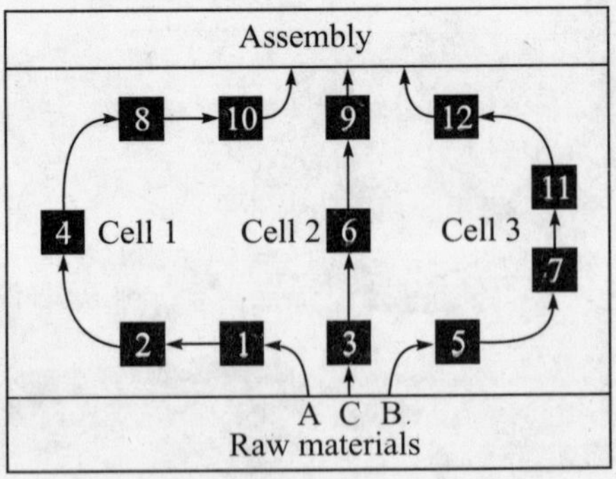

Figure 2.3 GT layout.

Benefits of cellular layouts:

These include:

- Reduced material handling and transit time
- Reduced set-up time
- Reduced Work In-Process (WIP) inventory
- Better use of human resources
- Easier control
- Easier automation.

Disadvantages:

Some of the disadvantages of cellular layouts are:

- Inadequate part families to make medium levels of product variety and volume
- Poorly balanced cells
- Expanded training and scheduling of workers
- Increased capital investment

2.2 BASIC GT LAYOUTS

The benefits of a good layout are numerous, and the hazards of a poor layout can be severe and costly. Good layouts should:

- minimize material handling costs,
- utilize space efficiently,
- utilize labour efficiently,
- eliminate bottlenecks,
- improve communications,
- reduce process and service time,
- eliminate wasted movements,
- facilitate heavy load movement,
- incorporate safety and security measures,
- promote product and service quality,
- encourage proper maintenance activities,
- provide visual control of operations or activities, and
- lend flexibility to adapt to changing conditions.

Layout issues are very important and require careful consideration. They arise in our personal lives as well as in our work. During one's work, there are three basic types of layouts:

- Process (functional) layouts
- Product (assembly line, sequential flow) layouts
- Fixed-position layouts

Three hybrid layouts are used to accommodate new technologies and production methods:

- Cellular layouts
- Flexible manufacturing systems
- Mixed-model assembly lines

Details for each layout type are discussed in the following sections.

2.3 PROCESS (FUNCTIONAL) LAYOUTS

Process layouts are also known as *functional layouts*. They group similar activities together in departments or work centres according to the process or function they perform. For example, all drills would be together in a work centre, and all mills would be together in another centre. Universities often use a process or functional layout by each customer.

Think about students in a University. Each one has a particular major background, academic or otherwise. Because they offer different courses, each student travels to different buildings and classrooms to attend classes. Often, the layout of a university creates a *jumbled flow* pattern (Of course, there are some universities like BHU, Nima University and institutes like the IITs and IIMs which have well-patterned layouts.) Products can also flow in a jumbled pattern in a job shop or batch process. Customer orders come in, and each order has products that require different tasks or work activities. A process planner must look at each order and write a plan that indicates which task must be done, in what order, and for how long. The product does not flow in a straight line from one work centre to another. It zigzags and backtracks throughout the entire factory. Figure 2.4 shows a typical process layout for manufacturing, as well as some jumbled flow patterns for a product.

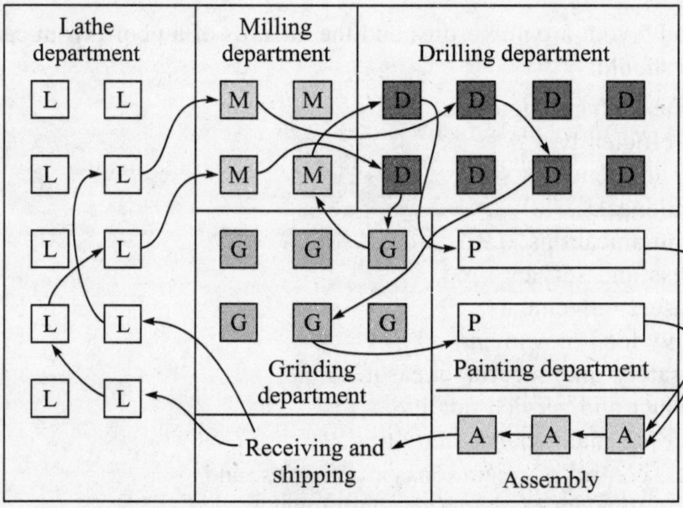

Figure 2.4 Process layout for manufacturing.

In summary,

1. Process or functional layouts are used for job shop or batch processes.
   - Equipment is arranged by function in departments or work centres.
   - Products flow in a jumbled flow pattern according to the custom needs of each customer order.
   - General-purpose equipment (can be used to make many different things) is used.
   - Workers tend to be highly skilled, but may be underutilized since some workstations are idle, and others are bottlenecked.

2. We know that products come with a variety of process requirements and are routed in different ways. The wide variety of products in a process layout tends to create:
   - Scheduling difficulties for each machine or work centre
   - Unbalanced work at each machine
   - Long queues and bottlenecks
   - Frequent set-ups, which tend to take a long time
   - High WIP inventory
   - Long product queues
   - Scheduling difficulties for customer orders, as orders tend to spend 80-90% of time in queues, or waiting to be worked on.

The major layout concern for a process layout is where to locate the work centres in relation to each other. Frequent travel and heavy loads between two work centres are a good indication that the distance between them should be minimized.

2.4 PRODUCT LAYOUTS (SEQUENTIAL-FLOW; ASSEMBLY LINE)

Product layouts, also known as assembly lines, arrange production activities in a line according to the sequence of operations that need to be performed on a particular product. Multiple assembly lines may exist in a single factory, and each assembly line would have a particular product associated with it. The characteristics of product layouts include:

- Flow is sequential, linear and predictable.
- Assembly lines are efficient.
- Assembly lines are inflexible. (Any attempt to pull a product off the line for adding a particular feature could cause major disruptions to the smooth flow of products.)
- Products tend to be standardized.
- Materials and products move in one direction and always in the same pattern.
- Conveyors are the major means for moving materials and products.
- Conveyors can be paced (automatically set to control the speed of work) or unpaced (stopped and started by the workers according to their pace).
- Assembly work can be performed on-line (on the conveyor) or off-line (at a workstation serviced by a conveyor).
- Product layouts tend to have large capital investment requirements; so a change in product volume could create difficulties in trying to generate sufficient cash flow to pay-off long-term debt. The difficulty associated with a change in volume is an inflexibility associated with product layouts.

A major concern of product layouts is balancing the line so that workstations along the line are using about the same amount of time to complete the necessary work before the next unit comes along. If a conveyor moves units every 30 seconds, and a worker at one station gets his work done in 10 seconds, while a worker at another station gets his work done in 28 seconds, the line is not well balanced.

2.5 COMPARISON OF PRODUCT AND PROCESS LAYOUTS

Table 2.1 summarizes the differences between product and process layouts.

TABLE 2.1 Difference between Product and Process Layouts

	Product layout	*Process layout*
Description	*Sequential arrangement of activities*	*Functional grouping of activities*
Type of process	Continuous, mass production and mainly assembly	Intermittent, jobshop, batch production mainly fabrication
Product	Standardized and made to stock	Varied, made to order
Demand	Stable	Fluctuating
Volume	High	Low
Equipment	Special purpose	General purpose
Workers	Limited skill	Varied skill
Inventory	Low in-process and high finished goods	High in-process, low finished goods
Storage space	Small	Large
Material handling	Fixed path using conveyors	Variable path using fork lift
Advantage	Efficiency	Flexibility

Fixed-Position Layouts

Fixed-position layouts are typical for projects in which the product manufactured is too fragile, bulky, or heavy to move. Because the product must remain stationary, equipment, materials, workers, and other resources are brought to the site where the product is manufactured, and not to a centralized factory. Ships, houses, and office buildings use fixed-position layouts. The characteristics of fixed-position layouts include:

- Scheduling problems that are associated with sequencing of work activities at the construction site to avoid labour and material shortages
- Highly skilled workers, often temporary workers
- Tendency to rent equipment rather than to buy it
- Tendency to subcontract specialized work requirement
- High variable costs (to cover labour and leasing costs) and low fixed costs (no factory and little permanent equipment).

Projects are concerned with schedules, budgets, and new product performance.

2.6 DESIGNING PROCESS LAYOUTS

The basic design principle for designing process layouts is to minimize travel distance between points that have frequent travel or points that transfer heavy loads. Intuitively, most people are good at planning a layout for the places with which they are familiar.

Sometimes, our logic or judgment is not sufficient to design a good layout. If data from similar processes are available to estimate travel and material-handling loads, two operations management techniques can be used, namely **block diagramming** and **relationship diagramming**. The techniques are *heuristics*, or methods of using judgment to develop a layout, then testing it and trying another layout until a satisfactory answer is obtained. Heuristic techniques do not guarantee optimum result, but they do provide a good method for using judgment to come up with a satisfactory answer.

2.6.1 Block Diagramming

To use block diagramming, an analyst needs certain information. The first thing needed is a general idea of the sections for each work area. For example, a building may be sectioned into a 2 × 4 grid. An aisle may run through the middle to create two horizontal sections, and each side may have four sections in it. The major problem with block diagramming is representing the work areas on a two-dimensional grid. The second set of information needed by the analyst is the traffic flow and/or material loads between rooms represented by the grid. Traffic flows and material weight may be combined by multiplying "weight" by "number of trips", or number of trips could be used alone to compute traffic *loads*.

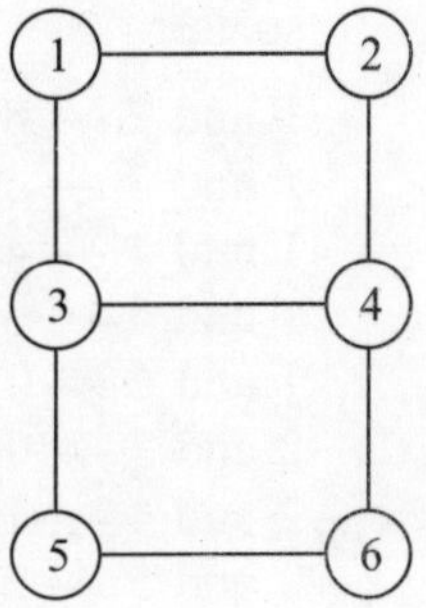

Figure 2.5 Block Diagram

Example: Some architects are designing a new building for themselves. Currently, they are located in a cramped multistorey building with the drafting department on one floor, architects on another, blueprint files on another, and so on. Earlier, they rented space as a small company, and as they grew, "temporarily" moved various departments around to allow for growth. Now, they want to design a one-floor, rectangular building to allow efficient movement of employees and supplies. For this purpose, they recorded movements in their existing office building and compiled the data to show how much movement occurs between each department.

These architects anticipate that they need six work areas, and they have created a 3 × 2 grid to begin design work. There is no aisle through the centre to separate one side from the other, and people can move freely from one work area to another. The movements between departments are listed on the Load Summary Chart.

TABLE 2.2 Load Summary Chart

From:			*To:*			
DEPARTMENT	1	2	3	4	5	6
1	–	3000	1000	4000	3500	0
2	5000	–	0	0	1800	5000
3	2000	0	–	300	500	0
4	700	1000	300	–	1500	0
5	2500	500	200	500	–	0
6	0	0	0	0	0	–

Read the chart from left to right by looking at Department 1 in the first column. Read across to the number just below Department 2. The number is 3000 and represents the travel load from Department 1 to Department 2. Now, look at Department 2 listed in the first column. Read across to the number listed under Department 1. The number is 5000. To get the total number for "travel load, to-and-fro" between Departments 1 and 2, the numbers are added together: (3000 + 5000) = 8000.

EXERCISES

1. What is the to-and-fro travel load between Departments 3 and 4?

   Answer

   To-and-fro travel loads are listed for each pair of work areas exactly as we have done so far:

   1 and 2 — 8000
   1 and 3 — 3000
   1 and 4 — 4700
   1 and 5 — 6000
   1 and 6 — 0
   2 and 3 — 0
   2 and 4 — 1000
   2 and 5 — 2300
   2 and 6 — 5000
   3 and 4 — 600
   3 and 5 — 700
   3 and 6 — 0
   4 and 5 — 2000
   4 and 6 — 0
   5 and 6 — 0

The loads are now placed on the grid diagram. Straight lines mean the departments are adjacent to each other. Dashed, curved lines mean the departments are not adjacent:

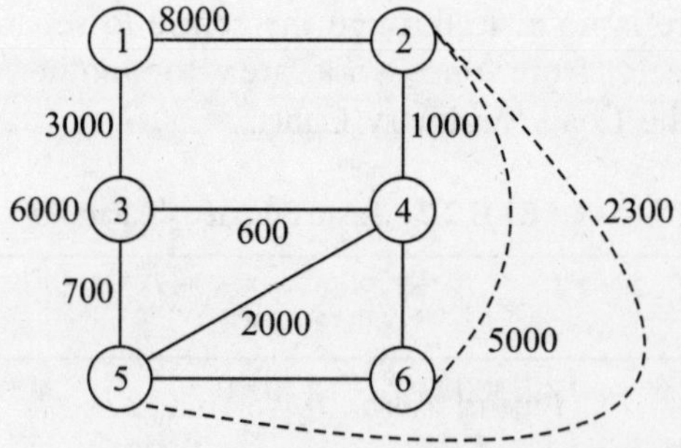

Figure 2.6(a) Grid diagram.

To evaluate our layout, we next calculate the total movements associated with non-adjacent departments:

1—5 : 6000
2—5 : 2300
2—6 : 5000
Total — 13,300

The most undesirable layout is one in which high loads are non-adjacent. A review of the grid diagram shows that very large loads exist between non-adjacent departments. If the grid is changed so that departments 5 and 6 are transposed with departments 3 and 4, we may improve the layout:

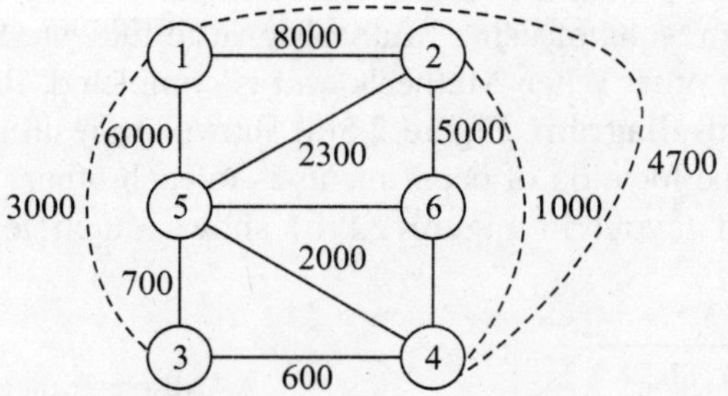

Figure 2.6(b) Grid diagram.

The new score for non-adjacent loads is:

$$\begin{gathered} 1-3 : 3000 \\ 1-4 : 4700 \\ 2-4 : 1000 \\ \text{Total} - 8700 \end{gathered}$$

The new layout is a dramatic improvement over the original one. The number of non-adjacent trips saved is (13,300 – 8700) = 4600.

2. What is the score for non-adjacent departments for the diagram below?

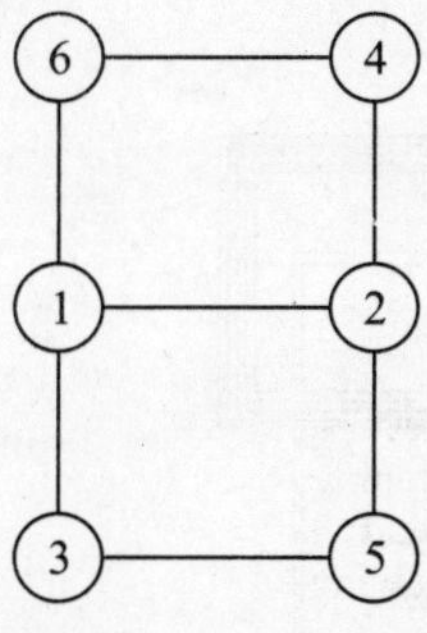

Figure 2.6(c) Block diagram.

2.7 RELATIONSHIP DIAGRAMMING

Relationship Diagramming is similar in concept to block diagramming, but it uses subjective judgment instead of numbers to evaluate the distances between departments. In relationship diagramming, departments are listed, then they are linked by a diagonal grid as shown in

Figure 2.7. Subjective judgments are used to determine if the relationship between two departments should be rated A, E, I, O, or U. The letters A, E, I, O, U stand for the ratings shown in the key. Ratings are read by looking at the intersection between two departments. For example, the rating between Production and Stockroom is A, which is absolutely necessary. The name of the grid used for relationship diagramming is **Muther's Grid**.

Figure 2.8 shows the results of using Muther's grid to indicate how important the location of departments is to each other. When Muther's grid is completed, the information can be used to construct a **relationship diagram**. Figure 2.8(a) shows the results of using Muther's grid to indicate how important the location of departments is to each other. Thicker lines mean greater importance. The revised layout in Figure 2.8(b) shows much less travel between critical departments.

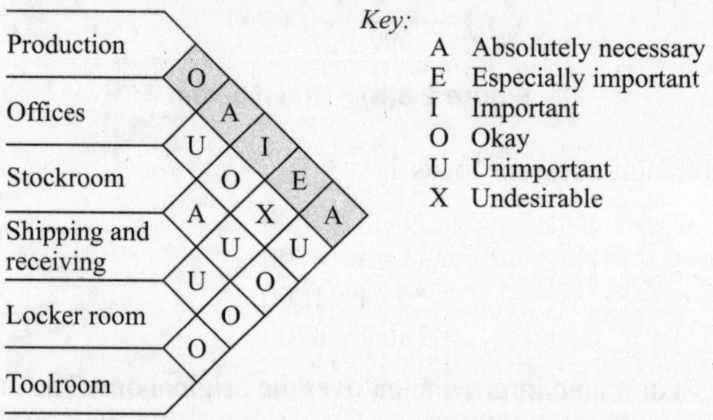

Figure 2.7 Relationship programming.

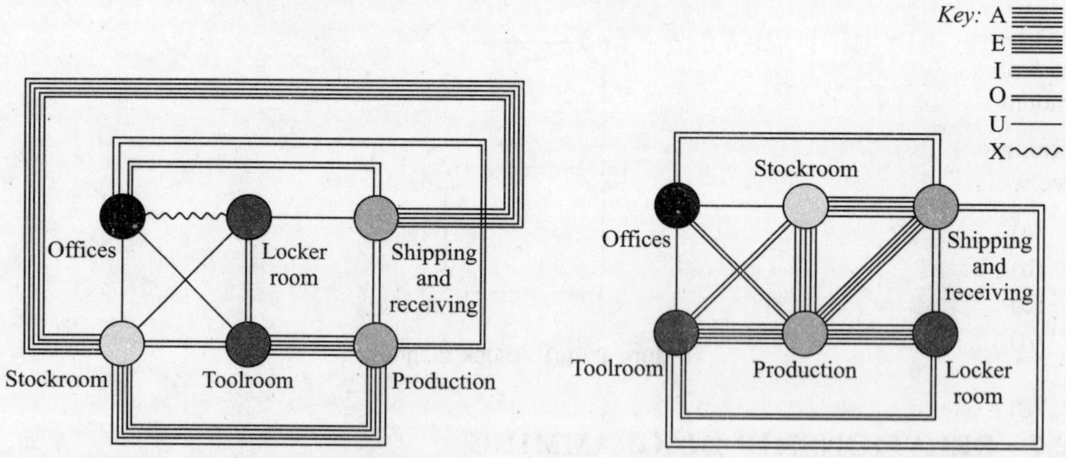

(a) Relationship diagram of original layout

(b) Relationship diagram of revised layout

Figure 2.8 Relationship diagram.

Computerized layout solutions

When process layouts are too complex or large for humans to analyze, computer packages like CRAFT, PREP, and ALDEP can be used.

2.8 SERVICE LAYOUTS

Most services use process layouts because of the variability of customer requests for service. A few services, however, use an assembly line approach. *Quick lube* service centres, which provide only lube, oil, and filter services, usually have assembly line layouts. Cafeterias also have assembly line layouts. On the whole, however, services use process layouts. The goal of service process layouts could be similar or the exact opposite of the goals for a manufacturing process layout. The goal of a manufacturing process layout is to minimize travel and transportation of heavy loads. The goals of service layouts are:

- Minimize the distance travelled by employees, paperwork, or customers. (Such travel would make the service more efficient for better customer satisfaction.)
- Maximize the distance travelled by customers when exposure to a large variety of retail goods is desired.
- Account for technical needs that restrict layout options.
- Make the layout aesthetically pleasing – this is very important in services where customers are exposed to work areas.

2.9 DESIGNING PRODUCT LAYOUTS

The main concern about designing product layouts is balancing the workload at each work station. The technique for doing so is called **line balancing**.

2.9.1 Line Balancing

Line balancing is used to group *work elements* (task activities) into each *workstation* (often the work area of a single worker) so that all workers in each workstation are active about the same amount of time.

Example: Consider an assembly line that assembles better quantity ballpoint pens. The line has five workers that complete the activities in the times listed below:

(a) The first worker fills the ink cartridge with ink and seals the cartridge (8 seconds).
(b) The second worker gets the front casing and inserts the ink cartridge into it (4 seconds).
(c) The third worker slips a spring around the ink cartridge (2 seconds).
(d) The fourth worker gets the back casing, places it over the ink cartridge, and screws it into the front casing (8 seconds).
(e) The fifth worker places the pen in a gift box with indentations for holding the pen, then closes the box (6 seconds).

Solution: Assume that the conveyor moves goods at the rate of 1 pen every 8 seconds. Put yourself in the position of worker no. 1 or worker no. 4. They are working non-stop, without even enough time to catch a sneeze. If they could sneak a peek toward the centre of the line, they would see that worker no. 2 is idle more often than working. All workers are getting paid the same, yet one works only 25% of the time. The efficiency of the system can be computed as follows:

- If each worker works the entire 8 seconds that the product is at his workstation, the system efficiency would be 100%.
- In the current system, two workers are working at 100%, one at 75%, one at 50%, and one only at 25% efficiency. But these efficiencies represent individual efficiencies, not system efficiency.
- To compute system efficiency, we note that the total time available in the system is 5 stations times 8 seconds = 40 seconds. But the time actually used for work activity is 8s + 4s + 2s + 8s + 6s = 28 seconds. The ratio of time actually used to time available is the system efficiency: 28/40 = 0.70 or 70%. The factory is paying workers for the total time they are there, yet they are only working for 70% of the time they are there! Mathematically, this concept is expressed by the following equation:

$$\eta = \frac{\sum t_i}{nc}$$

η = system efficiency
t_i = actual work times at each workstation
n = No. of workstations
c = the cycle time, or rate at which the conveyor moves units from one station to another

- The remaining (balance) delay is the percentage of time that workstations are idle. Balance delay = 1 – efficiency

For the above example (pen),

$$\text{Balance delay} = 1 - 0.70 = 30\%$$

- The number of pens produced per day can also be determined if the cycle time is known. The cycle time for pen production is one pen every 8 seconds. By converting seconds to hours, the number of pens produced in an 8-hour day can be computed:

$$\text{No. of pens produced/day} = \frac{1 \text{ pen}}{8 \text{ s}} \times \frac{60 \text{ s}}{\text{min}} \times \frac{60 \text{ min}}{\text{hr}} \times \frac{8 \text{ hr}}{\text{day}} = 3600 \text{ pens/day}$$

Exercise On observation, a consultant notices that a company manufacturing ballpoint pens can get rid of one worker by combining the activities at workstation 2 with those at workstation 3. The new line has four workers and moves units at the rate of 1 every 8 seconds. The first worker uses 8 seconds of the time available, the second uses 6 seconds (combining a 4-second activity with a 2-second activity), the third uses 8 seconds, and the fourth uses 6 seconds. What is the new system efficiency?

Answer: Let us look at line balancing from a different perspective. Suppose a company knows how many units of a particular product are needed each day, but they don't yet have work elements assigned to each worker. They would be interested in grouping work elements in an efficient manner and assigning the group of work elements to workers so that an adequate number of units is produced to meet the demand.

Example: The pen manufacturer experiences a sudden increase in demand. He must produce 4800 pens per day to meet the expected demand. The work elements must be regrouped so that enough pens can be produced. The number of workstations and workers will be determined by these requirements. Work elements are broken down as shown in Table 2.3.

TABLE 2.3 Work Element Table

Element No.	*Work element*	*Immediate predecessors*	*Time t_i (seconds)*
0	Start	-	0
1	Fill ink cartridge	0	4
2	Seal ink cartridge	1	4
3	Get front casing	0	2
4	Insert cartridge into front casing	2,3	2
5	Slip spring around ink cartridge	4	2
6	Get back casing	0	2
7	Insert cartridge into back casing	4,6	3
8	Screw back casing to front casing	7	3
9	Place pen in gift box	8	4
10	Close and seal gift box	9	2
Total			28

A trial-and-error method of line balancing uses the following steps to arrive at a solution:

1. Draw and label a precedence diagram.
2. Calculate the desired cycle time required for the line.
3. Calculate the theoretical minimum number of workstations.
4. Group elements into workstations, recognizing cycle time and precedence constraints.
5. Calculate the efficiency of the line.
6. Determine if the theoretical minimum number of workstations or an acceptable efficiency level has been reached. If not, go back to step 4.

Precedence diagram

The purpose of the precedence diagram is to systematically define the sequential order of activities. The diagram shows what must be completed before the next work activity can be completed. For example, the ink cartridge in the pen example cannot be sealed until the cartridge is filled with ink. A precedence diagram uses circles or nodes to represent work elements and arrows to indicate the direction of work flow. A precedence diagram for the pen example is shown below:

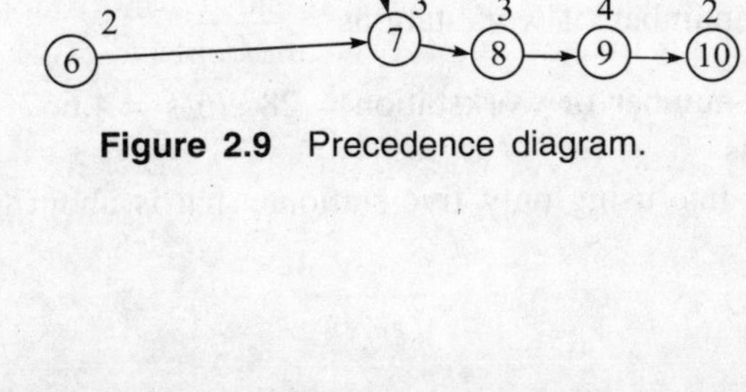

Figure 2.9 Precedence diagram.

One way of determining which work elements can be grouped together without violating the sequencing of activities is to list each element in the order of predecessor requirements. For example, element 3 has no predecessor on the diagram. Element 4, on the other hand, has the following predecessors: Elements 1, 2, and 3, for a total of 3 predecessors. The work elements with the smallest number of predecessors are listed first, then the remaining work elements are listed in the ascending order of number of predecessors:

TABLE 2.4 Work Element Table

Element No.	*Work Element*	*No. of predecessors, including "Start"*	*Time* t_i (seconds)
0	Start	–	0
1	Fill ink cartridge	0	4
3	Get front casing	0	2
6	Get back casing	0	2
2	Seal ink cartridge	2	4
4	Insert cartridge into front casing	3	2
5	Slip spring around ink cartridge	4	2
7	Insert cartridge into back casing	6	3
8	Screw back casing to front casing	7	3
9	Place pen in gift box	8	4
10	Close and seal gift box	9	2
Total			28

Desired cycle time

The desired cycle time can be computed from the new demand requirement of 4800 pens/day:

$$C = \frac{\text{Production time available}}{\text{Desired units of output}}$$

$$= \frac{(8 \text{ hours} \times 60 \text{ minutes/hr} \times 60 \text{ seconds/min})}{4800 \text{ units}} = 6 \text{ seconds}$$

The cycle time represents the speed of the conveyor. Each worker must get all assigned tasks completed in a 6-second timeframe. We can use this number to group work activities for each workstation.

Theoretical minimum number of workstations

Before grouping work elements and assigning them to workstations, we can compute the theoretical minimum number of workstations. The minimum number of workstations is a good figure for comparison purposes. The goal is to get as close as possible to the theoretical minimum:

$$\text{Theoretical minimum number of workstations} = \frac{\sum_{i=1}^{1} t_i}{C}$$

Theoretical minimum number of workstation = 28 s/6 s = 4.667

Round up to 5 stations

If we can balance the line using only five stations, that is about the best we can do.

Grouping work elements

To group work elements, we can go back to the table listing work elements in order of ascending predecessors. Sequential work elements that add up to 6 seconds or less can be grouped together. If the addition of an element creates a time longer than 6 seconds, the grouping cannot be used and the element must be separated into its own workstation or into a new grouping with a subsequent element after it. Groupings should add as close to 6 seconds as possible without going over:

TABLE 2.5 Work Element Table

Element No.	*Work element*	*No. of predecessors*	*Time t_i* (seconds)	*Grouped elements*	*Sum of times for grouped elements* (seconds)
1	Fill ink cartridge	0	4	–	
3	Get front casing	0	2	1, 3	6
6	Get back casing	0	2		
2	Seal ink cartridge	2	4	6, 2	6
4	Insert cartridge into front casing	3	2		
5	Slip spring around ink cartridge	4	2	4, 5	4
7	Insert cartridge into back casing	6	3		
8	Screw back casing to front casing	7	3	7, 8	6
9	Place pen in gift box	8	4		
10	Close and seal gift box	9	2	9, 10	6
Total		28			28

If we look at the table, the first two groupings, (1,3) and (2, 6) add up to 6 seconds, meaning operations at the first workstations will have an efficiency of 100%. The next grouping, (4, 5) cannot include work element no. 7. If it did, the time required to complete all activities would add to 7 seconds, which is more time than the conveyor allows. The workstation combining elements 4 and 5 has an efficiency of 4/6 or 66.7%. Each of the last two stations operate at 100% efficiency each, with work elements summing up to 6 seconds.

If you like working with graphic representations rather than lists, you can make the combinations from the precedence diagram. Always be sure your groupings add to the cycle time or less, and do not violate the precedence requirements:

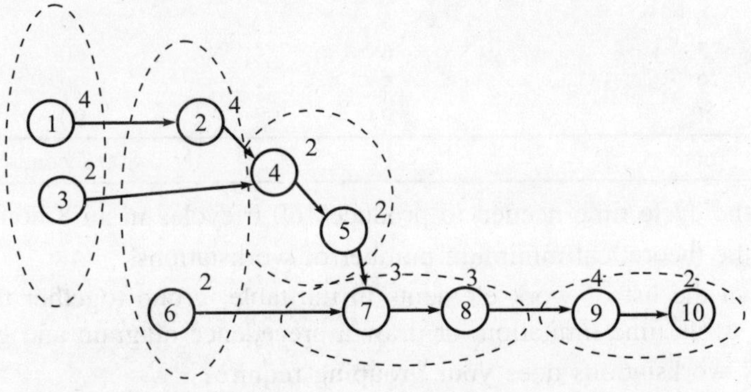

Figure 2.10 Modified precedence diagram.

The results can be summarized as in the following diagram:

Workstation 1	Workstation 2	Workstation 3	Workstation 4	Workstation 5
1, 3 6 s	6, 2 6 s	4, 5 4 s	7, 8 6 s	9, 10 6 s

Efficiency

The efficiency of this line can now be computed. There are five workstations, the cycle time is 6 seconds, and the total number of seconds for all work elements sums up to 28 seconds:

$$\eta = \frac{\sum t_i}{nC}$$

$$= (28)/(5 \times 6)$$

$$= 28/30$$

$$= 93.33\%, \text{ (Not bad at all)}$$

Comparison to the theoretical limits

The number of workstations needed for the line we just designed is 5. That is also the theoretical minimum number of workstations. The efficiency of the line is 93.33%. It appears that the line is well-balanced and meets the limits as closely as possible. If it did not, a trial-and-error approach could be used until the line balance is satisfactory.

The following table shows the work elements associated with assembling tricycles. Work elements are listed along with their predecessors. The tricycle manufacturer must make 160 tricycles in an 8-hour day.

TABLE 2.6 Work Element Table

Element No.	*Immediate predecessors*	*Time t_i (second)*
1	0	100
4	0	50
7	0	90
2	1, 4	70
3	2	140
5	3, 7	50
6	5	60
8	6	60
9	8	90
10	9	90
Total		800 seconds

(a) Calculate the cycle time needed to produce 160 tricycles in an 8-hour day.

(b) Calculate the theoretical minimum number of workstations.

(c) Going down the list of work elements in the table, group together the elements that satisfy the cycle time limitation, or draw a precedence diagram and group elements.

(d) How many workstations does your grouping require?

(e) At the second workstation, what is the time to complete the assigned activities?
(f) Calculate the line system efficiency.
(g) Is the balance adequate?

Computerized line balancing

When many work elements are needed on an assembly line, the complexity can be too time-consuming for a human to handle. Programs like IBM's COMSOAL and GE's ASYBL can work out such complex systems using heuristics.

2.10 CODING SYSTEM

Coding systems for manufacturing usually have **both product and process information.** The analyst works with both. If there is a great deal of inconsistency in existing processes, the analyst can revert to the more fundamental characteristics such as size, shape and material.

Parts with similar physical characteristics should usually have similar processes. The coding system allows an analyst **to standardize process routings and then group the parts into families.**

This avoids detailed design for many parts and avoids many tooling costs.

Existing parts have a cost history that helps product costing. Where a new part is required, designers learn from examining current drawings, tooling and cost history. They then incorporate this knowledge in the new design. Group technology thus institutionalizes organizational learning.

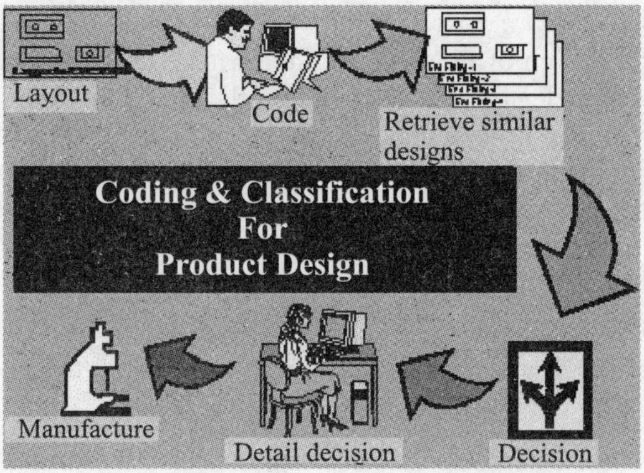

Figure 2.11 Product design.

Each digit position represents a **particular part characteristic**. For example, the third position indicates the finished weight of the casting. Each character in the code **carries information about the product characteristic assigned to that digit position.** For example, a one-third character position might indicate a finished weight between 1 kg and 5 kg.

Coding and classification present some major challenges which are not usually evident to an inexperienced practitioner. This is especially true of large databases. But then, large projects have correspondingly large returns. A successful project requires experience and judgment in coding system design, initial coding, and family development. It is not a task for the novice.

Many parts classification and coding systems are available in the market. Some are custom oriented while others are general. Custom engineered classifications are due to the fact that while the one system adapts best for one company, it may not be so for other companies.

2.10.1 Opitz Classification System

The Opitz coding system uses the following digit sequence:

12345 6789 ABCD

The basic code consists of nine digits, which can be extended by four more digits. The first nine digits are intended to convey both design and the manufacturing data. The general interpretation of the nine digits is indicated in Figure 2.9. The first five digits, 12345, are called the "form code" and describe the primary design attributes of the part. The four digits, 6789, constitute the "supplementary code". They indicate some of the attributes that would be of use to manufacturing (work material, raw workpiece shape, and accuracy). The extra four digits "ABCD", are referred to as the "secondary code" and are intended to identify the production operation type and sequence. The secondary code can be designed by the firm to serve its own particular needs.

In the form code, the first digit identifies whether the part is a rotational or a non-rotational part. It also describes the general shape and proportions of the part. Figure 2.8 shows the specification scheme. For the rotational workpieces, the coding of the first five digits is given in Figure 2.9.

Example Given the part design of Figure 2.10, the form code for this part is discussed below:

The overall length/diameter ratio, L/D = 1.6, so the first code = 1. The part is stepped on both ends with a screw thread on one end. So the second digit code would be 5, and the third digit code is 1 because of the through hole. The fourth and fifth digits are both 0 since no surface machining is required and there are no auxiliary holes or gear teeth on the part. The complete form code in the Opitz system is "15100". To add the supplementary code, we have to properly code the sixth through ninth digits with data on dimensions, material, workpiece shape, and accuracy.

2.10.2 The MICLASS System

The MICLASS classification number can range from 12 to 30 digits. The first 12 digits from the universal code that can be applied to any part "Lap to 18 additional digits can be used to code data that are specific to the particular company or industry. For example, lot size, piece time cost data, and operation sequence might be included in the 18 supplementary digits.

The component attributes coded in the first 12 digits of the MICLASS number are as follows:

1st digit	Main shape
2nd and 3rd digits	Shape elements
4th digit	Position of shape elements

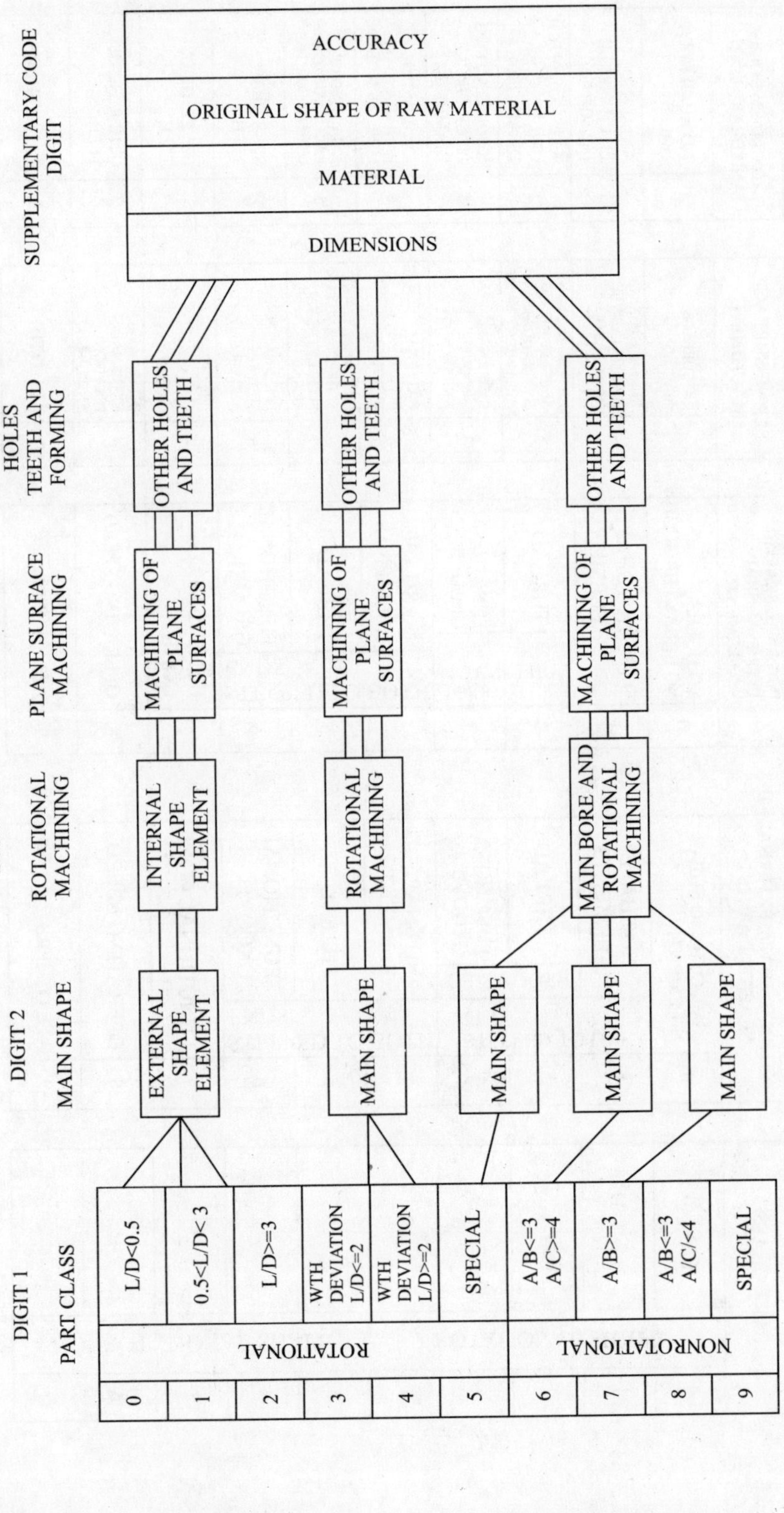

Figure 2.12 Opitz System.

DIGIT 1

PART CLASS		
0	ROTATIONAL SHAPE	L/D<=0.5
1	ROTATIONAL SHAPE	0.5<L/D<3
2	ROTATIONAL SHAPE	L/D>=3
3	ROTATIONAL SHAPE	
4	ROTATIONAL SHAPE	
5	NONROTATIONAL SHAPE	
6	NONROTATIONAL SHAPE	
7	NONROTATIONAL SHAPE	
8	NONROTATIONAL SHAPE	
9	NONROTATIONAL SHAPE	

DIGIT 2

EXTERNAL SHAPE, EXTERNAL SHAPE ELEMENTS			
0	SMOOTH, NO SHPAE ELEMENTS		
1	STEPPED TO ONE END		NO SHAPE ELEMENTS
2	STEPPED TO ONE END	OR SMOOTH	THREAD
3	STEPPED TO ONE END	OR SMOOTH	FUNCTIONAL GROOVE
4	STEPPED TO BOTH ENDS		NO SHAPE ELEMENTS
5	STEPPED TO BOTH ENDS		THREAD
6	STEPPED TO BOTH ENDS		FUNCTIONAL GROOVE
7	FUNCTIONAL CONE		
8	OPERATING SPEED		
9	ALL OTHERS		

DIGIT 3

INTERNAL SHAPE, INTERNAL SHAPE ELEMENTS		
0	NO HOLES NO BREAKTHROUGH	
1	SMOOTH OR STEPPED TO ONE END	NO SHAPE ELEMENTS
2	SMOOTH OR STEPPED TO ONE END	THREAD
3	SMOOTH OR STEPPED TO ONE END	FUNCTIONAL GROOVE
4	STEPPED TO BOTH ENDS	NO SHAPE ELEMENTS
5	STEPPED TO BOTH ENDS	THREAD
6	STEPPED TO BOTH ENDS	FUNCTIONAL GROOVE
7	FUNCTIONAL CONE	
8	OPERATING SPEED	
9	ALL OTHERS	

DIGIT 4

PLANE SURFACE MACHINING	
0	NO SURFACE MACHINING
1	SURFACE PLANE AND/OR CURVED IN ONE AROUND A CIRCLE
2	EXTERNAL PLANE SURFACE RELATED BY GRADUATION AROUND A CIRCLE
3	EXTERNAL GROOVE AND/OR SLOT
4	EXTERNAL SPLINE (POLYGON)
5	EXTERNAL PLANE SURFACE AND/OR SLOT EXTERNAL SPLINE
6	INTERNAL PLANE SURFACE AND/OR SLOT
7	INTERNAL SPLINE (POLYGON)
8	INTERNAL AND EXTERNAL POLYGON AND/OR SLOT
9	ALL OTHERS

DIGIT 5

AUXILIARY HOLES AND GEAR TEETH	
0	NO AUXILIARY HOLES
1	AXIAL, NOT ON PITCH CIRCLE DIAMETER
2	AXIAL ON PITCH CIRCLE DIAMETER
3	RADIAL NOT ON PITCH CIRCLE DIAMETER
4	AXIAL AND/OR RADIAL AND/OR OTHER DIRECTION
5	AXIAL AND/OR RADIAL AND/OR OTHER DIRECTION
6	SPUR GEAR TEEH
7	BEVEL GEAR TEEH
8	OTHER GEAR TEEH
9	ALL OTHERS

Figure 2.13 Coding the first five digits.

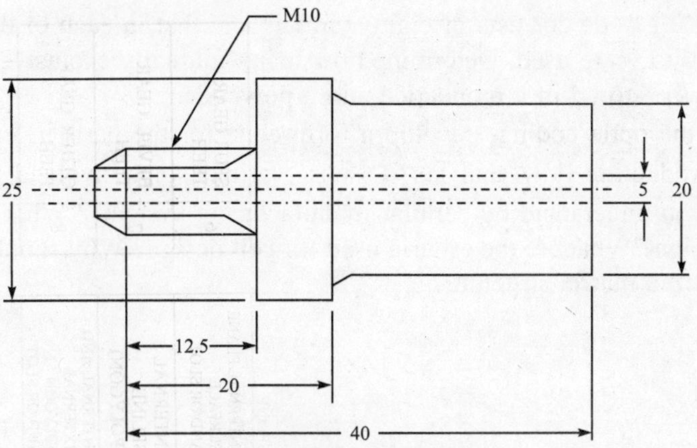

Figure 2.14 Part design.

5th and 6th digits	Main dimensions
7th digit	Dimension ratio
8th digit	Auxiliary dimension
9th and 10th digits	Tolerance codes
11th and 12th digits	Material codes

One of the unique features of MICLASS system is that parts can be coded using a computer interactively. To classify a given part design, the user responds to a series of questions asked by the computer. The number of questions depends on the complexity of the part. For a simple part, as few as seven questions are needed to classify the part. For an average part, the number of questions ranges between 10 and 20. On the basis of responses to its questions, the computer assigns a code to the part.

- Selection of the machining parameters also affects the time and cost required to produce components.
- Several machinability systems are currently marketed that recommend sets of parameters which either optimize machining cost, time or production rate, or simply retrieve data table or calculated values.

One extensive data system is CUT DATA from Machinability Data Systems of USA.

FAST system (table value retrieval system).

Recently, machinability systems have become computer based to take advantage of computational power to include optimized models.

REVIEW QUESTIONS

1. What do you understand by the term, *Group Technology* (GT)? Discuss the uses of GT in various functional departments of an organization.
2. Discuss with examples the following: Classification and coding systems, monocode, polycode, and mixed code.

3. Assume that a code consists of eight symbols and that in each of the eight code fields the digits 0 to 9 are used. Determine how many mutually exclusive characteristics can potentially be stored in a monocode and a polycode.
4. Based on the optiz coding, develop a form code for the parts in Figure 2.14.
5. What is production flow analysis? Discuss various steps involved in PFA.
6. What do you understand by cellular manufacturing systems? What do you understand by cell design? What are the criteria used for cell design? Why should you want to have a diagonalized matrix structure?

Chapter 3

FLEXIBLE MANUFACTURING SYSTEMS

3.1 INTRODUCTION

Globalization has created new demands for manufacturers to produce a wide range of quality products at lower prices, and shorten their manufacturing lead times. In an environment of fierce competition, industry can't afford large amounts of waste—in time, material or production capacity. Cost, quality and responsiveness are three requirements the industry must be aware of if it hopes to survive. This requires a new type of manufacturing system, the flexible or reconfigurable manufacturing system.

There are a number of machining systems used in industry such as dedicated machining systems, flexible manufacturing systems (FMS), cellular manufacturing systems, and the soon-to-be available reconfigurable machining system. Each system is designed for a particular family of parts and each has specific advantages.

3.2 DEDICATED MANUFACTURING SYSTEMS

Specialized and single-purpose machines, or dedicated machines, were developed in the early 1900s for mass production of identical parts. Many different transfer-type machines, each designed to produce a specific product or perform a specific machining operation, were used in manufacturing. This process was not very flexible, and as many as 150 different machines were required to produce a limited number of finished products. When the product mix changed, the machines had to be changed, which was a costly, time-consuming process.

3.3 FLEXIBLE MANUFACTURING SYSTEMS

A flexible manufacturing system or FMS generally consists of a number of CNC machine tools and a materials handling system which is supervised (controlled) by one or more dedicated computers. A typical flexible manufacturing system can completely process the members of one or more part families on a continuing basis without human intervention. FMS brings flexibility to manufacturing so that a part can be produced when the market requires it. The system is flexible enough to suit changing market conditions and product type without buying other equipment.

3.3.1 Cellular Manufacturing Systems

The main purpose of cellular manufacturing systems is to improve manufacturing productivity, by having machining processes, machinery, and people, all grouped together to manufacture a specific family of parts. The cells are designed to handle all the operations of a family of parts, ranging from receiving raw material to producing the finished product within the cell. Cellular manufacturing provides the economic advantages in high-variety, low-volume production, which are normally associated with low-variety, high-volume production of flexible manufacturing systems.

The Engineering Research Center of the University of Michigan, Ann Arbor, USA, along with more than 30 industrial partners, is working to develop a new generation of manufacturing systems that will be responsive to customer demands and changes to product mix. The Reconfigurable Machining System (RMS) is designed with rapid and constant changes in mind. The Research Center's goals are to develop a new factory that will:

1. reduce lead time and ramp-up time for new and reconfigured manufacturing systems,
2. be flexible enough to adapt to changes in product and market demands,
3. allow the reconfiguration of existing manufacturing and machining systems, and
4. allow new software technology to be incorporated into existing systems.

What is holding back the development of reconfigurable systems is the lack of a scientific approach to designing systems that include reconfigurable hardware and software. The Michigan Research Center is developing a Reconfigurable Machining System that will allow for rapid, cost-effective changes, regardless of the production requirements. This potential can only be realized when CNC control can be combined with the high production rate of dedicated machines by a systematic method. The goals of RMS research include developing methodologies for:

1. Computer-Aided Design (CAD) for manufacturing systems of part families.
2. Design of a new generation of reconfigurable machine tools along with open architecture controllers. The machine tools will consist of replaceable modules, any of which can be changed, to suit the product to be manufactured.
3. Rapid, systematic ramp-up after configuration to reduce costs. Ramp-up periods need to be as brief as possible in any manufacturing system and developing methodologies and reconfigurable tools will shorten manufacturing downtime.

Reconfigurability can be defined as the ability to adjust the production capacity and operation to new circumstances through the rearrangement or change of the system's components. Components could be machines and conveyors in the whole system, mechanisms in individual machines, and new sensors or a new controller. The two classes of machine tools being developed are the Adjustable Capacity RMT (a simpler machine with enough functionality to produce just a part family) and the Adjustable Functionality RMT (the capacity and functionality to produce a part family). New circumstances may be changing product demand, producing a new product on an existing system, or integrating new process technology into existing manufacturing systems.

Reconfigurable manufacturing systems will be open-ended so that they can be improved and upgraded rather than replaced. They will allow flexibility, not only in the products they can handle, but also in changing the system itself. The key to such a system is that it must be designed to be reconfigurable at the beginning, and must be created from basic hardware and software modules that can be rearranged quickly and reliably.

RMS technology requires a systems approach to creating new design knowledge and the modification of the manufacturing system. The system must contain the following elements, some of which have already been developed:

1. A system-level process planner given a part family, volume and mix, can suggest different production systems and choose the best system for production purposes.
2. A life cycle economic modelling methodology that can blend dynamic programming to recommend the most profitable system.
3. A reconfigurable machine tool, consisting of numerous modules that can be replaced on the basic machine to allow it to be modified quickly to suit the job.
4. A control system for sequencing and coordinating large manufacturing systems. The Engineering Resource Center's RMS Open Architecture Control System allows for the rapid reconfiguration of manufacturing systems by having a common Human Machine Interface (HMI) for operating machines with different control designs, and produced by different vendors, a stream-of-variations that can blend control theory with in-process statistics for ramp-up time reduction.
5. A measurement scheme selection that ensures the highest quality and productivity.

Multitasking machines available

Future machine tools will combine many operations performed by individual machines into a single machine tool. Some of these machines are already available today in the form of multitasking machines.

An example of a multitasking machine is the Mazak Integrex Model 200 III that combines a turning centre and a machining centre into one machine. It contains two turning spindles and a separate vertical machining spindle, and can perform all the operations normally performed by two or more machines required to complete a complex workpiece. Equipped with the proper tools, it can turn, machine, drill, tap, grind and cut gears on a complex part.

Multitasking machines, such as the Integrex, combine IT (Information Technology) and MT (Manufacturing Technology) in one machine to allow an expanded range of complex workpieces to be processed. Tool data is tied to a Tool Identification chip which is mounted on each tool's retention stud. Once all the tool data are registered, the information is automatically downloaded over the network when the tool ID chip is read.

When a tool is removed from a machine magazine, the tool ID number is read and all current data for that tool is automatically transferred to the Cyber Production Centre tool database for central management.

'Computer Integrated Manufacturing' is the phrase used to describe the complete automation of a manufacturing plant, with all processes functioning under computer control and digital information tying them together. It was promoted by machine tool manufacturers in the 1980s and the Society for Manufacturing Engineers (CASA/SME). Quite often it was mistaken for the concept of a "lights out" factory. It includes CAD/CAM (Computer-Aided Design/Computer-Aided Manufacturing), CAPP (Computer-Aided Process Planning), CNC (Computer Numerical Control), machine tools, DNC (Direct Numerical Control), machine tools, FMS (Flexible Machining System), ASRS (Automated Storage and Retrieval System), AGV (Automated Guided Vehicle), use of robotics and automated conveyance, computerized scheduling and production control, and a business system integrated by a common database.

The heart of computer integrated manufacturing is CAD/CAM. Computer-aided design (CAD) and computer-aided manufacturing (CAM) systems are essential to reducing cycle times in the organization. CAD/CAM is a high technology integrating tool between design and manufacturing. CAD techniques make use of group technology to create similar geometries for quick retrieval. Electronic files replace drawing rooms. CAD/CAM integrated systems provide design/drafting, planning and scheduling, and fabrication capabilities. CAD provides the electronic part images, and CAM provides the facility for toolpath cutters to take on the raw piece.

The computer graphics that CAD provides allow designers to create electronic images which can be portrayed in two dimensions, or as a three-dimensional solid component or assembly which can be rotated as it is viewed. Advanced software programs can analyze and test designs before a prototype is made. Finite element analysis programs allow engineers to predict stress points on a part, and the effects of loading.

Once a part has been designed, the graphics can be used to program the tool path to machine the part. When integrated with an NC postprocessor, the NC program that can be used in a CNC machine is produced. The design graphics can also be used to design tools and fixtures, and for inspections by coordinate measuring machines. The more downstream use that is made of CAD, the more time that is saved in the overall process.

Generative process planning is an advanced generation of CAD/CAM. It uses a more powerful software program to develop a process plan based on the part geometry, the number of parts to be made, and information about facilities in the plant. It can select the best tool and fixture and it can calculate cost and time.

Flexible manufacturing systems are extensions of group technology and cellular manufacturing concepts. Using integrated CAD/CAM, parts can be designed and programmed in half the time it would normally take to do the engineering. The part programs can be downloaded to a CNC machining centre under the control of an FMS host computer. The FMS host can schedule the CNC and the parts needed to perform the work.

Computer integrated manufacturing can include different combinations of the tools listed above. One of the key issues regarding CIM is equipment incompatibility and difficulty of integration of protocols. Integrating different brand equipment controllers with robots, conveyors and supervisory controllers is a time-consuming task with a lot of pitfalls. Quite often, the large investment and time required for software, hardware, communications, and integration cannot be financially justified easily.

Another key issue is *data integrity*. Machines react clumsily to bad data, and the costs of data upkeep as well as general information systems departmental costs are higher than in a non-CIM facility.

Another issue is the attempt to program extensive logic to produce schedules and optimize part sequence. There is no substitute for the human mind in reacting to a dynamic day-to-day manufacturing schedule and changing priorities.

Just like anything else, computer integrated manufacturing is no panacea, nor should it be embraced as a religion. It is an operational tool that, if implemented properly, will provide a new dimension to competing: quickly introducing new customized high quality products and delivering them with unprecedented lead times, swift decisions, and manufacturing products with high velocity.

3.4 MAJOR ELEMENTS OF FMS

Each of the major subsystems in an FMS performs a number of functions and is dependent on others to make the entire system work. The functions will vary, depending on the type of equipment and manufacturing operations involved.

3.4.1 Production Equipment

The production equipment used in FMS depends on the product manufactured.

(i) ***FMS for sheet metal work:*** The work centres used in sheet metal FMS include turret punch presses, laser machining centres, press brakes, guillotines, etc. A typical FMC consists of sheet stacking system, sheet unloading device, sorting conveyor, turret punch press, right angle shear, loading device, and automatic storage.

(ii) ***FMS for machining:*** This type of FMS typically has a number of machining centres and/or turning centres to provide general-purpose machining capabilities. Machining centres offer the greatest flexibility since they can perform many different machining operations (e.g. milling, drilling, and boring). This is made possible by a tool-changing system that is either built into or supports the machining centre. A part can therefore undergo multiple machining processes at a single workstation. Special purpose machines may also be included in the FMS to perform operations which are unique or require more efficiency (e.g. turning, grinding). Washing machines and inspection machines also form the equipment of FMS.

The family of parts which the FMS is designed to produce will determine the capabilities required from the machine tools (e.g. accuracy, size, power).

Support Systems

Automated machine tools typically require several systems to support their operation. The tools required to perform the multiple processes of a machining centre or a turning centre may be stored in magazines at each machine or in a central tool storage. Local magazines provide fast access as well as backup capability, but in a large FMS, a central tool facility may be more efficient. Centralization not only permits the total number of tools to be minimized, it also provides the opportunity to perform automatically additional functions such as:

1. Measuring tools for wear
2. Tool pre-setting
3. Tool regrinding, repair and maintenance
4. Replacement of broken or worn tools

Many automated machine tools have built-in systems to monitor tool wear and detect tool breakage. They may use probes or noncontact techniques such as acoustic emission for this purpose. When a tool needs replacement, the machine can signal the tool room for the delivery of a replacement. This may be performed by an AGV or gantry set-up or RGV.

Elaborate tool management support is an integral part of the FMS software. With this software, the operating personnel can have effective centralized control of a large tool inventory.

Automated machining operations also need to have the chips cleaned off the workstation and the workpiece. This may be performed by robots or special washing stations. Cleaning may involve turning the workpiece over, vacuuming and washing.

3.5 MATERIALS HANDLING SYSTEM

1. An FMS typically needs several materials handling systems to service the machines.
2. A transport system to move workpieces into and out of the FMS (e.g. overhead conveyors, gantry systems, AGVs, RGVs).
3. A buffer storage system for queues of workpieces at the machines (e.g. pallets).
4. A transfer system to load and unload the machines (e.g. robots, transfer fixtures).

For these systems to work effectively, they must be synchronized with the machine operations. The location and movement of workpieces must be tracked automatically. This is done by using sensors on the materials handling system and workstations.

They may be either contact devices (e.g. switches) or noncontact devices (e.g. optical or proximity devices).

Advantages

- Faster, lower cost changes from one part to another
- Lower direct labour costs
- Reduced inventory
- Consistent, and perhaps better quality

Disadvantages

- Limited ability to adapt to product or product mix changes
- Substantial preplanning and capital expenditure
- Technological problems of exact component positioning and precise timing
- Tooling and fixture requirements

3.6 PROBLEMS WITH FMS

- Determining if FMS is the best production system for your company (economically and socially)
- Possible expansion costs associated with implementing FMS
- Day-to-day maintenance of FMS operations

3.7 ENTERPRISE RESOURCE PLANNING (ERP)

Packaged business software systems that allow companies to

1. automate and integrate the majority of their business processes;
2. share common data and practices across the entire enterprise; and
3. produce and access information in a real-time environment.

Advantages

- Provides integration of supply-chain, production and administrative processes.
- Creates commonality of databases.
- Can incorporate improved, redesigned, "best processes".
- Increases communication and collaboration worldwide.
- Helps integrate multiple sites and business units.
- Is packaged with a software core that is off-the-shelf coding.
- Can provide a strategic advantage over competitors.

Disadvantages

- Is very expensive to purchase, and even more costly to customize.
- Requires major changes in the company and its processes to implement.
- Is such a complex program that many companies cannot adjust to it.
- Involves an ongoing process for implementation, which is often never completed.
- Expertise in ERP is limited, with staffing an ongoing problem.

3.8 MIXED-MODEL ASSEMBLY LINES

Some manufacturers make different models of a product in high volumes. The traditional way of manufacturing is to make a large volume of one model, then shut down and change over to a different model. The long set-up times require enough volume to be produced to cover the cost of making a very long changeover.

Example. XYZ Company makes five models of vacuum cleaners. The demand for the month for each model is:

Mode	*Demand*
A	200
B	400
C	800
D	400
E	400

A traditional manufacturer may choose to add the monthly demands for each model to get the total needed for the year. The manufacturer would then produce the amount of each model needed for the entire year in batches. For example, assuming that the demand for Model A is the same each month, the manufacturer has an annual demand of 2400 units (200/month × 12 months). The manufacturer would produce all 2400 units, then change over to the production of Model B.

Now, consider ModernXYZ Company. They have the same number of models and demands as TraditionXYZ Company. Their daily production is scheduled to produce the following models:

A BB CCCC DD EE A BB CCCC DD EE A BB CCCC DD EE . . .

How can this be? This concept is totally counter-intuitive. Why would anybody do this?

The answer to the question in the example is that a great deal of process management must take place before a manufacturer can mix models on the assembly lines as shown by ModernXYZ company. Techniques for getting to a mixed model layout include the following:

1. Many organizations start by using group technology to get families of parts.
2. Changeover and set-up times are reduced, often by thoughtful and creative use of mechanical and electrical devices that quickly move workpieces.
3. Workers are trained to perform a variety of tasks and work at several workstations — U-shaped lines are used; a flexible workforce is necessary.
4. Layouts are changed to facilitate the needs of all the models.
5. Lines are balanced using average times for all models.
6. Models are sequenced according to demand patterns and to take into account the time needed at each station to get an efficient line balance.

While sequencing of models may not use such small lot sizes as shown in the ModernXYZ example, mixed model assembly lines do not give any incentive to manufacturing products in large volumes. Think about the benefits of changing from one model to another over and over again. First, the operator becomes very good at making changes. Second, the finished goods inventory is reduced for certain models, and finished goods for other models become available in adequate amounts. In the old system, the manufacturer may not be adept at Model E, but the next two months of production are scheduled for Model B. With mixed-model assembly, each model is produced daily or weekly. A mixed-model production schedule allows quick change to different volume levels if the market suddenly shifts. The added flexibility allows better customer service.

3.9 CELL TECHNOLOGY AND FLEXIBLE MANUFACTURING SYSTEMS

Changes associated with JIT include the adoption of cell technology and flexible manufacturing systems. These improve operational flows.

3.9.1 Process/Production Flows and Cell Technology

Factory layout must contribute to simpler and improved process and production flows. Change to process or functional layouts is based on cells or groups so that each work-centre specializing in one or a narrow family of processes can produce benefits.

A cell is a self-sufficient unit, in which all operations to make a 'family' of parts, components or complete products can be carried out. The cell is like a mini-factory within the factory which a cell team can manage as their own operation—a server to other operations. Thus, client-server relationships can be emphasized. The ways in which a cell gives support to other cells downstream and receives support, e.g. from engineering or computing specialists, can be reviewed.

Cell layouts can be U-shaped or a segment of a line (a product or sub-assembly stage) allowing a self-organizing, multi-skilled group of fewer people to manage the operation. Shorter processing times, better team attention to quality problems, reduction of work in progress, lower handling costs, and simpler scheduling can be achieved.

3.9.2 Teams and Cells

Cells form an easily identifiable, self-reliant unit. Authority and responsibility, accountability and more clearly targeted rewards for performance can be better defined and team development strengthened. Cell teams can:

- buy-in technical expertise to supplement their own know-how;
- program their own continuous quality improvement;
- reap the benefits of their own achievements;
- become their own quality circles, make their own problems visible and apply problem solving quickly to sort them out continuous improvement means taking steps to prevent things from going wrong;
- be a value-added centre if information systems can be so geared up to provide feedback data.

3.10 AUTOMATIC GUIDED VEHICLES (AGVs)

AGV is one of the widely used types of material handling device in an FMS. These are battery-powered vehicles that can move and transfer materials by following prescribed paths around the manufacturing floor. They are neither physically tied to the production line nor driven by an operator like forklift. Such vehicles have on-board controllers that can be programmed for complicated and varying routes as well as load and unload operations. The computer for the materials handling system or the central computer provides overall control functions, such as dispatching, routing, traffic control and collision avoidance. AGVs usually complement an automated production line consisting of conveyor or transfer systems by providing the flexibility of complex and programmable movement around the manufacturing shop.

3.10.1 Advantages of Using AGV systems in FMS

(i) ***Flexibility:*** The route of the AGVs can be easily altered, expanded and modified, simply by changing the guide path of the vehicles. This is more cost-effective than modifying fixed conveyor lines or rail guided vehicles. It provides a direct access materials handling system for loading and unloading FMS cells and accessing the automated storage and retrieval system.

(ii) ***Real-time monitoring and control:*** Because of computer control, AGVs can be monitored in real-time. If the FMS control system decides to change the schedule, the vehicles can be re-routed and urgent requests can be served. AGVs are usually controlled through wires implanted on the factory floor using a variable frequency approach.

Radio control, an alternative to in-floor mounted communication lines, permits two-way communication between the on-board computer and a remote computer, independent of where the vehicle is, i.e. whether it is in the parking place or in motion. To issue a command to a vehicle, the central computer sends a bit stream via its transmitter using frequency shift keying methods to address a specific vehicle. The signal transmitted from the base station is, therefore, read by the appropriate vehicle only. The vehicle is also capable of sending signals back to the remote controller, to report the status of the vehicle, vehicle malfunction, battery status, and so on.

(iii) ***Safety:*** AGVs can travel at a slow speed but typically operate in the range 10–70 m/min. They have on-board microprocessor control to communicate with local zone controllers which direct the traffic and prevent collisions between vehicles as well as the vehicle and other objects. A bumper is attached to some designs of AGVs to prevent collision.

AGVs may also incorporate warning lights, fire safety interlocks, and controls for safety in shops. During design, the use of simulation can help detect whether there are enough vehicles to perform the necessary load, unload and transportation tasks, and thus optimize the utilization of the AGV system. Because these vehicles have to work in tandem with highly organized FMS cells as well as with automated warehouses under computer control, their level of performance will affect the entire efficiency of the FMS. Several Indian companies have reported the development of automatic guided vehicles. Some of them are:

BHEL R & D

Premier Automobiles Ltd.

TVS.

HMT uses a rail guided vehicle system in their flexible manufacturing cells.

3.11 AUTOMATED STORAGE (AS) AND RETRIEVAL SYSTEMS (RS)

A key part of any materials handling system is **storage**. Major advances have been made in recent years to automate the storage and retrieval of product and materials by employing sophisticated materials handling machines, high-density storage techniques, and computer control. Such systems come in a variety of forms and sizes, depending on the materials handling and storage job that has to be done. They often take the form of automated warehouses which use automatic storage and retrieval systems, conveyors and computers to control the materials handling machines and to track and control the inventory. The characteristics of such warehouses include:

- High density storage (in some cases, large, high-rise rack structures)
- Automated handling systems (such as elevators, storage and retrieval carousels and conveyors).
- Materials tracking systems (using optical or magnetic sensors).

In such a storage system, the computer can keep track of a large number of different parts, products and materials, and can assign bin locations to optimize the use of storage space. When such a system is connected to the production control system, parts and materials can be replenished as they are consumed on the factory floor, keeping the WIP to a minimum.

The various categories of AS/RS are:

The automatic storage and retrieval system can be classified into several types. Some of them are:

1. Unit load AS/RS
2. Mini load AS/RS
3. Man-on-board AS/RS
4. Automated item retrieval system
5. Deep lane AS/RS

The basic components of AS/RS are:

1. Storage structure
2. Storage and retrieval machine
3. Storage modules
4. Pick-up and deposit stations

The special features of AS/RS are:

- Aisle transfer cars
- Full/empty bin detectors
- Sizing stations load identification stations

3.11.1 Buffer Storing of Parts

In an FMS, parts go from one work cell to another where the various processing tasks are performed. Because of the almost random production facilities of FMS, the destination cell might not always be ready to accept the incoming part and the part has to wait in a buffer store. These and other aspects of materials handling problems can be successfully detected by simulation.

Buffer stores for parts will always be required, because they are the dislocations of the in-process materials handling and transportation system. Figure 3.1 shows a typical FMC cell layout where buffer stores are used as an integrated part of the cell as well as the overall materials handling system.

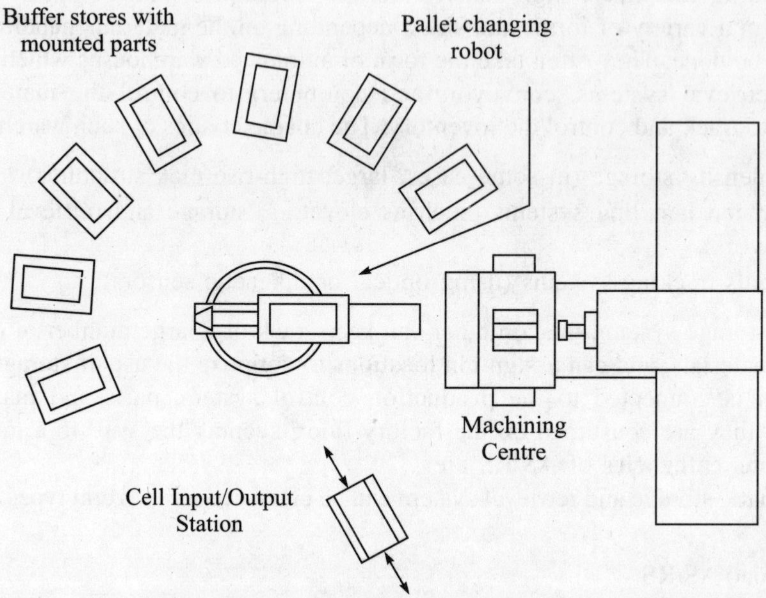

Figure 3.1 Typical FMS cell layout.

In the case of turning centres, the machining time may be of the order of a few minutes. A gantry robot is used for loading and unloading the component. It is better to present the raw workpieces in a pallet to the gantry. Finished workpieces can be deposited in another pallet. The empty raw material pallet and the filled finished part pallet will be transported by the AGV. Buffer store is also recommended for sheet metal items. Machining centres with multiple pallets (2, 4, 8 or more) incorporate adequate buffer capacity to last several hours.

3.11.2 Chip Removal and Washing Stations

Workpiece cleaning is important, especially before the part goes to the inspection station or assembly station, because unremoved swarf can cause problems during the inspection cycles or assembly.

The swarf removal is done at the washing station of the FMS. The pallet with fixtured part is loaded on to the washing station, where it is located as if it was a table of any other machining station. It is tilted by a hydraulic mechanism while being rinsed under high pressure coolant or pressurized air supply. Then, while reverting to its load/unload position, the pallet is blown clean with compressed air. Once the part is clean, it can be taken away by a robot or AGV, together with its pallet.

3.11.3 Computer Control System

The computer control system of an FMS integrates several subsystems including:

1. CNC systems
2. Support system controllers
3. Materials handling system controller
4. Monitoring and sensing devices
5. Data Communication system
6. Data Collection system
7. Programmable logic controllers
8. Supervisory computer

This control system must also integrate other computer systems if these are installed in the factory. The FMS system must also communicate with the following systems:

- The CAD/CAM system which generates the CNC programs for the machine tools
- The shopfloor control systems which schedule loading and routing of the work
- The MIS system which provides management with reports on the performance of the system

The various controllers and computers can be arranged in the form of a LAN for this purpose. The type of supervisory computer depends on the size of an FMS. A powerful minicomputer (preferably RISC based) should be adequate as a control computer.

3.12 OPTIMIZATION OF FMS

An FMS requires considerable investment. Thorough planning and analysis should precede the purchase of an FMS as the FMS should be designed to provide efficient operation. The following are some approaches which should be considered in order to optimize the overall efficiency and effectiveness of an FMS:

(i) *Minimizing the process cycle time:* The process must be designed to minimize machining and handling.

(ii) *Maximizing the utilization of each machine:* This can be done by balancing the work load in the system and real-time scheduling.

(iii) *Use of automated storage systems to keep work ready for machines to process:* The raw workparts must be replenished as and when needed to avoid starving the work centres.

(iv) *Provision of adequate sensors for the detection of errors or problems:* This includes the detection of the presence or absence of parts, jamming, tool wear, machine failures, and so on. This can be done with the use of vision systems, limit switches, proximity switches, etc. In some cases, special sensors like tool monitoring systems are used.

(v) *Back-up capabilities:* Redundancy is important in ensuring trouble-free operation of the FMS. The system should be able to run even when failures occur (e.g. use spare tools, provision to isolate defective machines, supply of alternative materials and transport paths, and additional machine capacity).

(vi) *Incorporation of in-process or post-process measurement and inspection techniques:* These assure product quality and reduce scrap and rework.

(vii) *Use of identification marking techniques:* Bar codes are now popular for identifying products as well as components. This permits automatic tracking of workpieces and tools.

A great deal of effort is required to implement FMS. These systems are complex and require careful planning and thorough design. Some of the tasks in selecting an FMS are:

1. Selecting a family of parts, which is both similar in design and business needs (Group technology concepts can be used for this purpose.)
2. Specifications of the capabilities and performance requirements of the subsystem and the total system
3. Benchmarking the performance of the alternative proposals
4. Economic justification of the system
5. Determining the size of the system
6. System simulation for optimization
7. Selection of the equipment for the FMS
8. Design of the control systems
9. Selection and training of the personnel to run the system

3.12.1 Operational Elements of a Typical Flexible Manufacturing Cell

The various functional elements of the FMC are discussed below:

The FMC software

The software system for the flexible manufacturing cell forms the basis for accommodating whichever software function modules the FMC plant requires. The basic system offers standard interfaces to various software functions. It also handles communication between the individual software modules and between the software modules and any peripherals (printers, other computers).

The CNC controls associated with the cell are accessed via appropriate programs. The basic system also supports or performs the following functions:

1. System generation and parameterization
2. System initialization
3. Collection and display of error messages
4. Log functions

3.12.2 Types of Data Associated with the Flexible Manufacturing Cell

A typical FMC system handles four different types of data: Master data, control data, status data, and general management data. These data are generated from CAD, CAP and CAM functions.

(i) **Master data:** Master data have to be set up only once when the flexible manufacturing cell is put into operation and this comprises:

- *System-specific data:* related to the architecture of the system
- *Resource master data:* resources master data comprise:
  - Tool master data
  - Workpiece carrier master data (data relating to the workpiece carriers and clamping fixtures)

(ii) **Control data:** Control data are product-specific data.

- CNC programs, tool layouts and work schedules (technical control data)
- Manufacturing orders (organizational control data)

(iii) **Status data:** Status data describe the current situation with regard to resources.

- Plant status data
- Resources data
- Workpiece carrier data
- Tool data
- Workpiece data

(iv) **Log data:** All operational data and machine data required for later analysis and diagnostics are recorded, evaluated by the software function modules, and filled with details of data and time of day. Such data include:

- *Machine specific messages* (e.g. from CNC, PLC handling devices, transport system)
  - Status and operational messages (e.g. NC start, NC end, NC program run time)
  - Alarms (e.g. machine fault)
- *Tool specific messages* (e.g. tool break, end of tool life)
- *NC messages* (e.g load NC program)
- *Entry and description of fault by way of a dialogue*

There are system statues in which the data are not defined. These are

— Metal start (all data),

— Restart after complete or partial loss of data

3.12.3 Job Scheduling

Manufacturing orders are entered into the cell computer by the operation interactively. In the input routine, the operator enters the job number, the parts number, the production quantities and deadlines. Manufacturing orders can be completed on a batch basis or may involve a parts mix. Figure 3.2 shows the various FMC functions.

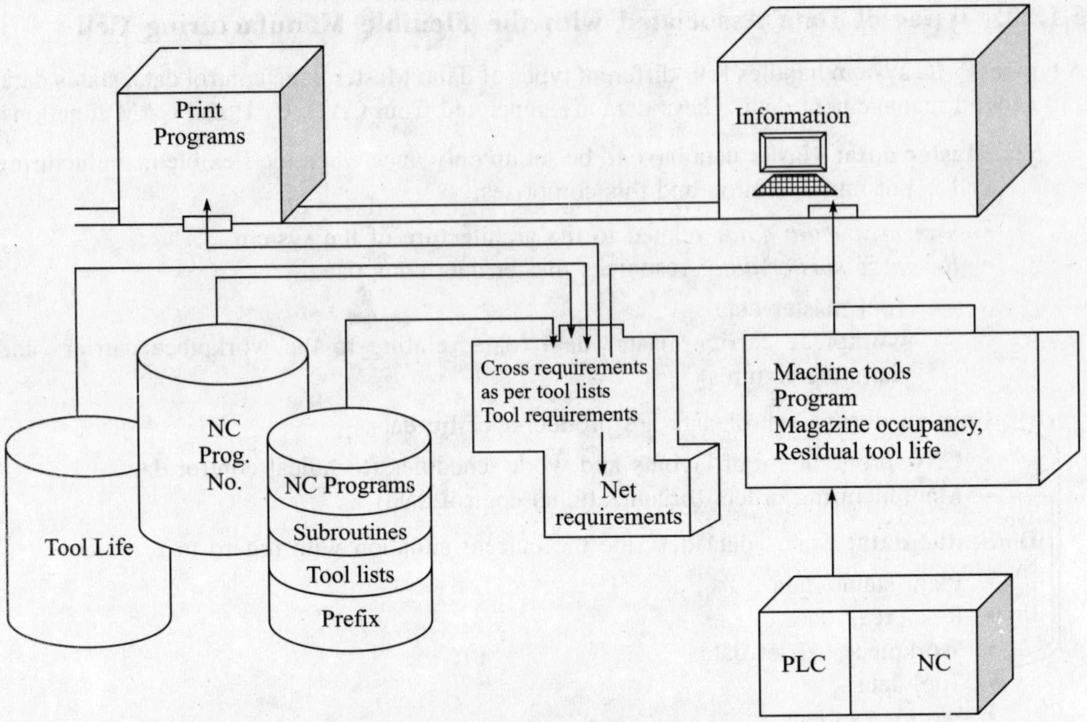

Figure 3.2 Data requirements.

By setting appropriate identifiers, it is possible to link manufacturing orders in such a way that parts are finished in the correct order for assembly. The fundamental planning method used is such that, with as few workpiece carriers as possible in the cell, the machines can be well utilized, and throughput times can be kept to a minimum.

Manufacturing orders with the highest priority are the first to be taken up for manufacture.

During the scheduling processes, a check is carried out to establish whether the necessary NC programs, workpiece carriers and master data are available (see Figure 3.2). If the operator informs the system that the missing resources can be procured in time, the order is then considered to be accepted and scheduled. This scheduling routine is repeated for as long as there is still free manufacturing capacity available. The availability check is repeated prior to the start of manufacture.

3.12.4 Tool Requirements

Tool requirements planning (Fig. 3.3) for scheduled manufacturing orders is carried out on two separate occasions:

- Immediately after job scheduling
- Prior to the start of manufacturing

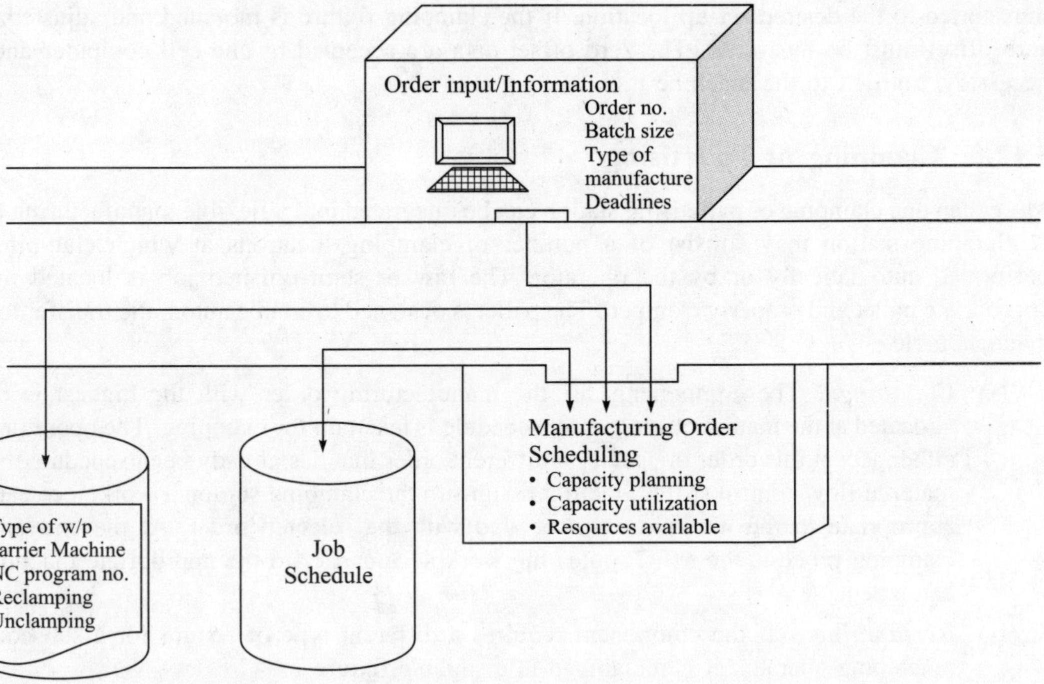

Figure 3.3 Tool data requirements.

(i) ***Preliminary planning phase:*** Tool identification numbers and all the intervention times for each tool application must be taken from the tool lists of the NC programs. The tool lives for all the tools used in the system are stored in the tool reference data management files. By linking all the relevant data, it is possible to ascertain the gross tool requirements; these requirements are notified to the operator (display, printout). The operator confirms that the tools can be obtained in good time.

(ii) ***Final planning phase:*** Before manufacturing starts, requirement analysis is performed, which takes into account the residual tool lives of the tools accommodated in the machine magazine. The net tool requirements list and the list of tools due for release in the machine magazine are printed in the form of tool loading and unloading.

A tool setting unit may be connected to the cell computer as an option. The setting unit is provided with the reference data for the tools to be set. The actual tool data or the tool correction data determined by the unit are transferred on-line to the cell computer. For changing the tools, the relevant data are available for the machine tool PLC.

3.12.5 Setting up Workplace Carrier

A workpiece carrier consists of a pallet and the workpiece-specific clamping fixtures. At the set-up location, the operator is informed of the set-up procedures to be performed for the particular workpiece. If there is no separate setting up location in the system itself, setting up takes place at the clamping location. Prior to setting up, the appropriate pallet is selected, blocked, and transported to the desired set-up location. If the clamping fixture is mounted and adjusted, the zero offset must be measured. The zero offset data are accepted by the cell computer and, if necessary, notified to the machine tool.

3.12.6 Clamping or Palletising

More than one clamping or palletising station can be integrated into a flexible manufacturing cell. A clamping station may consist of a number of clamping locations at which clamping is performed automatically or by the operator. The raw or semi-finished job is located in an appropriate pallet and properly clamped. The pallet is designed to enable automatic transfer to the machine table.

(i) ***Clamping:*** The components for the manufacturing order with the highest priority allocated at the manufacturing order schedule is taken up for clamping. The operator can either accept this order or specify a different order that has already been scheduled. The material flow control (MFC) facility assigns to the clamping station a workpiece carrier appropriate to the workpieces associated with the selected order. At the end of the clamping process, the MFC routes the workpiece carrier to the appropriate machining equipment.

(ii) ***Reclamping:*** If the component requires a different type of fixture for a subsequent machining operator, it is reclamped in a suitable fixture.

(iii) ***Unclamping:*** The MFC facility routes the finished workpiece to the clamping station where the pallet is fitted with a new workpiece appropriate to the manufacturing status. After all the clamping operations have been completed, the workpiece status data are updated. For reclamping and unclamping, "Remachining necessary", or "Reject" may be signalled as quality criteria with respect to the workpiece.

3.12.7 Material Flow Control

The material flow control (MFC) facility controls and supervises the transport of the workpiece carriers with the clamped workpiece in the flexible manufacturing cell. The workpiece carriers are transported by the transport system of the cell between a source station and a destination station. Source and destination stations include clamping locations, machine tools, washing measuring machines, etc.

The transport requests are specified by the station-specific programs, e.g. machine tool program, fixturing programme), to the MFC facility at the end of an operation and processed by the MFC facility according to the FIFO (First In First Out) principle. In addition to the stations, each cell normally has storage locations where the workpiece can be held if the destination station is occupied. If that happens, the transport job, together with the destination and the priority, is preserved. The source station identifier is replaced by the storage location number. Information on the source and destination is obtained by interpreting the workpiece-specific schedule, in which, among other things, the sequence of operations is defined. During transport, the status data of the transport system, source and destination stations, workpiece carrier and workpiece are updated by the MFC facility in real-time.

3.12.8 Machine Tool Program

In the FMC system, the machine tool program undertakes all the tasks arising at the interface to the machine tool. These tasks are online functions; so each of the machine tools integrated into the FMC system is assigned its own machine tool program.

Machine tool programs process

(i) NC functions

(a) Maintaining a list of the NC programs available in the CNC
(b) Loading a program by preparing the CNC for loading and activating program transfer via the NC program supply module
(c) Erasing the CNC memory

(ii) Tool data

(a) Reading the tool magazine and forwarding the contents to the tool requirement module
(b) Forwarding the tool offset data on loading the tool magazine alarms and status messages which lead to the cancellation of automatic processing of the machine tool

Alarms and status messages are logged by the basic software system. The relevant status data are updated.

3.12.9 Synchronization

Synchronization must be carried out in the event of a restart for a machine tool in the FMC.

(i) Resetting the PLC:
   (a) Interrogate alarms and statues.
   (b) Cancel NC programs.
   (c) Read tool magazine data.
(ii) Forwarding the information to the relevant program modules
(iii) Resetting the NC program supply module

3.12.10 CNC Program Supply

With the aid of this module (Fig 3.4), we can see the following CNC programs:

1. Programs transferred to the CNC of the cell
2. Programs retransferred from a CNC (updated, optimized programs)
3. Programs read from the floppy disk or via a host computer interface into the cell computer
4. Programs deleted from the NC program library

The CNC programs are held ready by a machine tool program at the interface to the CNC and transferred if the CNC signals that it is ready to receive. Retransmission of updated NC programs are initiated by means of operator inputs on the cell computer.

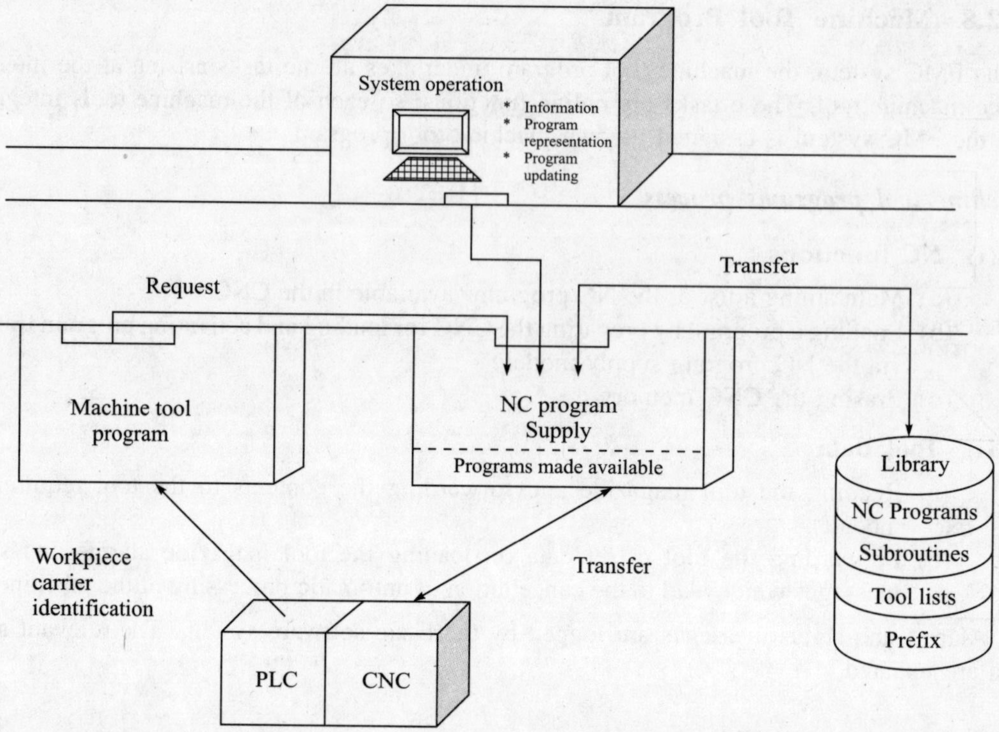

Figure 3.4 CNC program Management.

When CNC programs are transferred to the cell computer, they are read into an input file, provided with the necessary prefix and management data obtained by the operator interacting with the computer are transferred to the library. A particular CNC program number and references are given to subroutines and tool lists. The status of the CNC program indicates whether the program is blocked, released for production or it is an updated version by entering appropriate requests on the operator console lists, information relating to

- all blocked diagrams
- all programs with updated versions, etc.

3.12.11 Tool Flow Control

Figure 3.5 shows a schematic diagram of tool flow control. The tool flow control facility processes the bidirectional transfer of tools between the cell magazines. For each machine tool, the tool planning function supplies the tool flow control facility with lists containing the identification numbers of the tools required and the tools that are available.

The ruling factor is the time of requirement or the time of tool release for dismantling. Before a workpiece arrives at the machining location of the machine tool, a check is made to establish whether

(i) all the tools required are available in the machine tool magazine, or
(ii) the tools not in the machine tool magazine are available in the cell magazine.

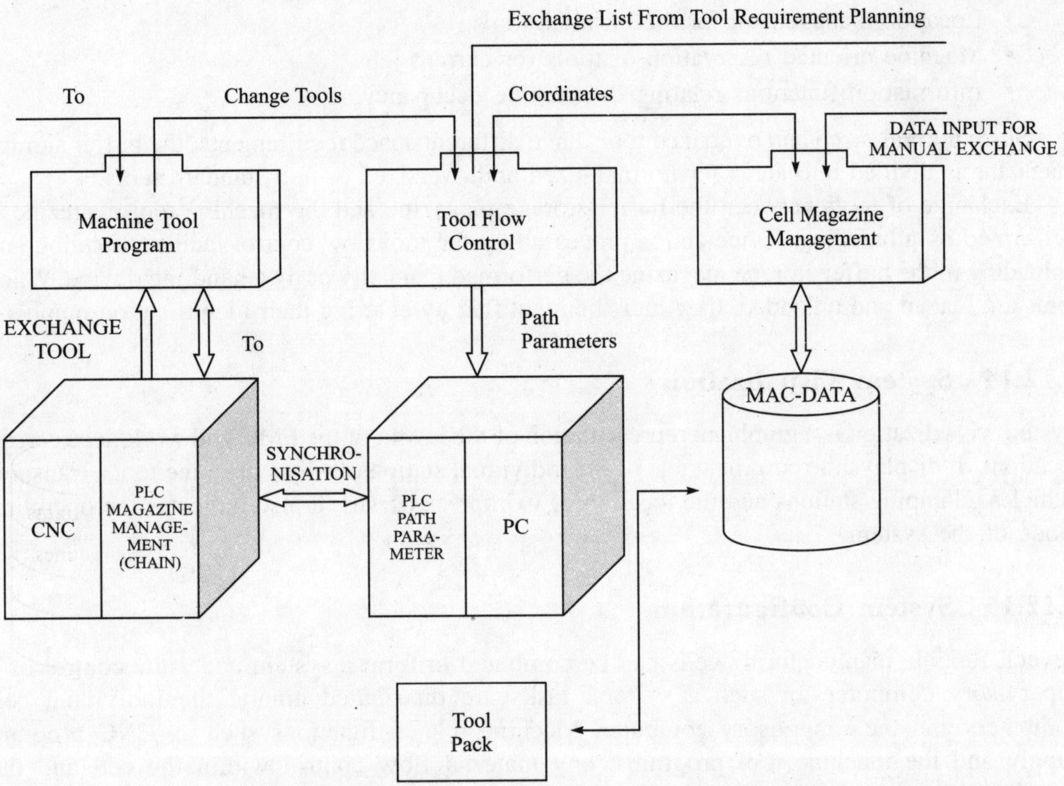

Figure 3.5 Schematic diagram of tool flow control.

If the necessary tools are not available, the workpiece is rejected. If the tools are in the cell magazine, a change of tools is initiated. The time of change of tools is notified to the relevant machine by the PLC as follows:

From flags set in the NC programs, the machine tool PLC knows if there is enough time available for a change of tools. If a flag is set, the tool flow control facility is informed.

3.12.12 Tool Changing

The coordinates of the tool location are passed to the PLC of the handling device where they are converted into movement instructions. At the same time as the handling device is executing the movement instructions, the PLC of the machine tool positions the magazine for tool changing. Once the new tool has been loaded, the tool that is no longer required can be exchanged and transferred to the cell magazine. A free location is specified by the cell store management facility. Physical tool exchange is accompanied by a magazine management update.

3.12.13 Cell Magazine Management

This program performs the following tasks in the buffer storage magazine of the FMC:

- Location management for the installed tools
- Machine oriented reservation of tools for current jobs
- Information functions relating to machine occupancy.

Since normal-sized and oversized tools have different space requirements, the buffer storage magazine is divided into areas for normal-sized and oversized or non-standard tools.

Exchange of tools between the buffer storage magazine and the machine tool magazine is performed by a handling device and is processed by the tool flow control facility. Loading and unloading in the buffer storage magazine are performed manually or by a handling device. When tools are loaded and unloaded, they must be identified by entering their identification numbers.

3.12.14 System Visualization

System visualization is a graphical representation of the layout of the FMC and workpiece carrier locations. It displays the current status of the individual stations such as machine tools, transport vehicles, clamping stations and the location of workpiece carves. It also indicates the operating mode of the system.

3.12.15 System Configuration

Several flexible manufacturing cells can be combined to form a system under the control of a supervisory computer. In such a system, tasks are distributed among the individual cell computers and the supervisory computer. Machine related functions such as CNC program supply and the machine tool programs, any material flow control within the cell and the associated data storage facilities remain at cell level. Execution of the higher functions of job scheduling, tool requirement management, material flow control and palletising, etc. are the responsibility of the computer that coordinates.

Shop data requirements people–Availability

(a) Workstations
(b) Idle
(c) Absent

Machines–Status

(a) Idle
(b) Set-up
(c) Production
(d) Delay
(e) Downtime

Parts/Raw materials location

(a) Raw Stores
(b) Transit
(c) On-machine

(d) Waiting
(e) Finished stores
(f) Assembly
(g) Missing

REVIEW QUESTIONS

1. Discuss the need for FMC.
2. What is FMC?
3. What are the benefits of FMS?
4. Discuss the relevance of FMS from the point of view of work centre utilization.
5. What are the major elements of FMS?
6. Discuss the importance of materials handling devices used in an FMS.
7. Describe the types of materials handling devices used in an FMS.
8. How is an FMS optimized?
9. What are the different types of data associated with FMS?
10. How are jobs scheduled in FMS?
11. Discuss the importance of palletising.
12. How is NC program control effected in an FMS?
13. Classify production systems based on volume-variety considerations.
14. What do you understand by FMS? What are the components of an FMS?
15. Flexibility is not the only essential ingredient for flexible manufacturing system. Explain.
16. All flexible manufacturing systems have common characteristics, irrespective of their types. Identify and explain these characteristics.
17. Discuss important differences among the following:
    (a) Transfer line
    (b) Stand-alone NC machine
    (c) Manufacturing cell
    (d) Special manufacturing system
    (e) Flexible manufacturing system
18. Describe the physical subsystems of an FMS.
19. Discuss the following control subsystems of an FMS:
    (a) Part control system
    (b) Machine-tool control system including inspection machines
    (c) Tool control and tool management system
    (d) Material handling and storage control system
    (e) Maintenance control system
    (f) Management control system
    (g) Interfacing of these subsystems with the central computer

20. Discuss various types of flexibility:

(a) Machine flexibility
(b) Routing flexibility
(c) Mix flexibility
(d) Mix-change flexibility
(e) Production flexibility
(f) Expansion flexibility

21. The data in Table P3.1 are available on the use of tools for various part types. Using the Hwang model and revised model due to Stecke and Kim, determine the batches of the parts selected. Discuss the differences between the results obtained. Assume that each tool requires one slot of a tool magazine.

22. What impact does tool magazine capacity have on FMS productivity? Discuss the following tooling policies:

(a) Bulk exchange policy
(b) Tool migration policy
(c) Resident tooling policy
(d) Tool-sharing policy

23. Layout considerations and flexible manufacturing systems are different from those in other manufacturing systems. Discuss the following types of layouts in the design of flexible manufacturing systems:

Table P3.1 Part-Tool Requirements

Part types	*P1 P2 P3 P4 P5 P6 P7 P8 P9*
Tools required	t1 t2 t3 t1, t2 t2, t3 t4, t5 t6, t7, t8, t9 t6, t8, t10, t11, t1, t2, t6, t5

(a) Linear single-row machine layout
(b) Double-row machine layout
(c) Circular machine layout
(d) Cluster machine layout
(e) Loop layout

24. Describe the optimal cycle time as well as robot sequences for a two-machine robotic cell with the following data:

Processing time of machine M1 = 20.00 min
Processing time of machine M2 = 10.00 min
Robot ripper pickup time = 0.20 min
Robot gripper release time = 0.20 min
Robot move time between the two machines = 0.30 min

25. Describe the optimal cycle time and corresponding robot sequence for a three-machine robotic cell. The following data are given:

Processing time of machine M1	= 10.00 min
Processing time of machine M2	= 07.00 min
Robot ripper pickup time	= 04.00 min
Robot gripper release time	= 0.20 min
Robot move time between the two machines	= 0.30 min

Chapter 4

SHOP FLOOR CONTROL

4.1 INTRODUCTION

Shop floor control deals with *managing the work-in-process*. This consists of the release of production orders to the factory controlling the progress of the orders through the various work-stations, and getting the current information of the status of the orders. This can be shown in the form of a factory information system (Figure 4.1). The input to the shop floor control system is the collection of production plans. These can be in the form of master schedule, manufacturing capacity planning, and MRP data.

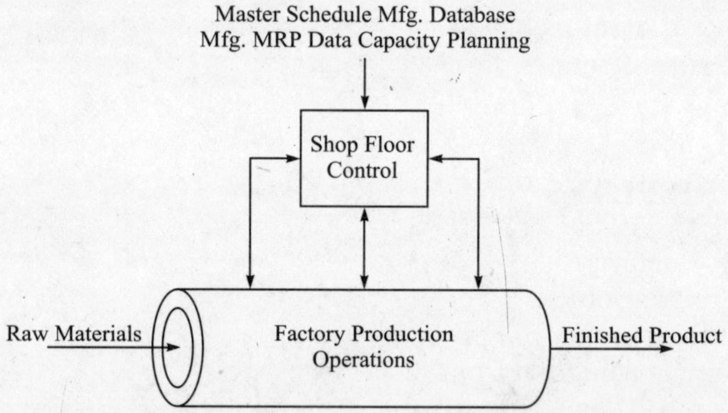

Figure 4.1 Factory information system.

The factory production operations are the processes to be controlled. A typical shop floor control system consists of three phases. In a computer integrated manufacturing system, these phases are managed by computer software. These three phases connected with the production management are shown in Figure 4.2. In today's implementation of shop floor control, these are executed by a combination of computers and human resources. The following sections describe the important activities connected with this task.

Benefits

The Shop Floor Control module provides the tracking mechanism that allows reporting of actual events from the production environment. By comparing planned (standard) events to actual events, variances can be identified. This variance reporting delineates inefficiencies and cost overruns in a timely manner, which allows corrective action to be taken immediately. The system provides management control over the material, capacity, costs, and productivity functions, which are critical to any manufacturing operation.

4.2 SHOP FLOOR CONTROL FEATURES

The Shop Floor Control module features comprehensive production management information on line. The information includes:

- Current status
- Material cost details
- Labour cost/time details
- Work centre cost, time and step details
- Miscellaneous cost details

The system maintains the materials ledger by lot (if desired) and provides for the reporting of partial or full completion of production, the reporting of work centre movement, the capture of work centre costs, the transfer of costs, the transfer of parts job-to-job, optional kitting of material to jobs or automated back flushing, the creation of pick lists, and the direct management of value-added costs.

The Shop Floor Control module is fully integrated with the General Ledger, Production Planning, Work Centres, Routings, Inventory Management, Purchasing, Payroll, and the Bill of Materials modules.

4.3 MAJOR DISPLAYS

- Job Status Summary
- Pegging for this Job
- Material Ledger
- Work Centre Loading and Utilization Percentages
- Labour Cost and Time Ledger
- Workcentre Cost Ledger
- Miscellaneous Cost Ledger
- Schedule and Routing Display
- Production Schedulers by Product Class

4.4 MAJOR REPORTS

- Production
- Traveller

- Pick List
- Detailed Job Cost Report
- Status Reports by Workcentre, Job, or Part
- Production Variance Reports (Detail or Summary)
- Pick List adjusted by Potency Factor

4.5 PHASES OF SFC

A typical shop floor control system consists of three phases. In a computer integrated manufacturing system, these phases are managed by computer software. These three phases connected with the production management are shown in Figure 4.2.

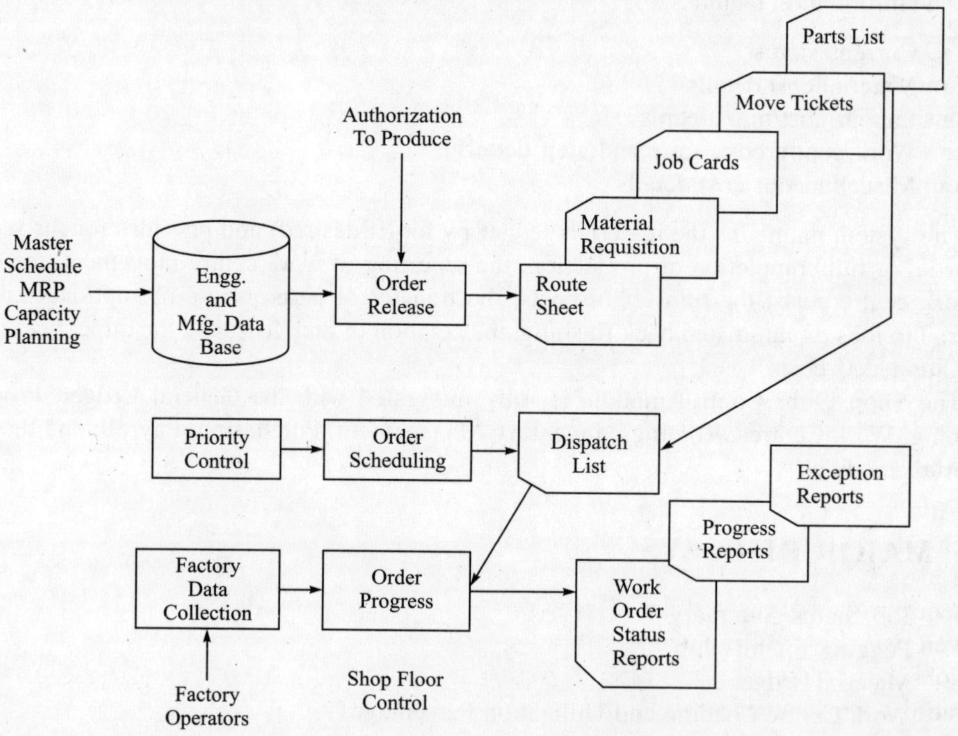

Figure 4.2 Phases in shop floor control.

4.5.1 Order Release

The order release in shop floor control provides the documentation needed to process a production order. The documents in the shop floor order may consist of the following documents:

1. Route sheet
2. Material requisition to draw necessary materials from the stores
3. Job cards or other means to report direct labour time given to the order

4. Instructions to materials handling personnel to transport parts between the work centres in the factory
5. Parts list for assembly, in the case of assembly operations.

In a typical factory which works on manual processing of data, these documents move with the production order and are used to track the progress through the shop. In a CIM factory, more automated methods are used to track the progress of the production orders.

The order release is connected with two inputs. Authorization proceeds through the various planning functions (MRP, capacity planning). These provide timing and scheduling information. The engineering and manufacturing database provides the product structure and process planning information needed to prepare the various documents that accompany the order through the shop.

4.5.2 Order Scheduling

This module assigns the production orders to various work centres, machine tools, welding stations, moulding machines, etc. in the plant. It follows directly from the order release module. Order scheduling executes the dispatch function in production planning and control. The order scheduling module prepares a dispatch list that indicates which production order should be accomplished at the various work centres. It provides the information on the relative priorities of the various jobs by showing the due dates for each job. By following the dispatch list in making work assignments and allocating resources to different jobs, the master schedule can be best achieved.

The order schedule module addresses two important activities in shop floor production control.

- Machine loading
- Job sequencing

Allocating the orders to the work centres is termed as **machine loading** or **shop loading**, which refers to the loading of all machines in the plant. In most cases, each work centre will have a queue of orders waiting to be processed. This queue problem can be solved by job sequencing which involves determining the order in which the jobs will be processed through a given work centre. To determine this sequence, priorities are given to jobs in the queue and the jobs are processed according to the priorities. Several queuing models are available in operations management to solve this problem.

Such control of priorities is an important input to the order scheduling module. Rules to establish the priorities are:

1. *Earliest due date:* These are given high priority.
2. *Shortest processing time:* Shorter processing time orders are given high priority.
3. *Least slack time:* Orders with least slack time are given high priority.

Fluctuations in market demand, equipment breakdown, cancellation of the order by customer, and defective raw material or delay in the receipt of materials affect the priority. The priority control plan reviews the relative priorities of the orders and adjusts the dispatch list accordingly.

4.5.3 Order Progress

The order progress module in the shop floor control system monitors the status of the various orders in the plant work-in-process (WIP) and other characteristics that indicate the progress and performance of production. The function of the order progress module is to provide the information that is useful in managing the factory, based on the data collected from the factory. The order progress report includes:

1. *Work order status reports:* These reports indicate the status of the production orders. Typical information in the report includes the current work centre where each order is located, processing hours remaining before completion of each order, whether the job is on-time or behind schedule, and the priority level.
2. *Progress report:* A progress report records the performance of the shop during the period of master schedule and reports the number of operations completed and not completed during the time period.
3. *Exception reports:* These reports bring out the deviations from the production schedule (e.g. overdue jobs).

The above reports are useful to production management in making the decisions about allocation of resources, authorization of the overtime hours, and other capacity issues, and in identifying areas of problems in the plant that adversely affect the implementation of the master production schedule

4.6 MANUFACTURING CONTROL

Manufacturing control provides information to efficiently manage flow of materials, effectively utilize human resource and equipment, coordinate internal activities with those of suppliers, and communicate with customers about market requirements. Another key use of CIM in the manufacturing enterprise is in **Manufacturing Control**. This includes all aspects of communication within the enterprise, which are necessary to control the manufacturing processes:

- Planning and supervizing activities at the upper (management) levels of the organization
- Process monitoring and control and the lowest (production) levels of the organization

Communication at and between these levels require different hardware, software and communication protocols to operate effectively. What they have in common is the requirement for open communication standards, generic hardware that is easily interconnected and scaled, user interaction and intervention, data collection, and system failure protection.

At the factory management level, communication typically consists of exchanging files and packets of information within relatively close proximity. Communication standards such as Manufacturing Automation Protocol (MAP) and Technical Office Protocol (TOP), and token protocols (such as CSMA/CD), to name the most common, operate over a variety of Local Area Network (LAN) configurations. While this does not need to be real-time data exchange, it does operate under heavy and continuous traffic and must be fast and reliable.

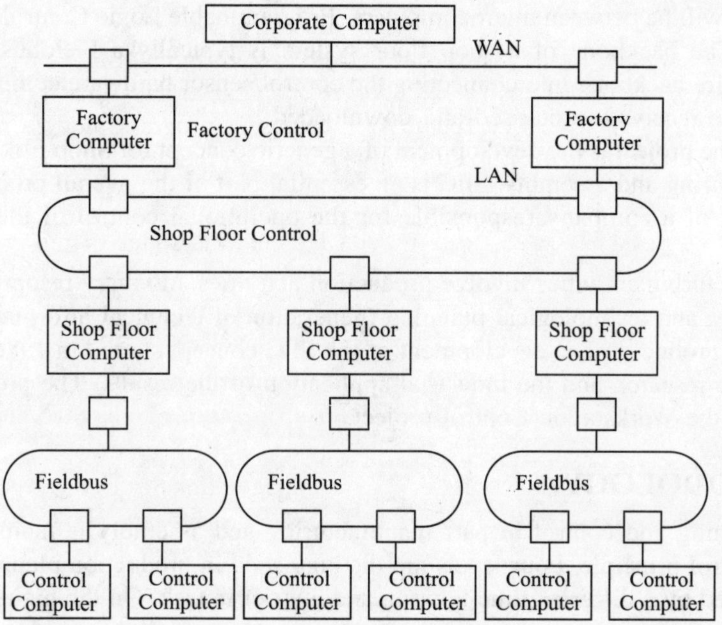

Figure 4.3 The structure of a fully integrated computerized manufacturing enterprise.

At the lowest level of manufacturing control is the shop floor. Shop floor control requires real-time communication that is fast, reliable, and can handle a wide variety of hardware and software. There are a variety of proprietary communications standards for shop floor, but the common ones including Profibus, InterBUS-S, and Manufacturing Message Service (MMS).

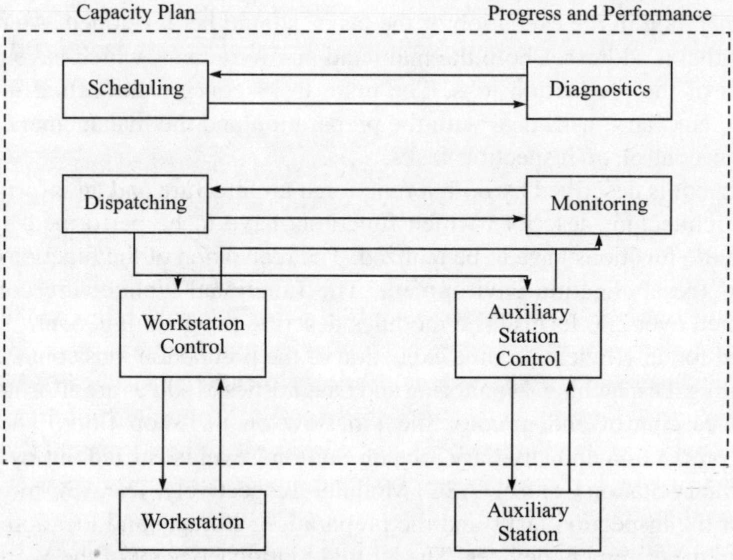

Figure 4.4 Functional architecture for shop floor control.

Communication will be between microcomputers, Programmable Logic Controllers (PLCs), and smart sensors. The backbone of a shop floor system is typically a Fieldbus, a closed-loop, twisted-paired wire backbone interconnecting the control/sensor hardware components allowing information to be received, processed and downloaded.

The aim of the project is the development of a generic concept for Shop Floor Control (SFC) in part manufacturing and assembly. SFC is an essential part of the overall production planning and control task of a company responsible for the operational control of the manufacturing related activities.

The manufacturing activities involve production activities, auxiliary resource management, quality assurance, and technological planning (generation of technical information, required to manufacture the products). The development of the SFC concept started in 1986 and is directed to both scientific research and the industrial application of the results. The project has a large interaction with the Workstation Control project.

4.7 METHODOLOGY

Production planning and control in part manufacturing and assembly is mainly based on the hierarchical control principle. Four levels are distinguished in production planning and control, each characterized by a different time horizon and a specific task. On the highest control level, the Company Level, the client order entrance function and master production planning are carried out. The next lowest level, Factory Level, deals, amongst others, with capacity planning on the basis of the available manufacturing resources. The following two levels are Cell and Station Levels, which concern the operational planning and execution of all manufacturing activities. SFC deals with Cell and Station Levels. Since these are the lowest levels in the hierarchy, the importance of SFC in the overall performance of production control is obvious.

The SFC concept is used to determine both how the manufacturing activities (which are performed by human beings and equipment) have to be organized and how SFC systems, which support the organization in the execution of the tasks, should be developed. An essential feature of the concept is that it addresses both the main and auxiliary tasks which have to be performed for the execution of the production jobs. The main tasks concern the actual manufacturing of the products. The auxiliary tasks deal with the preparation and the management of the auxiliary resources and the control of inspection tasks.

The SFC concept is described by both a functional architecture and an information structure. The functional architecture describes which functions have to be performed by SFC, without specifying how these functions have to be realized. The realization of the functions depends on the characteristics of the application environment. The functional architecture contains six basic modules distributed over 200 levels. The modules describe the planning, control and monitoring functions required for an efficient, in time execution of the production jobs contained in a capacity plan. The Scheduling, Dispatching, Monitoring and Diagnostics Modules are allocated on Cell Level.

These modules control and monitor the job flow on the shop floor. The control of the execution of the production and auxiliary jobs on Station Level is carried out by the Workstation Control and Auxiliary Station Control (ASC) Modules, respectively. The ASC module can be used for the control of the inspection tasks and the preparation, storage, and location management of tools, fixtures and measurement devices. The Virtual Station Concept of the Workstation Control and Auxiliary Station Control Modules enables the development of SFC systems which can be applied independently of the level of automation of the manufacturing equipment.

4.8 APPLICATION

A prototype SFC system, named FACT, has been implemented as an example of how such a system can be realized. The prototype has been installed in a machine shop for the evaluation of the functioning of the SFC concept, i.e. the control and monitoring of manufacturing activities based on a work plan. The evaluation has shown that the SFC system fulfils the requirements and criteria which are required for SFC in small batch part manufacturing.

The SFC system provides an information infrastructure on the shop floor which enables the in-time delivery of all the logistic and technical information required for the execution of the different manufacturing tasks. The SFC system provides both the adaptability to interface equipment controllers of different makes, and offers transparency to the users as regards the work that has to be carried out and the coordination between the different manufacturing activities.

The installation of the SFC system in the machine shop has led to a significant reduction of the throughput times and the work-in-progress, without a decrease in the utilization of the machine tools. In addition, it has been shown that the due date reliability has been improved and a significant reduction of the manufacturing costs can be achieved.

It has been proven that the functional architecture provides a structure which enables both an effective control of the manufacturing activities and an efficient use of the main and auxiliary resources. The real-world application of the SFC system has confirmed that SFC plays a crucial role in production planning and control in small batch part manufacturing. SFC enables companies to stay competitive in a market which has high demands as regards due date reliability, delivery times and prices.

Further developments are directed towards the extension of the functionality of the SFC concept, the integration with higher level production planning and control functions, the interaction with technological planning, and the application of the concept to other manufacturing types.

4.8.1 Simplified Manufacturing Planning and Control System

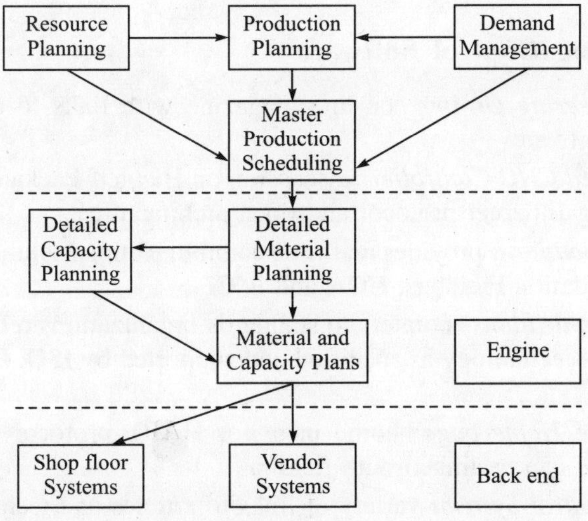

Figure 4.5 Manufacturing planning and control system.

4.8.2 Inputs Manufacturing Planning and Control System

I Product description

- Engineering drawings & specifications
- Bill of Materials
  - Components used to make product
  - Sub-assemblies at various stages of manufacture

II Process specifications

- Operations to make product
- Sequence of those operations
- Equipment & accessories required
- Standard labour and machine time required for each operation

III Manufacturing operations

- Time normally exceeds standards

IV Available facilities

V Quantities required

4.8.3 Manufacturing Planning Software

- CIRP Annals: articles on planning and planning software
- PART: Computer Aided Process Planning software created by Tecnomatix Machining Automation
- CIMx Home page: CAPP software
- Intergraph/CAPP: CAPP software from Sun Microsystems
- EASE CAPP Module: CAPP software from EASE Inc.
- Production Planning/CAP/Fine Planning: SAP's System R/3 software
- AMICS Shop Floor Control: AMICS manufacturing planning software

4.8.4 Manufacturing Control Software

- *Control Engineering On-line:* on-line magazine with links to manufacturing control articles and software
- *Hewlett-Packard CNC Controller:* discussion on Hewlett-Packard's latest "Open" CNC controller using Ethernet protocol and open architecture
- *Sisco Communications:* provides real-time communications solutions for MMS/Ethernet, Profibus, Foundation Fieldbus, UCA and ICCP
- *Standards Organizations:* pointers to standards organizations relating to Open Systems Interconnection technology as specified and supported by ISO, IEEE, MMS, and many others
- *HART Protocol Home page:* home page for HART protocol used for smart digital instruments and can be linked with fieldbus
- *Shop Floor Control System:* variety of links to articles and technical information

4.8.5 Cost of Inadequate MPC

- Poor customer service
- Excessive inventories
- Inappropriate equipment or labour utilization
- High rates of part obsolescence
- Large numbers of expediters dedicated to "fighting fires"

4.8.6 MPC System Activities

- Create the overall manufacturing plan.
- Perform detailed planning for material and capacity requirements.
- Execute those plans.
- MPC systems should support strategy and tactics.
- Different manufacturing processes often dictate the need for different MPC system designs.
- Effective MPC system can contribute to competitive performance by lowering costs and improving market responsiveness.
- Framework for MPC is general, but all three phases must be performed by manufacturing.
- MPC system's design will vary, depending on distinctive needs of firm and the state of its MPC system.

4.9 SHOP FLOOR DATA COLLECTION

A number of data collection techniques are available to collect data from the shop floor. These techniques require the employees to gather the data and later the data is compiled on a fully automated system that requires no human participation. These methods are identified by the term **DP Floor Data Collection Systems**.

These data collection systems consist of various paper documents, terminals and automated devices located through the plant. The shop floor data collection system serves as an input to the order progress module in shop floor. Examples of the data collection in shop floor are:

1. Supplying data to the order progress module in the shop floor control system.
2. Providing up to date information to the production supervisors and production control personnel.
3. Enabling the management to monitor implementation of master schedule.

To carry out these activities, the factory data collection system inputs the data to the computer system in the plant.

4.10 TYPES OF DATA COLLECTION SYSTEMS

The shop floor data collection systems can be classified into two groups:

- On-line data collection systems,
- Off-line data collection systems.

4.10.1 On-line Data Collection Systems

In an on-line system, the data are directly entered to the computer and are available to the order progress module. The advantage lies in the fact that the data file representing the status of the shop is always at the current state. As and when the changes in the order progress module are reported, they can be fed to computer and in turn to the status file. In this way, the production personnel are provided with most up to date information.

4.10.2 Off-line (batch) Data Collection Systems

In this the data are collected temporarily in a storage device or a stand-alone computer system to be entered and processed by plant computer in a batch mode. In this mode, there is delay in the entry and processing of the data. The delay may vary depending upon the situation. So this system cannot provide real-time information of shop floor status. The advantage of this system is that it is easier to install and implement.

4.11 DATA INPUT TECHNIQUES

As stated earlier, the data collection techniques include manual procedures and computer terminals located on the shop floor. The manual data collection methods require the production workers to fill out paper forms indicating the order progress data. These forms are compiled using a combination of clerical and computerization methods.

The manual data collection methods rely on the cooperation and accuracy of the employees to record a data properly on a paper. Errors may creep in this type of methods. The common form of errors that can be checked and rectified are wrong dates, incorrect numbers and incorrect operation numbers. These can be detected and corrected. There are, however, other errors which are difficult to identify.

Another problem is that there may be a delay in submitting the order progress for compiling. The reason is that there will be always a time between the time of occurrence of events and the recording of events.

These problems necessitate the location of the data collection equipment in the factory itself. The various input techniques include manual input by push-button pads or keyboards. Error checking routines can be incorporated to detect the syntax errors in the input. The data entry methods also include more automated technologies, such as bar code reader, and magnetic card readers.

An important type of equipment used in shop floor data entry is keyboard based system. There are various types of such systems. They are discussed in the following sections.

4.11.1 Centralized Terminal

A single terminal is located centrally in the shop floor. This requires the employees to go to the terminal and input the data. So employee's time will be wasted and in a big shop, this becomes inconvenient.

4.11.2 Satellite Terminals

These are multiple data collection centres located throughout the shop floor. In this arrangement a balance is to be struck between the minimization of the investment cost and maximization of the convenience of the employees in the plant.

4.11.3 Work Centre Terminals

The most convenient arrangement to the employees is to have a data collection terminal at each work centre. This reduces the time to go to me central terminal. This is possible when the amount of data to be collected is very large.

4.12 AUTOMATIC DATA COLLECTION SYSTEM

The recent trend in industry towards use of more automation necessitates automated data collection methods. Although the term automation is used, in many techniques the human participation is unavoidable.

The advantages of the automatic data collection methods are:

- The accuracy of data collected increases.
- The time required by the workers to make the data entry can be reduced.

The basic elements in a data collection systems are:

- Machine readable media
- Terminal configuration
- Software for data collection.

Machine readable media

- Typical machine readable media include
- Bar Code Technology
- Optical Character Recognition (OCR)
- Magnetic Ink Character
- Voice Recognition (VR)
- Magnetic Strip Technology
- Smart Cards

4.12.1 Bar Code Technology

Bar code technology is primarily an automatic identification technique. The data is simply reduced to a printed form, which consists of a symbol made of successive line segments. A bar code reader is used to illuminate the bar code symbol and examine successive segments of the symbol. The detected area may be a highly reflected area (space) or a non-reflective (bar) area. As the reader moves the reader over the bar code symbol, due to reflectivity and non-reflectivity, alternate transitions from light to dark and dark to light occur. These are detected and the time it will take will be converted to a digital representations of ones and zeros of the bar code message. Most commonly used bar codes are fixed beam reader and moving beam reader which are now described.

4.12.2 Fixed Beam Reader

Bar code readers are either fixture mounted or hand-held. The simplest form of bar code reader is a light pen. The tip of the light pen is moved in contact with the symbol and moves the tip from leading zone through trailing quiet in a smooth sweeping motion. The fixed beam light pen nearly or actually touches the symbol.

In fixture mounted reader the beam reader is fixed and the symbol moves. The reader is mounted on a conveyor or a transport system, observing a symbol while it passes through a reader beam. The fixed beam reader reads the symbol only once. In hand-held fixed beam reader, the symbol can be rescanned easily. In the fixture mounted bar code reader, some mechanism which moves the symbol towards the symbol is necessary; in other words, an intervention by an operator is needed.

In fixed beam reader, the contact with the symbol may erase the symbol and so it is less readable in subsequent attempts to read the symbol. Contact scanning of the symbol requires a smooth surface. So it is not suitable to read all the surfaces on which the symbols are printed.

Light pens interpret a narrow section of the symbol printed on a surface. This may cause the distortion of the image. Fixed beam reader takes more time to read the symbol. The speed of fixed beam readers is a function of conveyor speed and height of the bar code symbol. Faster scanners are required for shorter symbols and slower scanners are required for larger symbols.

4.12.3 Moving Beam Reader

Moving beam readers minimize limitations of the fixed beam readers, i.e. the intervention problem and contact with the symbol. The moving beam reader, as the name indicates, scans the symbol by a line of light emitted from the reader. This is actually a spot which moves at faster rates appearing as a straight line. A moving beam reader takes less time to scan the symbol depending on the type of equipment.

The light emitted from the moving beam reader can be drawn through the symbol in both the configurations, i.e. hand-held and conveyor configurations. Multiple viewing of the symbol provides a quick and correct transformation to convert the image to computer data.

REVIEW QUESTIONS

1. Discuss the need and importance of shop floor data collection systems. What are their functions?
2. List the major types of data collection systems.
3. What are the devices used for shop floor data collection?
4. How does a bar code reader work?
5. Define SFC.

6. Enumerate the benefits of SFC.
7. Name any three control features of SFC.
8. List the major display involved in SFC.
9. What are the phases of SFC?
10. Discuss the importance of order scheduling.
11. What do you mean by manufacturing control?
12. Compare MMS and SMS.
13. Briefly explain the methodology of SFC cell.
14. What are the types of data collections?
15. List the salient features of bar code technology.
16. Discuss the advantages of a moving beam reader.
17. Enumerate the to advantages of an integrated manufacturing planning and control system.
18. Describe the following elements of a manufacturing planning and control system.
    (a) Order release
    (b) Capacity planning

Chapter 5

PRODUCTION PLANNING AND CONTROL

Operations Management consists of all the activities, managers engage in to produce goods and services. A large number of activities are involved in operations management. First, an organization may undertake product development. Next, production must be planned. Finally, managers must concern themselves with the control of operations to ensure that the organization's goals are achieved.

5.1 OPERATION MANAGEMENT

1. It is also possible to define operations management as all those activities that managers engage in to produce goods and services. To produce a product or service successfully, a business must perform a number of specific activities. We discuss each of these activities in this chapter.
2. An operations manager is a person who creates and manages systems that convert resources and raw materials into goods and services.

5.2 THE CONVERSION PROCESS

A. Simply put, production is the **conversion** of *resources into goods and services.*

1. The purpose of this transformation of resources into goods and services is to provide utility to customers.
2. *Utility* is the ability of a good or service to satisfy a human need.
3. Although utility can be classified into four types—form, place, time, and possession—operations management focuses primarily on form utility.
4. *Form utility* is created by converting raw materials, people, finances, and information into finished products.

B. The **nature of the conversion**. The conversion of resources into products and services can be described in at least three ways:

1. *Focus*. The focus of a conversion process refers to the resource or resources that comprise the major input.
2. *Magnitude of change*. The magnitude of a conversion process is the degree to which the resources are physically changed.
3. *Number of production processes*. A single firm may employ one production process or many processes. In general, larger firms that make a variety of products use multiple production processes.

C. The **increasing importance of services**. The application of operations management to the production of services has coincided with a dramatic growth in the number and diversity of service organizations.

1. In 1900, only 28 per cent of American workers were employed in service organizations.
2. By any yardstick, service firms have become a dominant part of our economy.
3. A *service economy* is one in which more effort is devoted to the production of services than to the production of goods.
4. The production of services is uniquely different from the production of manufactured goods in four ways:
   (a) Services are consumed immediately and, unlike manufactured goods, cannot be stored.
   (b) Services are provided when and where the customer desires the service.
   (c) Services are usually labour intensive because the human resource is often the most important resource used in the production of services.
   (d) Services are intangible; therefore, it is more difficult to evaluate customer satisfaction.
5. Although it is often more difficult to measure customer satisfaction, today's successful service firms work hard at providing the services that customers want.

5.3 WHERE DO NEW PRODUCTS AND SERVICES COME FROM?

1. ***Research and development (R&D):*** For our purposes, *research and development (R&D)* is a set of activities intended to identify new ideas that have the potential to result in goods and services. There are three general types of R&D activities:
   (a) *Basic research* consists of activities aimed at obtaining new knowledge.
   (b) *Applied research* consists of activities geared to discovering new knowledge that has some potential use.
   (c) *Development and implementation* are research activities undertaken to put new or existing knowledge to use in producing goods and services.
2. ***Product extension and refinement:*** When a new product is first marketed, its sales start from zero base, but slowly they increase. If the product is successful, annual sales

increase more and more rapidly until they reach some peak. Then, as time passes, annual sales begin to decrease and they continue to decline until it is no longer profitable to market the product.

1. If a firm sells only one product, when that product reaches the end of its life cycle, the firm, too, will die. To stay in business, the firm must at least find ways to extend or refine the want—satisfying capability of its product.
2. Each refinement or extension results in an essentially "new" product whose sales make up for the declining sales of a product that was introduced earlier.

5.4 PLANNING FOR PRODUCTION

Only a few of the many ideas for new products, refinements, and extensions ever reach the production stage. For those ideas that do, the next step is planning for production. This step consists of design planning, facilities planning and site selection, and operational planning.

5.4.1 Design Planning

Design planning is the development of a plan for converting a product idea into an actual product. The major decisions involved in design planning are described as follows.

1. *Product line:* A *product line* is a group of similar products that differ only in relatively minor characteristics. Each distinct product within the product line must be designed. *Product design* is the process of creating a set of specifications from which the product can be produced.
2. *Required capacity: Capacity* is the amount of products or services that an organization can produce in a given period of time.
   - (a) Operations managers, together with marketing managers, determine the required capacity.
   - (b) If the facility has too much capacity, valuable resources will lie idle.
   - (c) If the facility has insufficient capacity, capacity will have to be added later, though doing so is much more expensive.
   - (d) The capacity of a service business is the number of customers it can serve at one time.
   - (e) Use of technology. Operations personnel must determine the degree to which automation should be used in place of human labour.
   - (f) A *labour-intensive technology* is a process in which people do most of the work.
   - (g) A *capital-intensive technology* is a process in which machines and equipment do most of the work.

5.4.2 Facilities Planning and Site Selection

Initially, managers must decide whether they will build a new plant or refurbish an existing factory. Generally, a business will choose to produce a new product in an existing factory as long as the existing factory has enough capacity to handle customer demand for both the new product and established products.

The cost of refurbishing an existing factory is less than the cost of building a new one. After exploring the capacity of existing factories, management may decide to build a new production facility. In determining where to locate production facilities, management must consider a number of variables, including the following:

1. Geographic locations of suppliers
2. Locations of major customers
3. Transportation costs to deliver finished products
4. The cost of both land and construction
5. Local and state taxes, environmental regulations and zoning laws
6. The amount of financial support offered by local and state governments
7. Special requirements, such as great amounts of energy or water
8. Quality of life for employees and management
9. Availability and cost of skilled and unskilled labour

Human Resources

The firm must recruit employees with the appropriate skills, develop training programs, or do both. It may even be necessary to transfer skilled workers to the new location or to train local workers.

Plant Layout

Plant layout is the arrangement of machinery, equipment, and personnel within the facility. Three general types of layout are used.

1. The *process layout* is used when different operations are required for creating small batches of different products.
2. The *product layout* is used when all products undergo the same operations in the same sequence, such as in an assembly line.
3. A *fixed-position layout* is used when a very large product is produced.

5.4.3 Operational Planning

Operational planning is the development of plans for using production facilities and resources. The objective of operational planning is to decide on a level of output for the facility.

Step 1: Selecting a planning horizon. A *planning horizon* is the period during which a plan will be in effect.

A common planning horizon for operational plans is one year.

- A planning horizon of one year is generally long enough to average out seasonal increases and decreases in sales.
- A planning horizon of one year is also short enough for planners to adjust production to accommodate long-range sales trends.
- Firms that operate in a rapidly changing business environment with many competitors may find it best to select a shorter planning horizon.

Step 2: Estimating market demand. The *market demand* for a product is the quantity that customers will demand at the going price.

Step 3: Comparing market demand with capacity. The third step in operational planning is to compare the projected market demand with the facility's capacity to satisfy that demand. One of three outcomes may result.

(a) Demand may exceed capacity.
(b) Capacity may exceed demand.
(c) **Capacity and demand may be equal.** If market demand and capacity are equal, the facility should be operated at full capacity. If they are not equal, adjustments may be necessary.

Step 4: Adjusting products or services to meet demand. When market demand exceeds capacity, several options are available to the firm.

(a) Output may be increased by operating the facility overtime with existing personnel or by starting a second or third work shift.
(b) Another response is to subcontract a portion of the work to other manufacturers.
(c) If the excess demand is likely to be permanent, the firm may expand the facility.
(d) When capacity exceeds market demand, workers may be laid off and part of the facility may shut down.
(e) The facility may be operated on a shorter-than-normal workweek for as long as the excess capacity exists.
(f) To adjust to a permanently decreased demand, management may shift the excess capacity to the production of other goods or services.
(g) The most radical adjustment is to eliminate the excess capacity by selling unused facilities.

5.5 OPERATIONS CONTROL

5.5.1 Purchasing

Purchasing consists of all the activities involved in obtaining required materials, supplies, and parts from other firms. The objective of purchasing is to ensure that required materials are available when they are needed, in the proper amounts, and at minimum cost.

Purchasing personnel should constantly be on the lookout for new or back-up suppliers, even when their needs are being met by their present suppliers.

The choice of suppliers should depend on careful analysis of price, quality, reliability, credit terms, and shipping costs.

5.5.2 Inventory Control

An inventory is a stock of goods and materials. Basically, there are three kinds of inventories to be managed.

1. The *raw materials inventory* consists of materials that will become part of the product during the production process.
2. The *work-in-process inventory* consists of partially completed products.
3. The *finished goods inventory* consists of completed goods awaiting shipment to customers.

Inventory control is the process of managing inventories to minimize inventory costs, including both holding costs and potential stock-out costs.

One of the most sophisticated methods of inventory control used today is materials requirements planning. *Materials Requirements Planning* (*MRP*) is a computerized system that integrates production planning and inventory control. A manager using an MRP system can arrange both order and delivery schedules so that materials, parts, and supplies arrive when they are needed.

An extension of materials requirements planning is known as Manufacturing Resource Planning, often called MRP II. The primary difference between the two systems is that materials requirements planning involves just production and inventory personnel, whereas the manufacturing resource planning system involves the entire organization.

Enterprise Resource Planning (*ERP*), is a second extension of MRP. This sophisticated software monitors not only inventory and production, but quality, customer satisfaction, and even variables at a supplier's location. Its use is not limited to the activities at one firm like MRP and MRP II. It can be used to monitor activities at more than one firm.

A *just-in-time inventory system* is designed to ensure that materials or supplies arrive at a facility just when they are needed so that storage is minimized.

5.6 SCHEDULING

Scheduling is the process of ensuring that materials and other resources are at the right place at the right time. These "materials and resources" may be raw materials, sub-assemblies, work in process, or finished goods. Both place and time are important to scheduling. The routing of materials is the sequence of workstations that the materials will follow. The timing function specifies when the materials will arrive at each station and how long they will remain there.

Scheduling through Gantt charts

A Gantt chart, developed by Henry L. Gantt, is a graphic scheduling device that displays the tasks to be performed on the vertical axis and the time required for each task on the horizontal axis.

Scheduling control through PERT

The Program Evaluation and Review Technique (PERT) is a technique for scheduling a process or project and maintaining control of the schedule.

1. The first step is to identify all the major activities involved in the project. The completion of each activity is called an event.
2. Then, the events are arranged in a sequence.

3. Thereafter arrows, representing activities, are used to connect events that must occur in sequence.
4. The sequence of production activities that takes the longest time from start to finish is called the *critical path*.

5.7 QUALITY CONTROL

Quality control is the process of ensuring that goods and services are produced in accordance with design specifications for the products. **Statistical Process Control (SPC)** is a system that plots data on control charts and graphs to see if the production process is operating as it should and pinpoint problem areas. **Statistical Quality Control (SQC)** is a set of specific statistical techniques used to sample both work in progress and finished products.

Increased effort is also being devoted to *inspection*–the examination of the quality of work in process. Inspections are performed at various times during production.

Improving quality through employee participation. Historically, quality and productivity were viewed as conflicting; one was increased at the other's expense. Over the years, more and more managers have realized that quality is an essential ingredient of the good or service being provided. Viewed in this light, quality becomes an overall approach to doing business and is the concern of all members of the organization.

Because of increased global competition, American manufacturers have set a goal that calls for better quality in their products. Total Quality Management (TQM) coordinates the efforts directed at improving customer satisfaction, increasing employee participation, strengthening supplier partnerships, and facilitating an organizational atmosphere of continuous quality improvement.

Another strategy employs the use of a *quality circle*, a group of employees who meet on company time to solve problems of product quality.

World quality standards

ISO 9000 and ISO 14000. As the number of companies competing in the world marketplace has increased, so has the seriousness of product quality. To deal with it, the International Organization for Standardization (a nonprofit organization in Geneva, Switzerland, with a membership of 95 countries) brought together a panel of quality experts to define what methods a company must use to produce a quality product.

This certification, issued by independent auditors, serves as evidence that a company meets the standards for quality control procedures in manufacturing design, production processes, product testing, training of employees, record keeping, and correction of defects. Although certification is not a legal requirement to do business globally, the organization's 95 member countries have approved the ISO standards. In fact, ISO 9000 is so prevalent in the European Community that many customers refuse to do business with non certified companies.

The International Organization for Standardization has also developed ISO 14000, a series of international standards for incorporating environmental concerns into operations and product standards.

REVIEW QUESTIONS

1. What is operations management?
2. What is production?
3. Write short notes on planning for production.
4. Define operation planning.
5. What do you mean by market demand?
6. What is the important role of purchasing?
7. Define MRP.
8. Define ERP.
9. Define the term "Just-in-Time inventory system".
10. Discuss the main function of scheduling.
11. Compare PERT and GANT charts.
12. Define critical path.
13. What is statistical process control?
14. What do you mean by statistical quality control?
15. What is the role of R&D in production?

Chapter 6

COMPUTER AIDED PROCESS PLANNING

6.1 AUTOMATED PROCESS PLANNING

"Process planning can be defined as an act of preparing processing documentation for the manufacturing of a piece, part or an assembly". Process plan can be prepared in any one of the two ways depending upon the work environment as:

- Rough process plan
- Detailed process plan.

If the process plan is carried out with the help of the computer, then we call it "Computer Aided Process Planning".

Components of process planning

1. Selection of machining operations
2. Sequencing of machining operations
3. Selection of cutting tools
4. Determining the set-up requirements
5. Calculation of cutting parameters
6. Tool path planning and generation of NC/CNC programs
7. Design of Jigs/Fixtures.

Requirements for process planner

1. Must be able to analyze and understand part requirements.
2. Have extensive knowledge of machine tools, cutting tools and their capabilities.
3. Understand the interactions between the part, manufacturing, quality, and cost.

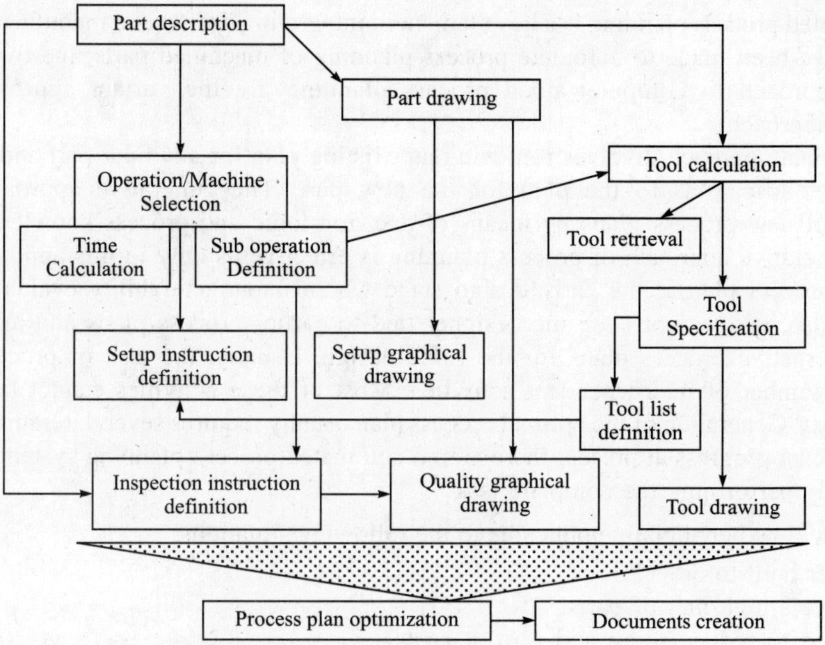

Figure 6.1 Process planning activities.

Some of the examples of process planning based on the work environment are as follows:

1. In tool-room type manufacturing "make part as per drawing" is sufficient.
2. In metal-forming type operations the process planning requirements are embedded directly into the die; process planning is fairly trivial.
3. Job-shop type manufacturing requires most detailed process planning.

Design of tools, jigs, fixtures and manufacturing sequence are dictated directly by the process plan.

Traditionally, process planning is done by

(a) Experienced-based and performed manually;
(b) variability in planner's judgment and experience can lead to differences in what constitutes best quality; and
(c) problem facing modern industry is the lack of skilled labour force today to produce machined parts as was done in the past.

6.2 COMPUTER-AIDED PROCESS PLANNING

Since the first Computer-Aided Process Planning (CAPP) system was developed in the early 1960s, there has been a proliferation of development that leads to various CAPP systems, from research and industrial prototypes to company and commercial packages, being tailored to different planning problems and offering a host of different solutions.

Automated process planning is a key element in integrating design and manufacturing. Many attempts have been made to automate process planning of machined parts; the two traditional types of approach to computer-aided process planning are the variant approach and the generative approach.

The variant approach involves retrieving an existing plan for a similar part and making the necessary modifications to the plan for the new part. The generative approach involves generation of new process plans by means of decision logic and process knowledge.

The generative approach of process planning is effective as only in this approach one can develop a process plan at the early design stage where manufacturability evaluation is most effective. Also, it does not keep the designer tied to earlier process plans and allows him to develop alternative process plans for the same design. Usually, the task of process planning involves a number of interdependent activities. Most of these activities cannot be performed independently. Generation of the optimal process plan usually requires several iterations, and even after significant progress at present there are no automated process planning systems capable of automatically performing the complete task.

CAPP will be beneficially applicable to the following situations:

1. Ever built to order?
2. Have a high mix of parts.
3. Need to reduce scrap and rework costs.
4. Ever done the wrong revision?
5. Need to improve productivity in process planning.
6. Have distorted capacity requirements.
7. Introduce a new part when a similar one exists.
8. Have inconsistent manufacturing instructions.
9. Need to increase tool and machine utilization.
10. Often have engineering changes.

Variant CAPP organizes standard process plans by category. When new plans are needed, we have to select the appropriate standard plan and modify it for the current product. With generative process planning, we can implement concurrent engineering by creating process plans using the company's business rules.

Top-down justification technique—the methodology

The evaluation of a CAPP system should be based on the current process planning methods and the level of financial detail required by management. As with any capital project, the evaluation should be in accordance with acceptable methodologies for the company, compared to competing projects, and should be based on a common yardstick, such as RoI or payback period.

The data to be collected and analyzed should be that which, in one's judgment, will be accurate and can be assembled in a reasonable time-frame. Economic benefits are usually calculated in three areas:

1. Tangible savings
2. Decision support or intangible enhancements
3. Revenue enhancements.

There are two fundamental approaches to any cost justification process: from the "bottom-up" or from the "top-down." The bottom-up approach starts with the assumption that there are no savings, and we attempt to build a case for justification, piece by piece, until we have combined all the elements into a total return. The bottom-up approach, while commonly used in financial justification efforts, tends to be tedious and time-consuming; the amount of detailed data required is often not available and must be accumulated over a long time-frame prior to even starting an evaluation.

To assist management team assess the benefits of a CAPP system and make an implementation decision in a reasonable time frame, we recommend the use of a worksheet for using the top-down method. By starting with known costs and evaluating the effect of reducing these costs with a good CAPP system, we will have a compelling reason to recommend the transition to a computer-aided planning process.

Most of the benefits derived from a CAPP system stem from standardizing the "best way" to manufacture company's products and eliminating those alternate method routings that are more expensive and less efficient. In a typical manufacturing environment, there is considerable variability in the routings selected by the process planners for identical parts. Since there can only be one "best way" to assemble an item, it is very likely that we can realize significant cost reductions simply by standardizing on the optimum methods of the most experienced process planners.

The Cost Justification Worksheet will take us through the top-down approach using company's costs and potential savings due to standardization. The Presentation Organization Worksheet is intended as an aid to organizing the results of the cost justification analysis for presentation to the senior management of the company.

Start the cost justification evaluation for the company by using the Cost Justification Worksheet and gathering all the cost data that are readily available. Complement the data available with the Industry data provided in this document. Organize the analysis for presentation by classifying the benefits that you anticipate and quantifying the payback period as suggested by the Presentation Organization Worksheet.

Cost justification worksheet

This worksheet will assist us to justify the cost in CAPP system for the company using the "top-down" approach.

Step 1: Estimate the annual manufacturing cost of goods sold

This can be estimated as a per cent of sales, or ideally, supplied directly by your Accounting Department. Some may argue that fixed costs, such as materials, depreciation and overhead, should be excluded from manufacturing cost. For this computation, it makes no difference as long as the same cost basis is used in the calculations for each process plan in averaging the variance in cost of a typical plan (Step 2). The effect of eliminating all fixed costs is simply to lower the manufacturing cost while increasing the average percentage variance of the individual process plans. If this is still unclear, try a few sample calculations with and without fixed costs included.

Step 2: Estimate the average variance in cost of a typical plan

A good way to determine this number is to ask a representative sample of process planners to plan the routing for a part (or assembly) of average complexity, determine the cost of each process plan; and then compute the difference between the average and the lowest cost plan. The ideal way to carry out this step would be to have the planners develop process plans for a random sample of parts using a variety of lot sizes since the routings can vary with the quantity produced. Similarly, there may be other factors that influence the choice of routings for any part, but the injection of too much sophistication may defeat the attractiveness of the top-down approach, namely the ease of generating a rapid estimate of the costs to be saved by implementing a state-of-the-art automated planning system.

Step 3: Multiply Step 1 by Step 2

The result of the product of Step 1 times Step 2 is a surprisingly good estimate of the annual cost reduction possible by delivering clear, consistent, unambiguous process instructions to the shop floor, which reflect the experience of the best manufacturing/industrial engineers in the company.

6.3 METHODS OF COMPUTER-AIDED PROCESS PLANNING

The ultimate goal of an automated process planning system is to integrate design and production data into a system that generates useable process plans. As already mentioned there are two approaches:

- Variant process planning
- Generative approach.

6.3.1 Variant Process Planning

A variant process planning system uses the similarity among components to retrieve the existing process plans. A process plan that can be used by a family of components is called a **standard plan**. A standard plans are stored permanently with a family number as its key. It follows the principle that similar parts require similar plans. Therefore, the process requires a human operator to classify a part, input part information, retrieve a similar process plan from a database (which contains the previous process plans), and edit the plan to produce a new variation of the pre-existing process plan. Planning for a new part involves retrieving of an existing plan and modification. In some variant systems, parts are grouped into a number of part families, characterized by similarities in manufacturing methods, and are related to group technology.

In comparison to manual process planning, the variant approach is highly advantageous in increasing the information management capabilities. Consequently, complicated activities and decisions require less time and labour. Also, incorporating a planner's manufacturing knowledge can standardize procedures and structure it to company specific needs. Therefore, variant systems can organize and store completed plans and manufacturing knowledge from which process plans can be quickly evaluated.

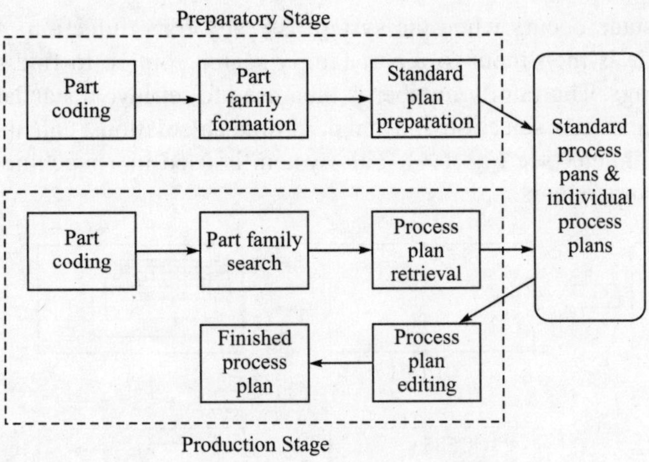

Figure 6.2 Standard plan.

There are difficulties, however, in maintaining consistency in editing practices adequately and inability to accommodate various combinations of geometry, size, precision, material, quality, and shop loading. The biggest disadvantage is that the quality of process plan still depends on the knowledge background of a process planner.

A family is represented by a family matrix which includes all possible members. The variant process planning system has two operational stages:

- Preparatory stage
- Production stage.

During the preparatory stage, the existing components are coded, classified, and subsequently grouped into families. The process begins by summarizing process plans already prepared for components in the family. Standard plans are then stored in a database and indexed by family matrices (Fig. 6.3).

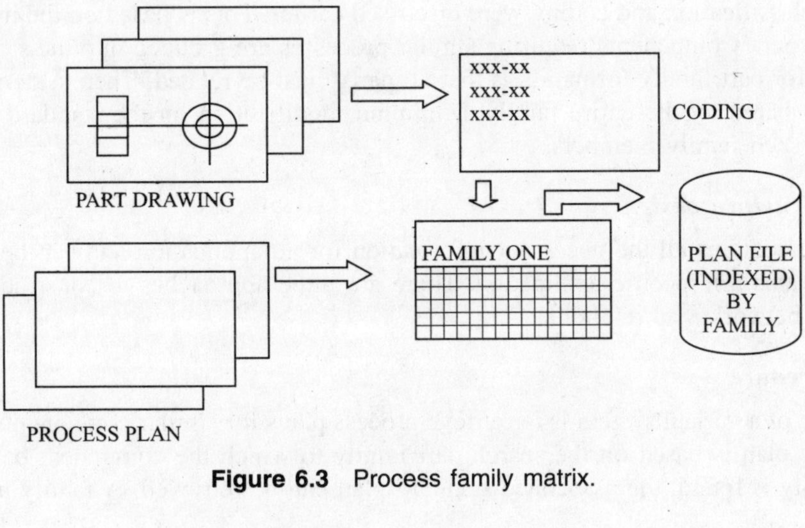

Figure 6.3 Process family matrix.

The operation stage occurs when the system is ready for production. An incoming part is first coded. The code is then input to a part family search routine to find the family to which the component belongs. The family number is then used to retrieve a standard plan. Some other functions, such as parameter selection and standard time calculations, can also be added to make the system more complete (see Fig. 6.4). This system is used in a machine shop that produces a variety of small components.

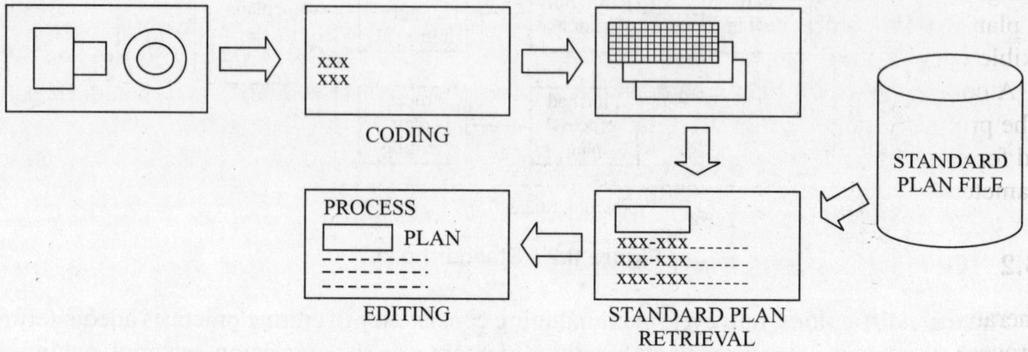

Figure 6.4 Part Search and retrieval.

Design of variable process planning system

The following are the sequences in the design of a Variant Process planning system:

1. Family formation
2. Database structure design
3. Search algorithm development and implementation
4. Plan editing
5. Process parameter selection/updating

Family formation

Part family classification and coding were discussed earlier. This is based on the manufacturing features of a part. Components requiring similar processes are grouped into the same family. A general rule for part family formation is that all parts must be related. Then, a standard process plan can be shared by the entire family. Minimum modification on the standard plan will be required for such family members.

Database structure design

The database contains all the necessary information for an application and can be accessed by several programs for specific application. There are three approaches to construct a database: hierarchical, network, and relational.

Search procedure

The principle of a variant system is to retrieve process plans for similar components. The search for a process plan is based on the search part family to which the component belongs. When the part family is found, the associated standard plan can be retrieved. A family matrix search can be seen as the matching of the family with a given code. Family matrices can be considered as masks. Whenever a code can pass through a mask successfully, the family is found.

Plan editing and parameter selection

Before a process plan can be issued to the shop, some modification of the standard plan may be necessary, and the process parameters must be added to the plan. There are two types of plan editing: One is the editing of the standard plan itself in the database, and the other is editing of the plan for the component. For editing a standard plan, the structure of the database must be flexible enough for expansion, addition, and deletion of the data records.

A complete process plan includes not only operations but also process parameters. The data in the process parameter files are linked so that we can go through the tree to find the speed and feed for an operation. The parameter file can be integrated into variant planning to select process parameters automatically.

6.3.2 Generative Process Planning

Generative process planning is a system that synthesizes process information in order to create a process plan for a new component automatically. In a generative planning system, process plans are created from information available in the manufacturing database without human intervention. Upon receiving the design model, the system can generate the required operations and operation sequences for the component. Knowledge of manufacturing must be captured and encoded into efficient software. By applying decision logic, a process planner's decision-making can be imitated. Other planning functions such as machine selection, tool selection, process optimization, and so on, can also be automated using generative planning techniques.

Generative process plans utilize decision logic, formulae, manufacturing rules, and geometry based data to determine the processes required to convert the raw materials into finished parts. They develop a new plan for each part based on input about the part's features and attributes. Due to the complexity of this approach, a generative CAPP system is more difficult to design and implement than a system based on the variant approach. But a generative CAPP system does not require the aid of a human planner, and can produce plans not belonging to an existing part family. It stores the rules of manufacturing and the equipment capabilities in a computer system. The generative approach is complex and a generative system is difficult to develop.

In comparison, the variant systems are better developed and mature than generative systems; they are suitable for planning processes in mass or large production volumes. For planning discrete processes of manufacturing products of great diversity, generative systems are much more suitable than variant systems. However, true generative systems are still to come although the earlier optimistic speculation were made by researchers. Most CAPP systems in use now are either variant systems or semi-generative systems (with some planning functions developed with variant approach, others with generative approach).

A proper combination of the two approaches can make an efficient CAPP system. First, the system will check, with the help of the variant approach, whether process planning is possible for a new part. If the variant system is unable to identify the part to be of a previous group or family it will use generative technique for process planning. So both the variant and generative process-planning approaches need further development in parallel.

The generative planning has the following advantages:

1. It can generate consistent process plans rapidly.
2. New components can be planned as easily as existing components.
3. It can be interfaced with an automated manufacturing facility to provide detailed and up to date control information.

The generative part consists of:

- Component representation module
- Feature extraction module
- Feature process correlation module
- Operation selection and sequencing module
- Machine tool selection module
- Standard time / cost computation module
- Report generation module

In order to generate a more universal process planning system, variables such as process limitations and capabilities, process costs and so on, must be defined in terms of the production stage. A wide range of methods have been used for generative process planning. A few methods that have been implemented successfully are:

1. Forward and backward planning
2. The input format
3. CAD models
4. Decision logic
   (a) decision trees
   (b) decision table
5. Artificial intelligence

Forward and backward planning

In generative process planning, when process plans are generated, the system must define an initial state in order to reach the final state (goal). The path taken represents the sequence of processes. For example, the initial state is the raw material and the final state is the component design. Then a planner works in modifying the raw workpiece until it takes on the final design qualities. This is called *forward planning.*

Backward planning uses a reverse procedure. Assuming that we have a finished component, the goal is to go back to the unmachined workpiece. Each machining process is considered a filling process.

Forward and backward planning may seem similar. However, they affect the programming of the system significantly. Planning each process can be characterized by a precondition of the surface to be machined and a post condition of the machining (it results). For forward planning, we must know the successor surface before we select a process because the post condition of the first process becomes the precondition for second process. Backward planning eliminates this conditioning problem since it begins with the final surfaces, and processes are selected to satisfy the initial requirements.

Figure 6.5 shows the modular structure of a generative CAPP system.

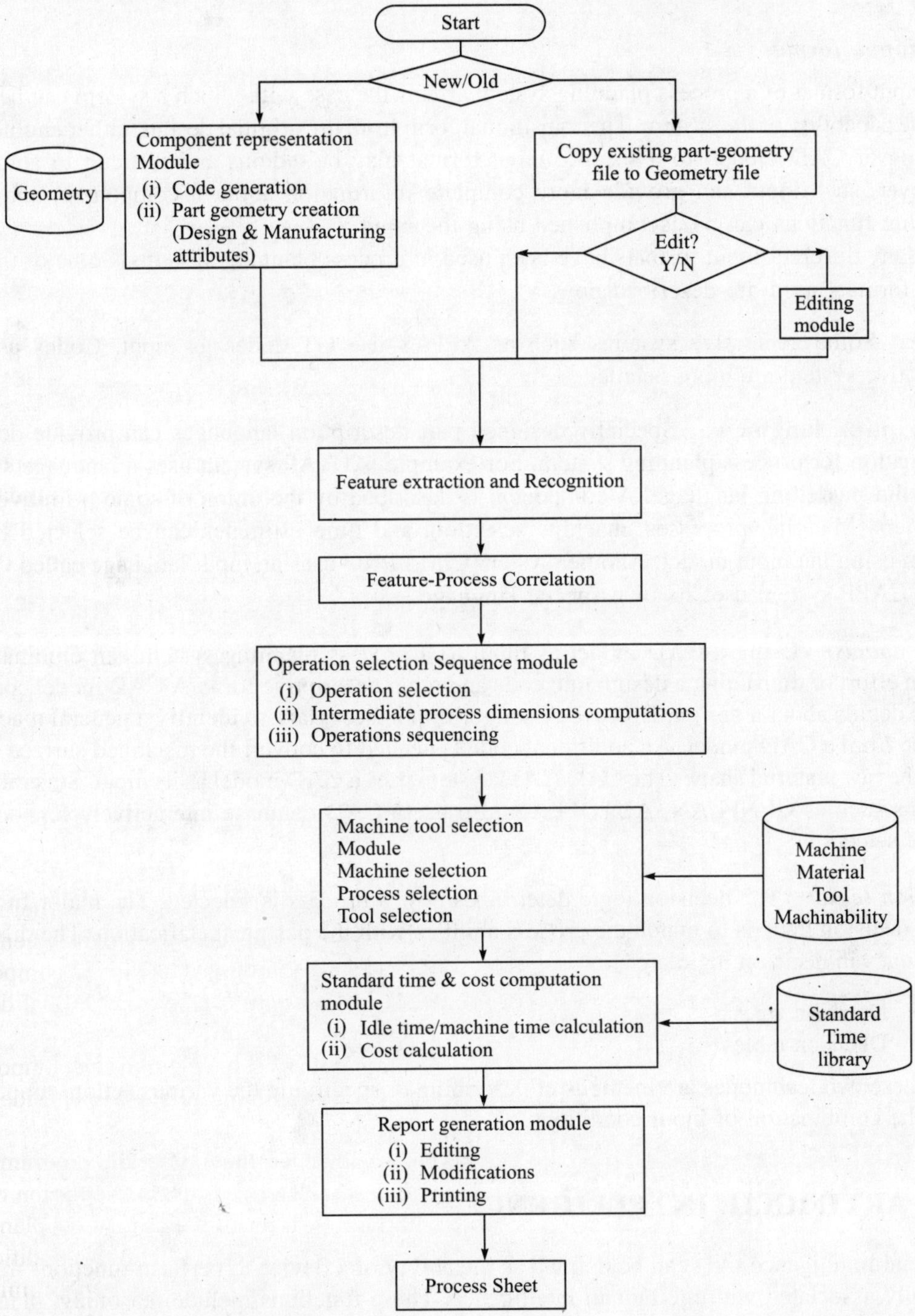

Figure 6.5 Flow chart generative CAPP system.

In forward planning, the objective surface must be taken to guarantee the result. On the other hand, backward planning starts with the final requirements and searches for the initial condition.

The input format

The input format of a process planning system affects the ease with which a system can be used, and the capability of the system. The transitional form from the original design (either engineering drawing or CAD model) to a specific input format may be tedious and difficult to automate. However, such input can provide more complete information about a component, and more planning functions can be accomplished using the input.

Many different input formats have been used in process planning systems. Some of the few input formats used are described now.

Code: Some generative systems such as APPAS use GT codes as input. Codes used in generative system are more detailed.

Description language: Specially designed part description languages can provide detailed information for process planning system. For example, AUTAP system uses a language similar to a solid modelling language. A component is described by the union of some primitives and modifiers. Material, processes, machine selection, and time estimates can be selected by the system using the input model. Another system CIMS/PRO uses an input language called CIMS/DEC. CAPP system uses its own special language.

CAD models: Using a CAD model as input to a process planning system can eliminate the human effort of translating a design into code and other descriptive form. A CAD model contains all the details about a design. However, an algorithm is necessary to identify a general machined surface from a CAD model. An additional code is needed to convert the machined surface shape from the raw material shape. The CAD/CAM system uses a CAD model as its input. Several other systems such as GENPLAN, AUTOPLAN also use a CAD database interactively for tool and fixture selection.

Decision logic: The decision logic determines how a process is selected. The major function of the decision logic is to match the process abilities with the design specification. The different techniques in decision logic are:

- Decision trees
- Decision tables

These two techniques are methods of describing or specifying the various actions associated with the combination of input conditions.

6.4 ARTIFICIAL INTELLIGENCE

Artificial Intelligence **(AI)** can be defined as the ability of a device to perform functions that are normally associated with the human intelligence. These functions include reasoning, planning, and problem solving. Applications for AI have been in natural language processing, intelligent database retrieval, expert consulting systems, theorem proving, robotics, scheduling, and perception problems. Process planning applications have been considered as part of an expert consulting system.

There are two types of knowledge involved in process planning systems: Component knowledge and process knowledge. *Component knowledge* defines the current state of the problem to be solved (declarative knowledge). On the other hand, the *process knowledge* defines how the component can be changed by processes (procedural knowledge).

There are several methods to represent declarative knowledge:

- First-order predicate calculus
- Frames and semantic networks

Procedural knowledge can be represented by **IF** (condition), **THEN** (action) statements which are similar to decision trees or decision tables. In AI such rules can be called *production rules.*

Even after the descriptive and procedural knowledge have represented, conclusions cannot be deduced because we do not have the mechanism to apply the appropriate rules to the problem. Control knowledge is similar to human knowledge in reasoning which deduces certain facts from the knowledge base concerning a problem. This can be a difficult task to program on a computer.

Computer Aided Process Planning (CAPP) has been an active area of research work in AI for many years. Feature recognition, feature process correlation, process sequencing, blank selection, cutting parameter selection, tool selection, etc. are the segments of CAPP where AI contribute to improve the quality of process plans.

Figure 6.6 shows the block diagrammatic representation of an expert CAPP system.

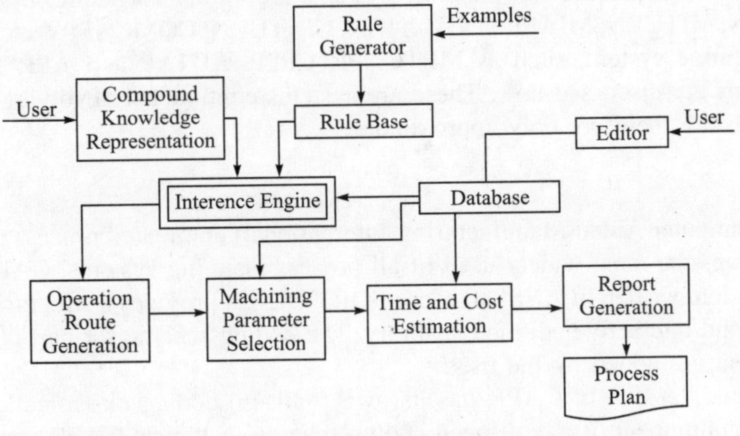

Figure 6.6 Block diagram of expert CAPP system.

6.5 IMPLEMENTATION CONSIDERATIONS

The process planning function is manufacturing system dependent. This implies that no one single process planning system can satisfy all the different manufacturing system needs. There are several factors that must be considered when one attempts to implement a process planning system. These include:

1. Manufacturing system components
2. Production volume/batch size
3. Number of different production families

For a moderate number of component families and many similar components in each family, a variant process planning system is usually economic automated planning alternative. Figure 6.7 illustrates regions for the different planning alternatives.

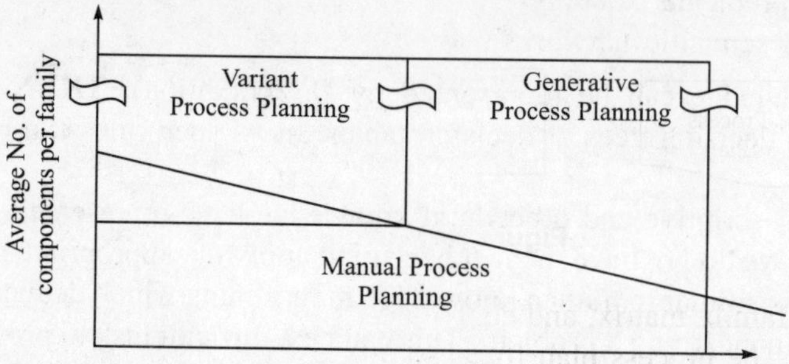

Figure 6.7 Economic regions for different types of process planning.

6.6 PROCESS PLANNING SYSTEMS

The majority of existing processes planning systems are variant in nature. Some of them are: CAPP, MIPLAN, MITURN, MIAPP, UNIVATION, CINTURN, COMCAPPV, etc. However, there are some generative system, such as METCAPP, CPPP, AUTAP and APPAS. Some of the planning systems are discussed now. These are systems continuously evolving in many cases. The descriptions are therefore only approximate.

CAM-I CAPP

The CAM-I (Computer Aided Manufacturing-International) automated process planning system (CAPP) is perhaps the most widely used of all process planning systems. CAPP is a database management system written in ANSI standard FORTRAN. It provides a structure for a database, retrieval logic, and interactive editing capability. The coding scheme for part classification and the output format are added by the user.

PI-CAPP, an extension of CAPP, has its own (built-in) coding classification system. This eliminates the requirement of a developed coding scheme. A typical CAPP system is shown in Figure 6.8.

MIPLAN and MULTICAPP

Both MIPLAN and MULTICAPP were developed in conjunction with OIR (Organization for Industrial Research). They are both variant systems that use the MICLASS coding system for part description. These are data retrieval systems which retrieve process plans based on part

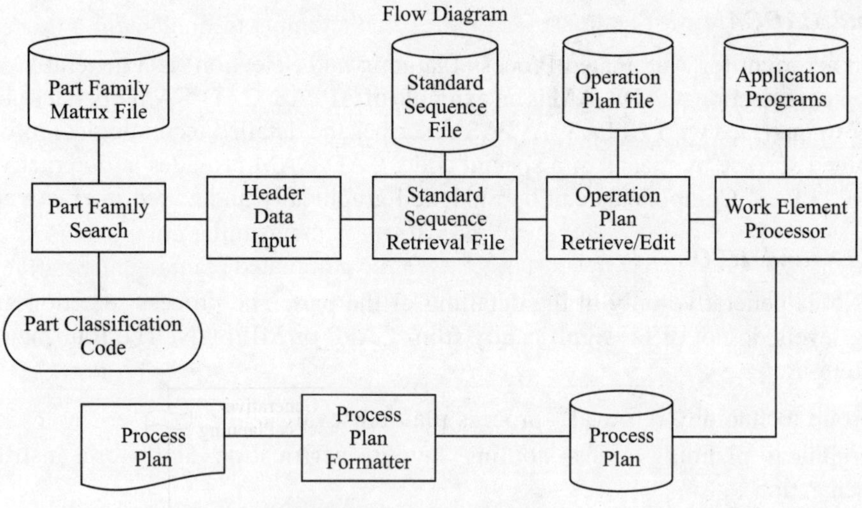

Figure 6.8 CAPP system.

code, part number, family matrix, and code range. By inputting a part code, parts with a similar code are retrieved. The process plan for each part is then displayed and edited by the user. A typical MULTICAPP system is shown in Figure 6.9.

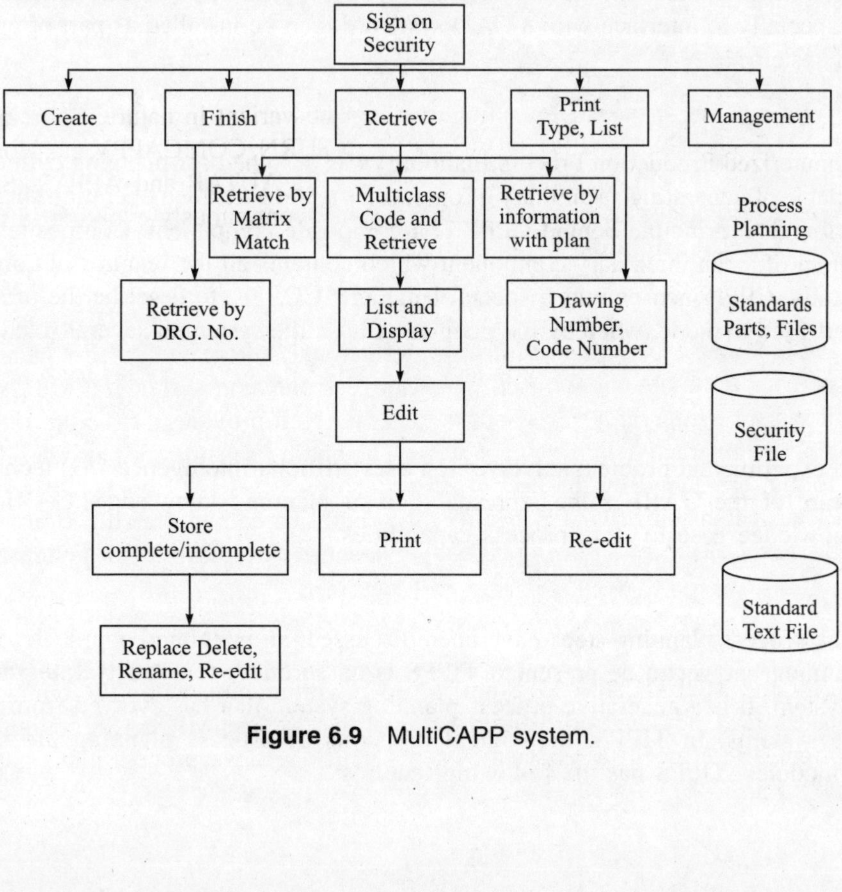

Figure 6.9 MultiCAPP system.

APPAS and CADCAM

APPAS (an acronym for Automated Process Planning and Selection) is a generative system for detailed process selection. CAD/CAM is an example of APPAS. CAD/CAM operates using a CAD "front end" to interface with APPAS. APPAS describes the detailed technological information of each machined surface by means of a special code. CAD/CAM provides an interactive graphics interface to APPAS. Components can be modelled graphically and edited interactively.

AUTOPLAN and RPO

AUTOPLAN is generative only in the detailing of the part. The process selection and process sequencing levels do not differ significantly from CAPP or MIPLAN. The four major modules of the system are:

1. Group technology retrieval—process plan retrieval
2. Graphical planning aides—tooling layout, verification and work instruction and preparation
3. Generative process planning
4. Process optimization.

AUTAP System

The AUTAP system is one of the most complete planning systems in use today. AUTAP uses primitives to construct a part similar to a constructive solid geometry (CSG). AUTAP is a system designed, especially to interface with a CAD system. It can be installed as part of an integrated CAD/CAM system.

CPPP

CPPP (Computerized Production Process Planning) was designed for planning cylindrical parts. CPPP is capable of generating a summary of operations and the detailed operation sheets required for production. The principle behind CPPP is a composite component. Composite component can be thought of as an imaginary component which contains all the features of components in one part family. CPPP incorporates a special language, COPPL, to describe the process. CPPP allows an interactive mode, whereby the planner can use the system at several fixed interaction points.

GARI

GARI is an experimental problem solver which uses artificial intelligence (AI) techniques. The unique feature of the GARI is the representation of planning knowledge. GARI employs a product, knowledge base to store process capabilities.

TIPPS

Although the process planning steps have been discussed, an integrated approach to generative process planning has yet to be presented TIPPS is an acronym for Totally Integrated Process Planning System. It is a generative process planning system that has evolved from the APPAS CAD/CAM systems. In TIPPS, the logical divisions of process planning are broken into functional modules. TIPPS has the following features:

- has a modular structure;
- can interact with a CAD system;
- allows for interactive surface identification; and
- contains a process/knowledge description language.

6.7 MRP (MATERIAL REQUIREMENTS PLANNING)

Production control systems often sound dauntingly technical and complicated. However, the need for a system like MRP can be explained with a simple analogy (Slack et al. 1995). Imagine that in four weeks time you are hosting a party for around 40 guests. You have decided to provide beer, wine and soft drinks plus sandwiches and savoury snacks to the guests. Putting all of this together on the night would involve some planning and production control. First, you would need some simple estimates of all the items you have to buy—for red and white wine, beer, hard drinks, soft drinks, fruit juice, etc. Before shopping for these items, you will take into account what you already have in stock. Again, shopping for the food at the party would involve identifying from your food recipes what ingredients are required to make up the various dishes (subtracting from your shopping list what you already have). Then, as well as specifying the quantity of your needs for materials and ingredients against the likely demand, you also need to think about the sequencing of the cooking and preparation. You can't do everything on the night so that you might choose, for example, to do some of the cooking in one or two days in advance, and then—put the items into the refrigerator so that they can be preserved. This might mean shopping for some ingredients first, leaving others for a later date. In short, to organize a party, you need to plan and control the acquisition of materials and the process of production. You need to make decisions about the quantity and timing of the purchasing and production of different ingredients, based on forecasts of the number of people coming to the party and your recipe for making the finished products.

Although the production process is much more complex than organizing a party, the production control systems such as MRP are designed to address similar problems. Figure 6.1 is a general model of a manufacturing planning and control system. It breaks activities into *a front end, an engine, and a back end.* The front end produces *the master production schedule* (MPS)—the equivalent of food and drink at the party. The MPS plans the production of the goods offered to customers over a given planning period. The back end handles factory scheduling and manages materials from suppliers (this equates with bringing home the shopping and planning, period cooking and preparation). *Material requirements planning* (MRP) is the core of the engine. It takes a period-by-period set of MPS requirements and (in the way our food recipes produce shopping lists) generates a related set of components and raw materials requirements. MRP is a detailed plan for the components required to enable the MPS to be fulfilled.

The MPS and MRP has two other inputs: A *Bill of Material* (BOM) which shows, for each part number, and the associated component part numbers. Thus, for a dining table, the BOM would show that a top assembly and four legs are required. The BOM for the top assembly would show that two end panels, a subframe, and two leaf inserts are required. The BOM for the legs would show that a solid timber stock and associated hardware kits (screws and castors) are required. *Inventory status data* (the third input into MRP) would indicate how many legs or leaf inserts, etc. were on hand, how many of those were already committed for production, and how many hardware kits had been ordered. This would then allow the requirements for further table production to be worked out.

MRP data thus make it possible to generate a *time-phased requirement record* for any part number. This data can also drive the *detailed capacity planning* modules. This is a massive computational task, only made possible by the use of modern computers.

MRP plays a central role in material planning and control. It translates aggregate plans for production into the individual steps necessary to realize those plans, and generates data to develop capacity plans. It also provides links to the shop floor and vendor systems that accomplish production. MRP is, therefore, the natural starting point for many companies to computerize their overall production control.

With sufficient computer power, sophisticated MRP systems can provide continual updates of the various component and materials requirements to match changing circumstances. This enables better priority-setting and fine-tuning of shop floor operations. Better capacity planning procedures can also be incorporated. As well as managing material flows, these enhanced systems can allocate resources such as machinery and personnel more efficiently. They can also include financial details. Simulation techniques allow the examination of various "what if" scenarios. Such enhanced systems are more broadly focused on manufacturing planning and control. They are clearly more company wide. Indeed, their scope is so much wider than the original concepts of MRP, that the Guru Oliver Wight coined a new term – 'MRP2'. This stands for "manufacturing resource planning".

Sophisticated, yet it is easy-to-use MRP's capabilities to plan and control material flow. MRP is widely recognized as one of the most valuable materials management tools a manufacturer can have. The MFG Material Requirements Planning subsystem translates your management, marketing, and production planning goals into detailed and coordinated schedules for purchasing and production. It takes into account a host of factors such as lead times, policies, shop calendar, and planned engineering changes, and rapidly generates a complete MRP report for your entire product line.

This application has been designed to coordinate activities of all subsystems to meet production goals. It will automatically manage special part calculations, open and blanket orders, and design changes. It has facilities for manual overrides, open order rescheduling, and firm planned orders. And it offers flexible reporting facilities allowing variable reporting periods, level-by-level pegging, and 'action-only' reporting Material Requirements Planning gives you the orderly, integrated materials management you need.

Material Requirements Planning (MRP) is widely recognized as one of the most valuable materials management tools a manufacturer can have. The unanswered question is: Which computer-based system will deliver effective MRP to your manufacturing operations? MFG's Material Requirements Planning subsystem, in conjunction with its companion subsystems, can deliver the orderly, integrated control you need. Equally important, its inherent speed, ease of operation, and concise reporting capabilities will enable you to use it efficiently and

- Print a net-change report
- Set an MRP-EXPLODE flag to either suppress or allow the explosion for a part

- Exclude certain types of parts from MRP processing
- Compensate for shrinkage, and
- Generate material requisitions.

Pegging

To solve a potential part requirement problem highlighted by the MRP run, a system-generated pegging report is offered. It identifies the source of the part's requirement from its own master schedule or dependent demand from other planned orders, work orders, or firm planned orders. Knowing the source, the planner is able to readily decide the best action.

Rescheduling open orders

Each time a Material Requirements Plan is run, the system checks whether orders should be rescheduled to an earlier (or later) period... determines the proper new period..., and prints a message specifying the appropriate moves. MRP does not actually move the orders, but pinpoints the need and leaves the decision to the planner's judgment and experience. Similarly, MRP won't arbitrarily generate planned orders to satisfy parts requirements, which can be more practically answered by rescheduling. MRP manages all the pertinent information for you, but leaves final control where it belongs in your hands.

Overriding the MRP

Occasionally, MRP logic will indicate that a requirement can't be met when, in fact, expediting an order or reducing a lot size may resolve the problem. The firm planned order facility allows the planner to override the MRP suggestion in such cases.

Tailoring the system to your preferences

MFG Material Requirements Planning subsystem is easily modified to match your needs. The MRP Report can display information customized to your needs. Further, you control the scope of the MRP. You can include all parts or omit those which do not need exact-quantity planning. And you choose the planning horizon most appropriate to your operation.

6.7.1 Manufacturing Resource Planning (MRP II)

This is a development that seeks to address some of the shortcomings of MRP. It includes all of the elements of MRP, it is based around the Bill of Materials, MRP II

- uses a Master Production Schedule (MPS) as its starting point, and
- uses the three steps of Explosion, Netting and Offsetting to create the initial schedule. However, MRP II includes the following four major developments from MRP.

Feedback

MRP II includes feedback from the shop floor on how the work has progressed, to all levels of the schedule so that the next run can be updated on a regular basis. For this reason, it is sometimes called 'Closed Loop MRP'.

Resource scheduling

There is a scheduling capability within the heart of the system that concentrates on the resources, i.e. the plant and equipment required to convert the raw materials into finished goods. For this reason, the initials 'MRP' now mean "Manufacturing Resources Planning". The advantages of this development are that detailed plans can be put to the shop floor and can be reported on by operation, which offers much tighter control over the plant. Moreover, loading by resource means that capacity is taken into account. The difficulty is that capacity is only considered after the MRP schedule has been prepared. It may turn out that insufficient time was allowed within the MRP schedule for the individual operations to be completed.

Batching

Rules: batching rules can be incorporated; indeed, they have to be if resource scheduling is to take place. Most software packages offer a variety of batching rules. Three of the more important are 'Lot for Lot', 'EBQ', and 'Part Period Cover'.

1. 'Lot for Lot' means "batches that match the orders". Therefore, if a company is planning to make 10 Nos. of Product A, followed by 20 Nos. of Product B, then the batches throughout the process will match this requirement. If both A and B require 2 Nos. of a certain sub assembly, then that will be made in quantities of 20 of A and 40 of B. It is the batching implicitly followed in basic MRP.
2. 'EBQ' stands for "Economic Batch Quantity". The batch size is calculated by a formula that minimizes the cost through balancing the set-up cost against the cost of stock.
3. 'Part Period Cover' means making batches whose size covers a fixed period of demand. A policy of making a week's requirement in one batch is an example.

Software extension programs

A number of other software programs are included in the MRP II suite. Some of these are further designed to help the scheduling procedure. The most important is Rough Cut Capacity Planning (RCCP), an initial attempt to match the order load to the capacity available, by calculating (using a number of simplifying assumptions) the load per resource. Overloads are identified and orders moved to achieve a balance. This has been described as "knocking the mountains (the overloads) into the valleys (periods of underload)".

Other additions are designed to extend the application of the MRP II package. For example, it may include an option for entering and invoicing sales orders (Sales Order Processing). Another common extension is into stock recording, and a third into cost accounting. A full MRP II implementation can, therefore, act as an integrated database for the company.

Data accuracy

This last development means that the company must lay great emphasis on data accuracy. Errors in recording in one part of the system will result in problems for all the users. The suppliers of such systems encourage users to aim at an accuracy ranging from 95% to 98%.

Manufacturing

The Manufacturing Department manages the manufacturing activities including bill of materials, routing, work orders, material requisitions, and production.

Table 6.1 Features of MRP II

Feature	Benefit
Creating Production Orders from Sales Orders	Items which are to be manufactured can be automatically transferred to production orders from sales orders entered by the sales department. Each item on the sales order may specify a different required date which can create separate production orders for shipments on different dates.
Bill of Materials	Create a multi-tier Bill of Material (BOM) for a finished item or sub-assembly. A BOM can contain sub-assemblies, which have their own BOM to create an unlimited number of levels. The BOM tracks all material costs, including shrinkage, required in producing an item.
Multiple BOM per Item	System allows multiple BOM for the same item to be stored for use in different production orders or for reference purposes.
Routing	Routing (manufacturing process) can be created with unlimited number of steps (sequences). Each step (sequences) usually represents the complete unit of work in the manufacturing process such as stamping a steel sheet or soldering an electronic board or assembling the computer. For each step you can define the work centre where the work will be done, set up and run labour required to perform the operation, other costs associated with the operation, set up and run machine time required for the operation. Overheads associated with the operation are calculated if the operation was meant for Labour Class and/or Machine Class.
Multiple Routings per Item	Routing can be common to multiple items. A single item can have multiple routings defined for use at different times.
Production Order Types	Different order types can be created. Estimated orders can be created to calculate estimated costs of production. Planned orders can be entered for purposes of what-if analysis and production schedule analysis. Firmed production orders record quantities, routings, and other information needed to initiate the manufacturing process.
Cost and Time Analysis	During and after production, planned costs and hours can be compared to actual costs and hours for each production order or a group of orders.
Flexible Issue to WIP	Items can be issued into Work-In-Process (WIP) based on all the components required for a sub-assembly or partial quantities for large production runs.
Sub-assembly Production	Sub-assemblies with an unlimited number of levels can be produced in parallel and combined to create finished items. Items can be assigned serial or lot numbers for tracking purposes during the production process.

Table 6.1 Features of MRP II (*Contd.*)

Feature	Benefit
Transferring W.I.P. to Inventory	Finished items can be transferred to various inventory locations. Raw materials which were not used can be transferred back to inventory at any time for use in the production of other items.
Integration with Inventory Warehouse Management	Inventory warehouse management can be used to issue items to Work-In-Process of specific production orders directly from warehouse bins.
Labour Time Tracking or Backflushing	Labour time can be accounted for by entering time transactions or importing the information from data collection devices located in the plant. Where the actual time does not need to be tracked, the labour time can be calculated with automatic backflushing based on the number of items manufactured. Labour tracking can be tracked by the employees.
Machine Time Tracking or Backflushing	Machine time can be accounted for by entering time transactions or importing the information from data collection devices located in the plant. Where the actual time does not need to be tracked, the machine time can be calculated with automatic backflushing based on the number of items manufactured. Machine time can be tracked at the machine level to determine appropriate maintenance intervals and machine utilization.
Other Costs	Capture miscellaneous manufacturing costs and sub-contracting costs incurred in the manufacturing process.
Overheads Calculations	Define overhead methods for use in production bill of materiais or production routing definitions. Overheads can be classified as fixed or variable depending on the analysis required and on which G/L accounts are charged. An overhead method can specify fixed, variable, or both. Overheads can be based on cost or quantity of items.
WIP Location Tracking	Once items are moved from inventory into Work-In-Process, their exact location in the production area or if items are sent to an outside contractor. Each movement can be tracked by location, time, and the person who moved the items.
Work Centres	The manufacturing area can be divided into logical work centres which can be used in the routing process. Each work centre can define standard production routing steps which can then be used when creating new routings or modifying existing routings.
Comprehensive Reporting	A collection of 20+ reports can be used to extract information quickly, modify existing reports or create new reports quickly.
Unlimited Transaction History	All transactions entered in the system are available for review at any time, performance of the system is not affected by the number of transactions.
No Month End or Year End Closing Process	Transactions can be entered for any period in any year, which has not been locked. Periods can be locked or unlocked at any time.

Latest Trends in CAPP

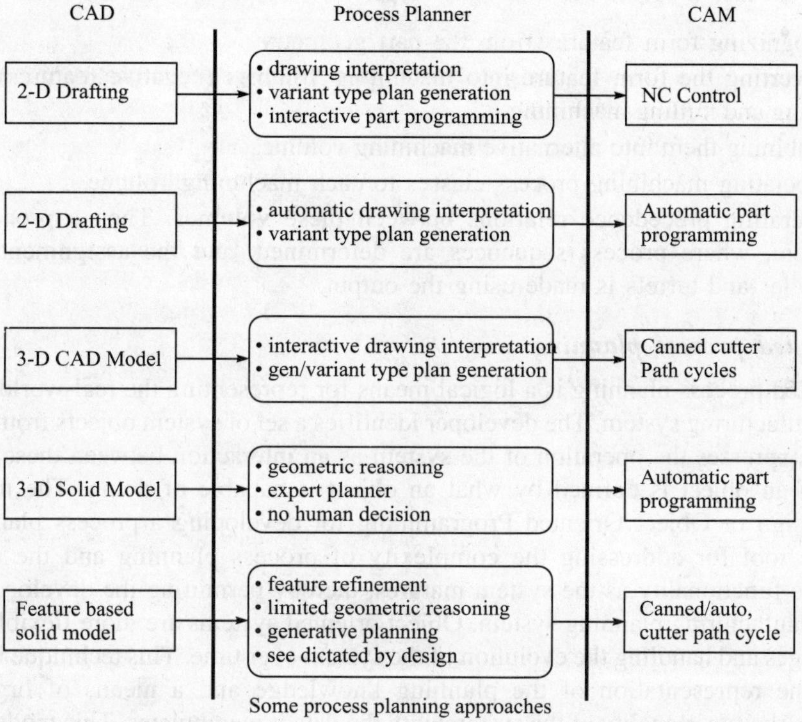

Figure 6.10 Process planning approaches.

Feature-based and solid model-based process planning

Solid model-based process planning uses solid modelling package to design a 3D part. In feature-based process planning systems, a part is designed with a design-oriented manufacturing feature or a feature extraction/feature recognition system is used to identify part feature and their attributes from the CAD file. A prototype **solid model based automated process planning system** for integrating CAD and CAM system. In this system, a three-dimensional (3D) finished part is built by using a solid modelling package. The primitive (cylinder, cone, block, wedge, sphere and torus) is used to define the removal volume. This system consists of three major sections: CAD interface, production knowledge and process planning. The CAD interface includes the finished part drawing. The finished part is built by using the AutoCAD Advanced Modelling Extension (AME) module in a PC. With the AME module, the user can create complex 3D parts and assemblies by using Boolean operations to combine simple shapes. The production knowledge is placed before the process planning procedure, where it accommodates the essential knowledge. It contains information about the machine tools, materials in stock, cutting parameters, and so forth. The process plan is then generated based on recognized solid primitives and production rules.

In the **prototype Feature Based Automated Process Planning** (FBAPP) system, features are recognized from the removable volume point of view rather than from the design part point of view. The entire process in FBAPP is naturally closer to the thinking of a human process planner. The process involves the following steps:

1. Recognizing form features from the part geometry
2. Converting the form feature into machining volumes (negative features) suitable for turning and milling machining
3. Combining them into alternative machining volumes
4. Associating machining process classes to each machining volume
5. Generating precedence relations between these volumes. Then a process planning system, where process sequences are determined, and the assignment to multiple spindles and turrets is made using the output.

Object oriented process planning

Object oriented process planning is a logical means for representing the real-world components within a manufacturing system. The developer identifies a set of system objects from the problem domain and expresses the operation of the system as an interaction between these objects. The behaviour of an object is defined by what an object is capable of doing. The use of Object Oriented Design or Object Oriented Programming for developing a process planning system provides the tool for addressing the complexity of process planning and the capability to increment the functionality as the system matures, thereby permitting the developer to create a complete manufacturing planning system. Object oriented systems are more flexible in terms of making changes and handling the evolution of the system over time. This technique is an efficient means for the representation of the planning knowledge and a means of organizing and encapsulating the functionality of the system with the data it manipulates. This modularity results in a design that can be extended to include additional functionality and address other processes. The design also expands on traditional piece part planning by extending the part model to support planning for end products composed of multiple parts and sub-assemblies.

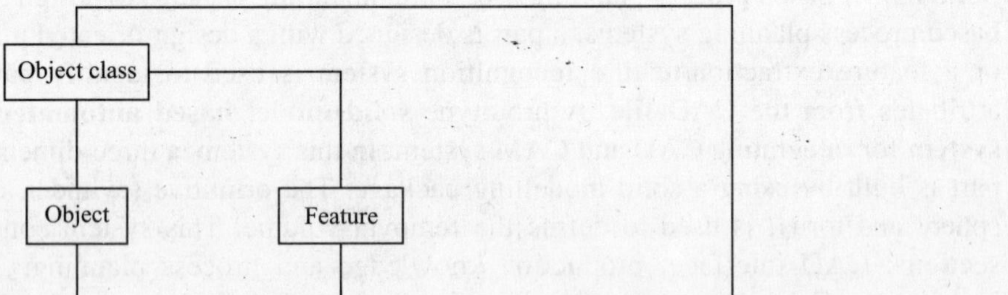

Figure 6.11 Partial view of some CAPP environment entities and their relationships.

AI techniques in CAPP

In the past, traditional computer programs have been used to solve formalized problems, where the statement and principles are well understood. The ill formalized and less understood problems have led to development of Artificial Intelligence (AI), particularly from the Knowledge Based Systems (KSB) or Expert System (ES). Knowledge based systems are developed from the field of artificial intelligence, which in turn is a branch of computer science. AI is developed in an attempt to simulate human intelligence in a computer. AI tries to adapt, learn, invent, and accumulate the combined wisdom of a profession. Currently some AI techniques such as Genetic Algorithms (GAs), Artificial Neural Networks (ANNs), and Fuzzy Logic are widely used for the CAPP system, but Expert System (ES), which is an AI tool, has been used in CAPP system for a long time. Many ES based CAPP systems have so far been developed and researchers are still trying to solve many problems encountered in practice.

The process planning domain is analyzed in terms of the problem specifications and knowledge characteristics. The expert system is divided into the following main components:

1. Knowledge base
2. Knowledge acquisition mechanism
3. A recognition/inference mechanism
4. A user interface mechanism.

Expert systems differ from databases in that their knowledge is not simply readable as data, but can also be used for solutions of different types of problems. Knowledge is thus distinguished from data in that it is integrated with instructions for its use. For this, the division of knowledge into as many fragments as possible makes a knowledge base more modular. Manufacturing knowledge in a CAPP system implies the following:

1. The design related manufacturing knowledge is employed, for instance, by the CAD expert module, which allows to extract/add process planning information from/to the part design.
2. The part process manufacturing knowledge associates the data content of the part model to the process model; it embodies the process selection expert module, which determines the different manufacturing steps to be undertaken on a certain part or feature and the sequencing relationships between those manufacturing steps.
3. The resource process manufacturing knowledge associates the data content of the available resources to the process model; it embodies, for example, the machine selection expert, which determines the candidate machine tools for the operation on a certain work piece.
4. The resource part manufacturing knowledge relates the part data with the resource data (e.g. selection of a tool/machine is influenced by the dimension of the part).
5. Process plan generation knowledge encloses the knowledge that brings all other knowledge sources together.

Recently, **Artificial Neural Network (ANN)** has been gaining attention among many researchers.

Neural networks are a class of computing systems which utilize a high parallel architecture. Problem solving tasks, such as process planning, may be considered as pattern classification tasks, which can be efficiently solved by neural networks. [36]. The process planner learns mapping between input patterns consisting of features and attributes of a part, and output patterns consisting of sequencing of machining operations to apply to these parts. Thus, the neural network offers a promising solution for automating of process planning knowledge. The network comprising large numbers of simple processing units, which communicate in parallel through weighted connections.

REVIEW QUESTIONS

1. What do you mean by CAPP?
2. Discuss the importance of variant CAPP.
3. What are the benefits of cost justification worksheet?
4. Name any two approaches involved in CAPP.
5. Discuss the salient features of process family matrix.
6. What do you mean by database structure design?
7. Discuss the importance of generative process planning.
8. Define code.
9. What are the CAD models involved in CAPP?
10. Briefly explain the advantages of an artificial intelligence.
11. Compare MIPLAN and MULTI CAPP.
12. What is GARI?
13. Define the term totally integrated process planning.
14. Compare MPS and MRP.
15. What is BOM?
16. Compare MRP and MRP II.

Chapter 7

JIT-LEAN PRODUCTION

7.1 HISTORY OF MRP, JIT AND LEAN PRODUCTION

MRP has been a viable planning and execution system for almost 40 years ever since the first computers were introduced into the manufacturing sector of the business. Dr. Joe Orlecky introduced the basic concepts of MRP in the early 1960s, and one of the first real software applications, Bill Of Material Processor (BOMP) was released by IBM in the year 1980. The basic assumptions of this technique was to plan all shop floor activities through a series of "planned" events associated with a set of operating parameters and conditions that would be monitored and executed by the software application. The basic driver of this approach lies in the concepts of a "plan-driven" philosophy. In the late 1970s and early 1980s, the influence of the Japanese approach, called **Just-In-Time (JIT)**, emerged as a dramatically new approach to executing all activities on the shop floor. The primary focus of JIT centred on a 'pull' approach where all upstream manufacturing activities were linked and controlled by the activities of the next downstream operation. The downstream operation would physically send distinct and direct signals to the prior upstream operation in the form of 'Kanban' or card signals, directing that operation to respond to its needs for product or service. In essence, this approach can be classified as a "process-driven" philosophy.

In the mid to late 1980s, a high-bred version of both systems emerged, which was called **Computer Integrated Manufacturing CIM**. It combined the best of both operational scenarios, with the strong influence of integration from both computer systems as well as automation in the form of robotics, automated processes, and integrated computer platforms and networks.

Starting in the late 1980s, the term 'Lean Production' was introduced in a book '*The Machine that Changed the World*' by James Womack et al. (1990) which focused on the comparisons of MRP, JIT, CIM, and the new emerging concept that known as the Lean Enterprise. It was at this point in the manufacturing cycle where dramatic changes began to evolve into what we now recognize as Lean Production. US manufacturers have always searched for efficiency strategies that help reduce costs, improve output, establish competitive position, and increase market share. Early process oriented, mass production manufacturing methods common before

World War II shifted afterwards to the results-oriented, output-focused, production systems that control most of today's manufacturing businesses.

Japanese manufacturers rebuilding after the Second World War were facing declining human, material, and financial resources. The problems they faced in manufacturing were vastly different from their Western counterparts. These circumstances led to the development of new, lower cost, manufacturing practices. Early Japanese leaders such as the Toyota Motor Company's Eiji Toyoda, Taiichi Ohno, and Shingeo Shingo developed a disciplined, process-focused production system now known as the **Toyota Production System**, or **lean production**. The objective of this system was to minimize the consumption of resources that added no value to a product.

The *lean manufacturing* concept was popularized in American factories, in large part by the Massachusetts Institute of Technology (MIT) study of the movement from, mass production to production as described in *The Machine That Changed the World*, referred to above, which discussed the significant performance gap between Western and Japanese automotive industries. This book described the important elements accounting for superior performance as lean production. The term *lean* was used because Japanese business methods used less human effort, capital investment, floor space, materials, and time in all aspects of operations. The resulting competition among US and Japanese automakers over the last 25 years has lead to the adoption of these principles within all US manufacturing businesses.

7.1.1 Automobile Manufacturing

Forty years ago, Peter Drucker dubbed automobile manufacturing "the industries of industries." Today, automobile manufacturing is still the world's largest manufacturing activity. After First World War, Henry Ford and General Motors' Alfred Sloan moved world manufacture from centuries of craft production (led by European firms) into the age of mass production. As a result, the United States soon dominated the world economy.

7.1.2 Production Methods

The craft producer uses highly skilled workers and simple but flexible tools to make exactly what the customer asks for—one item at a time. Few exotic sports cars provide current examples. We all love the idea of craft production, but the problem with it is obvious: Goods produced by the craft method—as automobiles once were exclusively manufactured—cost too much for most customers to afford. So mass production was developed at the beginning of the twentieth century as an alternative.

The mass-producer uses narrowly skilled professionals to design products made by unskilled or semi-skilled workers tending expensive, single-purpose machines. These churn out standardized products in very high volume. Because the machinery costs so much and is so intolerant of disruption, the mass-producer keeps standard designs in production for as long as possible. The result: The customer gets the product at lower costs but at the expense of variety and by means of work methods that most employees find boring and dispiriting.

The Toyota Motor Corporation, by contrast, combines the advantages of craft and mass production, while avoiding the high cost of the former and the rigidity of the latter. Toward this end, they employ teams of multi-skilled workers at all levels of the organization and use highly flexible and increasingly automated machines to produce volumes of products in enormous variety.

The Toyota Production System is also defined as Lean Production because it uses less of everything compared with mass production—half the human effort in the factory, half the manufacturing space, half the investment in tools, half the engineering hours to develop a new product in half the time. Also, it requires keeping far less than half the needed inventory on site, results in fewer defects, and produces a greater and ever growing variety of products.

Perhaps the most striking difference between mass and Toyota production system lies in their ultimate objectives. Mass-producers set a limited goal for themselves— "good enough," which translates into an acceptable number of defects, a maximum acceptable level of inventories, and a narrow range of standardized products. Lean producers, on the other hand, set their sights explicitly on perfection.

7.2 TOYOTA PRODUCTION SYSTEM

The pioneering of the concept of Toyota Production System saw the rise of Japan to its current economic pre-eminence subsequents, other companies and industries copied this remarkable system. Manufacturers around the world are now trying to embrace this innovative system, but they are finding the going rough. The companies that first mastered this system were all headquartered in one country-Japan. However, many Western companies now understand Toyota Production System, and at least one is well along the path of introducing it. Though superimposing this method on the existing mass production systems causes great pain and dislocation.

This chapter we believe is an effort to explain the necessary transition from mass production to the revolutionary production, called Toyota production System. By focusing on the global auto industry, this chapter explains in simple, concrete terms what the Toyota Production System is, where it came from, how it really works, and how it can spread to all corners of the globe for everyone's mutual benefit. The global adaptation, as it inevitably spreads beyond the auto industry, will change everything in almost every industry choices of customers, the nature of work, the fortune of companies, and, ultimately, the fate of nations.

What is Toyota Production System? Perhaps the best way to describe this innovative production system is to contrast it with craft production and mass production, the two other methods humans have devised to manufacture products.

The Toyota Production System is a technology of comprehensive production management the Japanese invented about a 100 years after opening up to the modern world. The basic idea of this system is to maintain a continuous flow of products in factories in order to flexibly adapt to demand changes. The realization of such production flow is called Just-In-Time (JIT) production, which means producing only necessary units in a necessary quantity at a necessary time. As a result, the excess inventories and the excess work-force will be naturally diminished, thereby achieving the purposes of increased productivity and cost reduction.

The basic principle of JIT production is rational, that is, the Toyota Production System has been developed by steadily pursuing the orthodox way of production management. With the realization of this concept, unnecessary intermediate and finished product inventories would be eliminated. However, although cost reduction is the system's most important goal, it must achieve three other sub-goals in order to achieve its primary objective. They include:

1. *Quantity control*, which enables the system to adapt to daily and monthly fluctuations in demand in terms of quantities and variety.
2. *Quality assurance*, which assures that each process will supply only good units to the subsequent processes.
3. *Respect-for-humanity*, which must be cultivated while the system utilizes the human resource to attain its cost objectives.

It should be emphasized here that these three goals cannot exist independently or be achieved independently without influencing each other or the primary goal of cost reduction. All goals are output of the same system, with productivity as the ultimate purpose and guiding concept, the Toyota Production System strives to realize each of the goals for which it has been designed. Before discussing the contents of the Toyota Production System in detail, an overview of this system is in order. The outputs or result side as well as the inputs or constituent side of the production system are depicted in Figure 7.1.

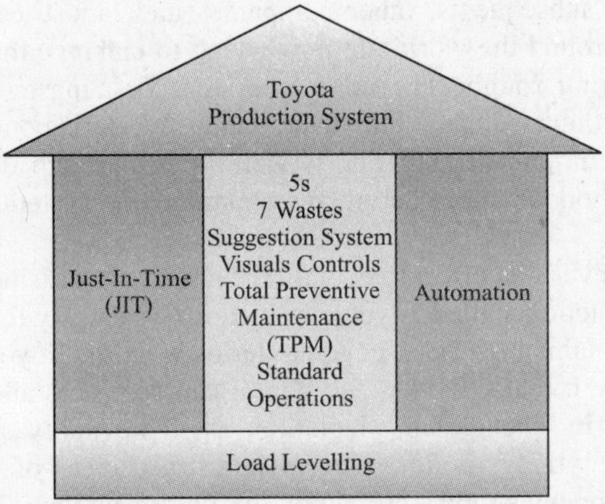

Figure 7.1 Lean manufacturing: Toyota Production System.

Achieving two key concepts creates a continuous flow of production, or adapting to demand changes in quantities and variety,: JIT and Automation. These two concepts are the pillars of the Toyota Production System.

Just-In-Time basically means to produce the necessary units in the necessary quantities at the necessary time. Autonamation may be loosely interpreted as autonomous defects control. It supports JIT by never allowing defective units from the preceding process to flow into and disrupt a subsequent process. Two concepts, which are also key to the Toyota production system, include **flexible work force,** which means varying the number of workers to demand changes, and **creative thinking,** or capitalizing on workers' suggestions.

To realize these four concepts, Toyota has established the following systems and methods:

1. Kanban system to maintain JIT production
2. Production smoothing method to adapt to demand changes

3. Shortening of set-up time for reducing the production lead time
4. Standardization of operations to attain line balancing
5. Machine layout and the multi-function worker for flexible work force
6. Improvement activities by small groups and the suggestion system to reduce the workforce and increase the worker's morale.
7. Visual control system to achieve the Autonamation concept
8. Functional management system to promote company-wide quality control.

Global adaptation: Since Toyota production system has been created from actual practices in the factories of Toyota, it has a strong feature of emphasizing practical effects, and actual practice and implementation over theoretical analysis. This system can play a great role in the task of improving the constitutions of the companies worldwide, especially those of the automobile industry.

7.3 JUST-IN-TIME PRODUCTION

The idea of producing the necessary units in the necessary quantities at the necessary time is described by the short term JIT. JIT means, for example, that, in the process of assembling the parts to build a car, the necessary kind of sub-assemblies of the preceding processes should arrive at the product line at the time needed in the necessary quantities. If JIT is realized in the entire firm, then unnecessary inventories in the factory will be completely eliminated, making stores or warehouses unnecessary. The inventory carrying costs will be diminished, and the ratio of capital turnover will be increased.

However, to rely solely on the central planning approach which instructs the production schedules to all processes simultaneously, it is very difficult to realize JIT in all the processes for a product like an automobile, which consists of thousands of parts. Therefore, in the Toyota system, it is necessary to look at the production flow conversely; in other words, the people of a certain process go to the preceding process to withdraw the necessary units in the necessary quantities at the necessary time. Then what the preceding process has to do is produce only enough quantities of units to replace those that have been withdrawn.

7.3.1 Kanban System

Many people think the Toyota Production System is a Kanban system: this is incorrect. The former is a way to make products, whereas the latter is a way to manage the JIT production method. In short, the Kanban System is an information system to harmoniously control the production quantities in every process. It is a tool to achieve JIT production. In this system, what kind of units and how many units needed are written on a tag-like card called *Kanban.* The Kanban is sent to the people of the preceding process from the subsequent process. As a result, many processes in a plant are connected with each other. This connecting of processes in a factory allows for better control of necessary quantities for various products. The following support the Kanban system:

- Smoothing of production
- Reduction of set-up time design of machine layout

- Standardization of jobs
- Improvement activities
- Autonamation

A Kanban is usually a card put in a rectangular vinyl envelope. Two kinds are mainly used: Withdrawal Kanban and Production-ordering Kanban. A Withdrawal Kanban details the kind and quantity of product which the subsequent process should withdraw from the preceding process, while a Production-ordering Kanban specifies the kind and quantity of the product which the preceding process must produce. The Withdrawal Kanban implies that the preceding process which makes this part is forging, and the carrier of the subsequent part must go to position B-2 of the Forging Department to withdraw drive pinions. The subsequent process is machining. The Kanban shows that the machining process SB-8 must produce the crank shaft for the car type. The crank shaft produced should be placed at store F26-18. These cards circulate within Toyota factories, between Toyota and its many cooperative companies, and within the factories of cooperative companies. In this manner, the Kanban can contribute information on withdrawal and production quantities in order to achieve JIT production.

Suppose we are making products A, B and C in an assembly line. The parts necessary to produce these products are a and b which are produced by the preceding machining line (Figure 7.4). Parts a and b produced by the machining line are stored behind this line, and the production-ordering Kanbans of the line are attached to these parts. The carrier from the assembly line making product A will go to the machining line to withdraw the necessary part 'a' with a withdrawal Kanban. Then, at store, he picks up as many boxes of this part as his withdrawal Kanbans and detaches the production-ordering Kanban attached to these boxes. He then brings these boxes back to his assembly line, again with withdrawal Kanbans. At this time, the production-ordering Kanbans are left at store 'a' of the machining line, showing the number of units withdrawn. These Kanbans will be the dispatching information to the machining line. Part 'a' is then produced in the quantity directed by that number of Kanbans. In this machining line, actually, parts a and b are both withdrawn, but these parts are produced according to the detached order of the production-ordering Kanbans.

7.3.2 Automation

In order to realize JIT perfectly, 100 per cent good units must flow to the prior process, and this flow must be rhythmic without interruption. Therefore, quality control is so important that it must coexist with the JIT operation throughout the Kanban system. Autonamation means to build in a mechanism a means to prevent mass production of defective work in machines or product lines. Autonamation is not automation, but the autonomous check of abnormality in the process.

The autonomous machine is a machine to which an automatic stopping device is attached. In Toyota factories, almost all the machines are autonomous so that mass production of defects can be prevented and machine breakdowns are automatically checked. The idea of Autonamation is also expanded to the product lines of manual work. If something abnormal happens in a product line, the worker pushes the 'stop' button, thereby stopping his whole line. For the purpose of detecting troubles in each process, an electric light board, called Andon, indicating a line stop, is hung so high in a factory that it can easily be seen by everyone. The Andon in the

Toyota System has an important role in helping this automatic check, and is a typical example of Toyota's "Visual Control System."

7.4 LEAN MANUFACTURING

Lean is simply the thought process, not a tool, you use to look at your business, whether it is manufacturing, service or any other activity where you have a supplier and a customer/receiver.

The origins of 'Lean' date back well before the twentieth century even though it wasn't called Lean. The word 'Lean' came from a student at MIT while he was doing research to discover the differences between some Japanese carmakers and the traditional North American carmakers.

The key thought processes within Lean are identifying 'waste' from customer's perspective, and then determining how to eliminate it the 'right' way. *Waste* is defined as the activity or activities that a customer would not want to pay for and/or that add no value to the product or service from the customer's perspective. "Doing it the right way" means that you or your group has not only identified 'waste' but has designed a plan to reach the Future State in an effective manner that encompasses the entire 'system'.

Most companies we encounter see 'waste' and then want to attack it. The problem they face is attacking of waste without a plan creates an environment where everyone is eliminating waste in their area or department without total consideration for the upstream or downstream activities. We call this 'point' improvements or 'exciting chaos'. The result of 'exciting chaos' is minimal bottom line or, worse still, detrimental effect on the bottom line with a negative impact on the culture. Excitement is created by everyone doing their own job and being disconnected from the 'system', but feeling good about their actions. So it is important to realize that leadership knowledge and passion without a plan will hinder your results and future progress.

Knowledge + Passion – A plan = Exciting chaos
Knowledge + Passion + A plan = Success

Lean is about being able to assess your organization from customer's perspective and knowing how to eliminate the waste properly knowing when and how to apply the tools (Value Stream Mapping, 5S, Six Sigma, SMED, Kanban, Cells, Supermarkets, etc.) to maximize results. If done properly, then you will decrease (a) your lead time, (b) your costs (bottom line), (c) defects, and increase your ability to respond to the ever-changing customer demands.

7.4.1 What is Lean Manufacturing?

Lean Manufacturing can be defined as: "A systematic approach to identifying and eliminating waste (non-value-added activities) through continuous improvement by running the product at the pull of the customer in pursuit of perfection."

The 10 rules of lean production can be summarized as follows:

1. Eliminate waste
2. Minimize inventory
3. Maximize flow
4. Pull production from customer demand

5. Meet customer requirements
6. Do it right the first time
7. Empower workers
8. Design for rapid changeover
9. Partner with suppliers
10. Create a culture of continuous improvement.

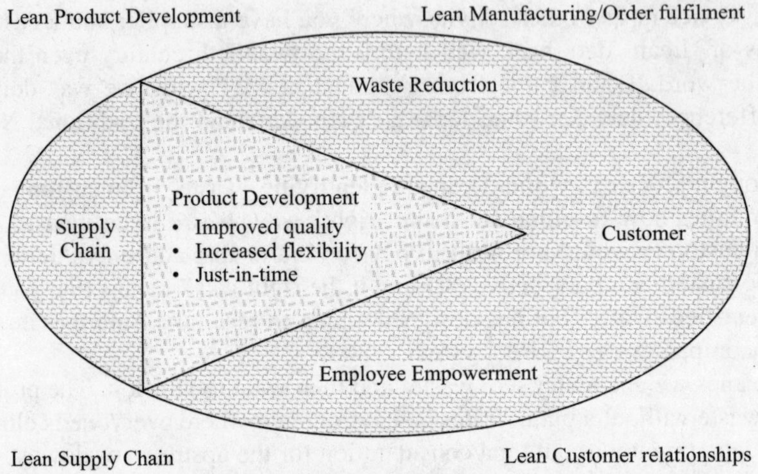

Figure 7.2 Lean production system.

7.4.2 Principle of Lean

A lean enterprise is defined as one whose aim is **"to create a new mechanism for looking at the whole, a channel for the value stream within the enterprise"**. Lean enterprise principles focus on five basic principles:

1. **Value** Specific product meeting specific needs at a specific time and specific price.
2. **Value stream** Transformation of product from design through production from order to delivery.
3. **Flows** A consistent and level flow of all value-stream steps linked together.
4. **Pull** Responding to the demand 'pull' from the customer flowing through the value stream.
5. **Perfection** Forever reducing wasteful practices to meet or exceed customer expectations.

These five principles form the basis for the Lean Enterprise concept of operations. The goal is to minimize or eliminate as many non-value-added steps in the process across the entire product supply, production and customer delivery cycle as possible. With this goal in mind, the remainder of this presentation will focus on the integration of MRP and JIT as they are executed in a Lean Enterprise Production System.

7.4.3 Some Basic Characteristics of Lean Manufacturing

- Elimination of waste
- Equipment reliability
- Process capability
- Continuous flow
- Material flows, one part at a time
- Less inventory required throughout the production process, raw material, WIP, and finished goods
- Reducting defects
- Reduction of lead time
- Error proofing
- Stop the line quality system
- Kanban systems
- Standard work
- Visual management
- In station process control
- Production level
- Task Time
- Quick changeover
- Teamwork
- Point-of-use storage

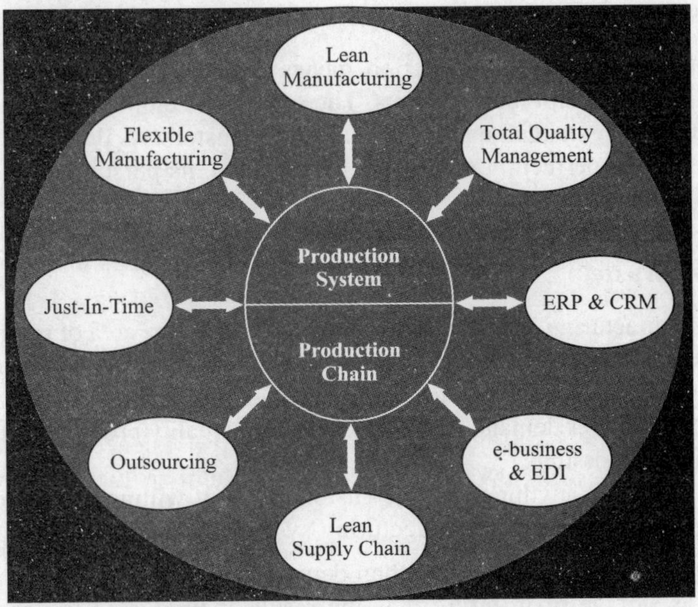

Figure 7.3 Production system and chain.

7.4.4 Value of Product

In lean production, the value of a product is defined solely by the customer. The product must meet the customer's needs at both a specific time and price. The thousands of mundane and sophisticated things that manufacturers do to deliver a product are generally of little interest to customers. To view value from the eyes of the customer requires most companies to undergo comprehensive analysis of all their business processes. Identifying the value in lean production means to understand all the activities required to produce a specific product, and then to optimize the whole process from the view of the customer. This viewpoint is critically important because it helps identify activities that clearly add value, activities that add no value but cannot be avoided, and activities that add no value and can be avoided.

7.4.5 Continuous Improvement

The transition to a lean environment does not occur overnight. A continuous improvement mentality is necessary to reach your company's goals. The term 'continuous improvement' means incremental improvement of products, processes, or services over time, with the goal of reducing waste to improve workplace functionality, customer service, or product performance (Suzaki, 1987). Continuous improvement principles, as practised by the most devoted manufacturers, result in remarkable improvements in performance that competitors find nearly impossible to achieve.

Lean production, applied correctly, results in the ability of an organization to learn. As in any organization, mistakes will always be made. However, mistakes are not usually repeated because this is a form of waste that the lean production philosophy and its methods seek to eliminate.

7.4.6 Perfection

The concept of perfection in lean production means that there are endless opportunities for improving the utilization of all types of assets. The systematic elimination of waste will reduce the costs of operating the extended enterprise and fulfils customer's desire for maximum value at the lowest price. While perfection may never be achieved, its pursuit is a goal worth striving for, because it helps maintain constant vigilance against wasteful practices.

7.4.7 Focus on Waste

The aim of Lean Manufacturing is the elimination of waste in every area of production, including customer relations, product design, supplier networks, and factory management. Its goal is to put in less human effort, less inventory, less time to develop products, and less space to become highly responsive to customer demand while producing top quality products in the most efficient and economical manner possible.

Essentially, a 'waste' is anything that the customer is not willing to pay for. Typically, the types of waste considered in a lean manufacturing system include the following:

1. *Overproduction:* To produce more than demanded or produce it before it is needed. It is visible as storage of material. It is the result of producing to speculative demand. Overproduction means making more than is required by the next process, making earlier

than is required by the next process, or making faster than is required by the next process. Causes for overproduction waste are, among others:

(a) Just-in-case logic
(b) Misuse of automation
(c) Long process setup
(d) Unlevelled scheduling
(e) Unbalanced work load
(f) Over engineered
(g) Redundant inspections

2. *Waiting:* Delay for a machine to process should be eliminated. The principle is to maximize the utilization/efficiency of the worker instead of maximizing the utilization of the machines. Causes of waiting (waste) include:

(a) Unbalanced work load
(b) Unplanned maintenance
(c) Long process set-up times
(d) Misuses of automation
(e) Upstream quality problems
(f) Unlevelled scheduling

3. *Inventory or Work in Process* (*WIP*): It is defined as usable but idle resources. Causes of excess inventory include:

(a) Protecting the company from inefficiencies and unexpected problems
(b) Product complexity
(c) Unlevelled scheduling
(d) Poor market forecast
(e) Unbalanced workload
(f) Unreliable shipments by suppliers
(g) Misunderstood communications
(h) Reward systems

4. *Processing waste:* It should be minimized by asking why a specific processing step is needed and why a specific product is produced. All unnecessary processing steps should be eliminated. Causes for processing waste are, among others:

(a) Product changes without process changes
(b) Just-in-case logic
(c) True customer requirements undefined
(d) Over processing to accommodate downtime
(e) Lack of communications
(f) Redundant approvals
(g) Extra copies/excessive information

5. *Transportation:* It does not add any value to the product. Instead of improving the transportation, it should be minimized or eliminated (e.g. forming cells). Some of the causes of transportation waste are:
   (a) Poor plant layout
   (b) Poor understanding of the process flow for production
   (c) Large batch sizes, long lead times, and large storage areas
6. *Motion:* Movement of the workers, machines, and transport (e.g. due to the inappropriate location of tools and parts) is a waste. Instead of automating wasted motion, the operation itself should be improved. Causes of motion waste include:
   (a) Poor people/machine effectiveness
   (b) Inconsistent work methods
   (c) Unfavourable facility or cell layout
   (d) Poor workplace organization and housekeeping
   (e) Extra 'busy' movements while waiting
7. *Making defective products:* It is pure waste. Prevent the occurrence of defects instead of finding and repairing defects. Causes of processing waste include:
   (a) Weak process control
   (b) Poor quality
   (c) Unbalanced inventory level
   (d) Deficient planned maintenance
   (e) Inadequate education/training/work instructions
   (f) Product design
   (g) Customer needs not understood.
8. *Underutilizing people:* Not taking advantage of people's abilities. Causes of people waste include:
   (a) Old guard thinking, politics, the business culture
   (b) Poor hiring practices
   (c) Low or no investment in training
   (d) Low pay, high turnover strategy

Nearly every waste in the production process can fit into at least one of these categories. Those that understand the concept deeply view waste as the singular enemy that greatly limits business performance and threatens prosperity unless it is relentlessly eliminated over time. Lean Manufacturing is an approach that eliminates waste by reducing costs in the overall production process, in operations within that process, and in the utilization of production labour. The focus is on making the entire process flow, not the improvement of one or more individual operations.

7.4.8 The Relationship of Waste to Profit

1. ***Customer Focus:*** *Waste factor. Zero customer dissatisfaction/Relationship to profit:* Customer input and feedback assures quality. Customer satisfaction supports sales.

2. ***Leadership:*** *Waste factor. Zero misalignment/Relationship to profit.* Direction and support for development improves cost, quality, and speed.
3. ***Lean organization:*** *Waste factor. Zero bureaucracy/Relationship to profit.* Team-based operations reduce overhead by eliminating bureaucracy and ensuring information flow and cooperation.
4. ***Partnering:*** *Waste factor. Zero stakeholder dissatisfaction/Relationship to profit.* Flexible relationships with suppliers, distributors, and society improve quality, cost, and speed.
5. ***Information architecture:*** *Waste factor. Zero lost information/Relationship to profit.* Knowledge required for operations is accurate and timely, thus improving quality, cost, and speed.
6. ***Culture of improvement:*** *Waste factor. Zero wasted creativity/Relationship to profit.* Employee participation in eliminating operations waste improves cost, quality, and speed.
7. ***Lean Production:*** *Waste factor. Zero non-value-added work/Relationship to profit.* Total employee involvement and aggressive waste elimination promote speedier operations and eradicate inventories.
8. ***Lean equipment management:*** *Waste factor. Zero failures, zero defects/Relationship to profit.* Longer equipment life and design improvements reduce cost. Meticulous maintenance and equipment improvements increase quality. Absolute availability and efficiency increase speed.
9. ***Lean engineering:*** *Waste factor. Zero lost opportunity/Relationship to profit.* Early resolutions of design problems with customers and suppliers significantly reducing cost, while improving quality and cycle time.

7.4.9 Four Functions of Lean Production

Lean production can be defined by reviewing Figure 7.2, Lean Enterprise Model, to understand the major components necessary to define and understand a lean manufacturing enterprise. There are four primary functions, represented by the four pillars of the model in Figure 7.4, that focus on specific concepts, attributes and objectives required in a lean production environment. These four functions include:

Production Flow	– Products, workflow, equipment and support
Planning	– Planning, scheduling, and control
Organization	– Accountability, responsibility, teaming and culture
Performance Measures	– Measurements, performance and rewards

The combination of these four functions is then integrated into the enterprise by defining various product families, production units and production cells that are focused on the needs and requirements of the customer.

7.4.10 Production Flow

Production flow is focused on defining products, equipment and workflows that can be produced in small lot production runs being 'pulled' by downstream customer demands. A key

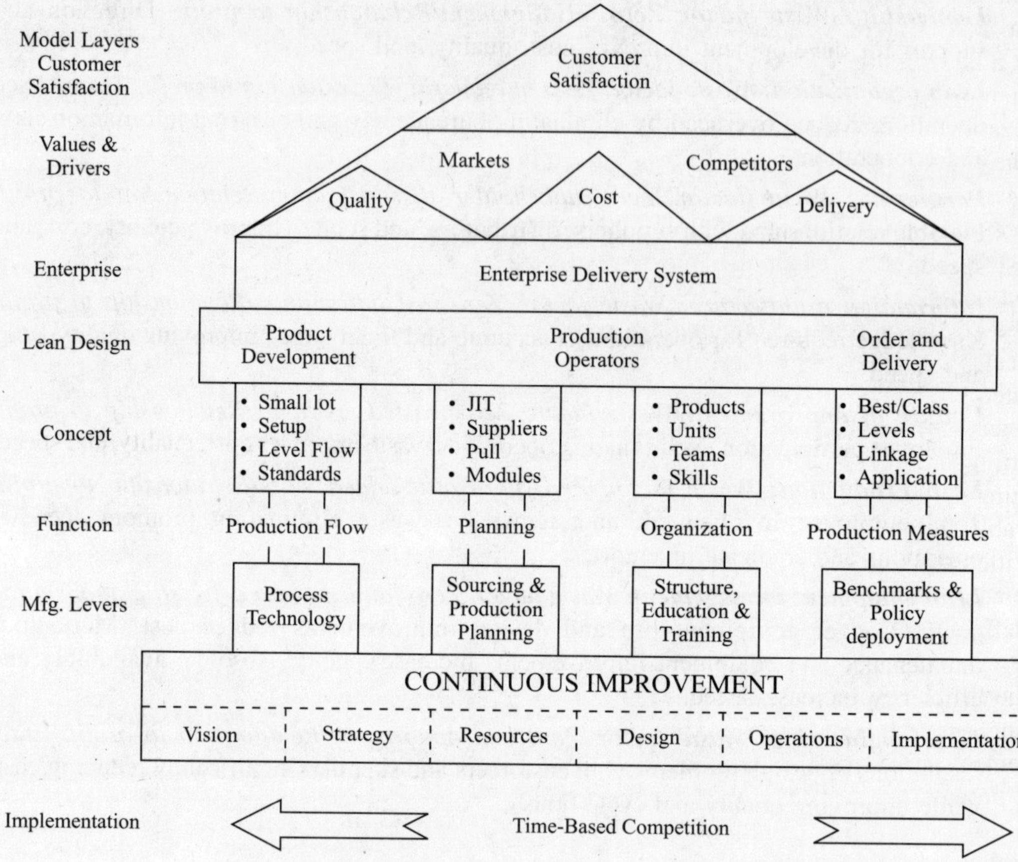

Figure 7.4 Lean enterprise model.

strategy in this approach is that of the Single-Piece Flow concept that emphasizes small lot production thinking. Enterprise design considerations include product design, production flow arrangement; and production units and cells that are oriented to getting the product to the customer in the shortest period of time and at the lowest cost with superior quality, providing customer satisfaction.

7.4.11 Planning

Planning involves scheduling and control. Key design concepts should include the separation of the planning (MRP System) functions from the execution, JIT and cellular manufacturing, functions. The role of MRP is to maintain a high-level profile for the development of plans, resources, financial goals and capacity analysis. It must also project intermediate and long-range procurement objectives to suppliers. MRP boundaries must be altered in a lean production environment to allow for the hand-off to process-oriented control systems that are directed by production units and cells. These units will operate from a different set of rules and controls.

The day-to-day and shift-by-shift operations within the manufacturing process will operate from a 'Pull' system concept. The physical facility and product configuration will be altered to reflect a new production flow and layout. A new plant-wide execution and control system will in turn, replace the 'traditional' MRP order-based systems, typically found in most MRP-oriented organizations that will assist the transition toward Lean production via JIT execution.

The primary concepts that must be defined and understood include:

Reconciliation: the process of high-level demand planning and scheduling

Regulation: the process of regulating command and control sequences within a pull-process environment capable of controlling 'workable work' for each production unit and production cell.

The key to successful planning and execution is ensuring that all production units, lines or work cells are loaded with workable work by the overall planning function. In some cases, a dedicated production unit or cell scheduler performs this activity as part of the production unit or cell team. In smaller work cells, the cell leader has responsibility for this activity. This entails converting production demands into a detailed plan, ensuring required materials are present, managing inventory and acting as a liaison with the *master scheduler* to confirm that the plan is achievable.

Within both processes, analysis must be performed to determine how small lot production can be achieved in a near continuous flow process that is physically and logically controlled by well-defined product groupings associated with focussed production units and cells all focused on the customer.

The major change to MRP and MRPII systems is determining where to stop the end of the planning process and when to let the physical process determine execution needs and requirements based on the pace, referred to as Take Time, flow and sequencing needs of the actual process.

In the traditional MRP environment, these decisions and actions are planned out through MRP-related activity modules and rules governed by inventory, firm and planned orders, and demand in the form of actual customer orders, production orders, forecasts, production plans, etc. Capacity for the most part in MRP is considered *infinite*, and great emphasis is laid on bills of materials, projected lead times and other factors which influence the development and issuance of manufacturing and purchase orders that drive the process.

In a *process-oriented* pulled manufacturing environment, actual production execution coupled to well-defined process flows, linked to JIT support, determine the rate, flow and finite control characteristics of the process. All activities in this type of execution process are linked through well-defined production flows that have synchronized operations, standardized work content, and built-in product and process quality control functionality to control and monitor the process. Thus, the entire process is linked as one operation at a time through all operations within the production unit or work cell. All operations have the capability to respond to the demand of the next "downstream" process when the proper signal is given to pull product into this area or function.

In many cases, the use of an "assembly broadcast system" approach is used. A given number of future assembly days will be issued electronically to suppliers. For the immediate rolling five-day assembly schedule, actual assembly model line sequence numbers will be issued for product configuration to assist in detailed assignment of component sequencing requirements for inbound material.

A real-time Plant Operations Control (POC) system should be incorporated into the systems design requirements to assist in performing real-time directives for each material move and be capable of giving back real-time status information on each component, operation or workstation. This can be accomplished by incorporating the use of bar code and radio frequency (RF) technologies working in unison with both automated and manual material handling systems if required by the nature of the product.

The design and use of information and control systems will be critical to the success of a well-defined manufacturing execution system. Information workstations should be defined at numerous locations throughout the process and in key production support offices that will be disbursed within the plant, Logistics planning communication will be made easier by decentralizing the planning and control activities from one physical administrative office to strategically placed offices by function and task within the plant and process. This will make daily and hourly communication needs more responsive to the manufacturing process and provide better feedback to the MRP-based planning systems to ensure that planning and execution will occur at all levels of the process.

Each area of the manufacturing process will need to define its operating requirements and translate these needs into well-defined workstation definitions. Workstations should be standardized as much as possible and built around a "work team" concept. Teaming areas should be physically designed into the process to allow each team to work out daily logistics, production and quality concerns by production unit, work cell or manufacturing function. The major emphasis of the entire execution system should be focused on implementing a 'Pull' system approach whereby the actual process directs actions and events that must occur to operate the production process.

These events should reflect Visual Factory Techniques where there is possible and should be as simple as for ease of use by the workers. The use of well-defined Work Breakdown Instructions by workstation and function should be planned for and designed into the entire production process. This is referred to as *standardized work.* Ultimately, all functions, systems, material movements and control methods should be integrated into one plant-wide operating system that can then interface to higher level MRPII-based planning systems.

Supply chain logistics completes the planning and execution function by having fewer supply partners that have more integration with the manufacturing organization on a long-term partnership basis. The partnership arrangement must have well-defined interfaces for information sharing, material handling and sequencing for JIT release notification, point-of-use delivery and change notification for both schedule and product engineering changes.

Suppliers must have the capability to deal with their downstream partners at two levels. The first level deals with high-level capacity, cost and resource planning commitments. This will be an active interface with the customer's ERP and MRPII planning systems. The second level will deal with weekly, daily or even hourly production releases for JIT deliveries in a proper sequence, production lot sizes, containers and product configurations that support the operations style of the customer's lean production operations strategy and execution plan. Directives and plans from both these systems must have full integration but distinct origins based on the 'Pull' system demands from the execution of the manufacturing process and planning system demands for high-level planning concerns.

7.4.12 Performance Measures

Performance measures must be developed in a hierarchy that is cascaded upward and downward within the organization. The measures should be understandable, simple to administer and enable management and team members to easily make decisions and take action when appropriate.

7.4.13 Organization

Organization focuses on a new organizational structure that is oriented to provide the necessary resources for each production unit and its associated work cells and teams that directly support products and services directly linked to the customer. This will require a dramatic change from typical pyramid-style organizations of the past where like job function and activities were grouped together. Lean production demands a rethinking of the structure, role and responsibility of each individual within the organization.

The formation of production units operating as 'business units' will become a new way of life for most organizations who truly understand the needs of their customers and organize all resources, products and capabilities to support and add value to its customers needs and expectations.

Transitioning to a lean production organization

Most organizations typically have numerous projects and initiatives ongoing simultaneously within the organization. This tends to cause confusion and conflict when it comes to fully implementing Lean production. This is especially true with the Planning function when large concept, procedure, operations directives and information software systems must be altered or changed to support Lean production strategies.

The organization should have a high-level champion in a senior-level position that will sponsor the transition to "Lean". There should also be a dedicated implementation team that comprised cross-functional representation from within the organization. The role of this team is to administer direction, focus and leadership that support the future vision and end-state look of the organization. All initiatives would then come under the direction and guidance of this group.

Lean Production is another integral step toward future development of the Lean Enterprise. It is one of the first steps that will dramatically change the way that the organization plans, schedules and operates its production environment. MRP, JIT, CIM and other organizational initiatives must be integrated into one common production system which is customized to the needs and requirements of each organization relative to the needs and requirements for its customers. Like the popular and most benchmarked production system in the world, the Toyota Production System (TPS), the new Lean Production System of the future will be another migration and transition along the path to world-class competitiveness for the future. It will only succeed when each and every individual within the organization embraces the new operating philosophy and becomes a strong proponent of its principles and practices.

7.4.14 The True Costs of Inventory

Reducing inventory is an important goal of the lean organization. Carrying inventory has many costs associated with it. Obvious costs include: capital tied up in inventory and the associated loss of interest on that capita., loss due to material handling damage, increased labour costs for material handling, and increased space and storage requirement. A cost from excess inventory which is not so obvious is quality. In fact, many companies have seen quality improvements resulting from inventory reductions while not focusing on quality. The reasoning is that if an upstream process is producing parts on a machine and defects occur halfway through the batch in an organization with low levels of inventory the next downstream process will discover the defects sooner. An organization with low inventory levels can stop the process when the defect is discovered, throw out the defective inventory, and request the previous process to start another batch. The organization with lower inventory levels will also be more effective at determining what caused the defect because the batch that the defect occurred in is fresh in the minds of both production and maintenance.

7.4.15 Production Operations

Production Operations can be further divided into two types: **Continuous** and **Discreet** production. *Continuous production* is characterized by those operations which produce 'streams' of products, such as laundry soap, food products, paper products, general materials production, and other products produced in bulk and made into discreet products during the packaging process. Continuous production may be further characterized by a relatively high-level of automation, although this isn't always the case. *Discreet production* is characterized by those operations which produce 'stand-alone' products, or individual components. Both continuous and discreet production systems will have different challenges and opportunities with respect to lean implementation.

As with all lean implementations, three things can be done to improve performance on the shop floor. These are: *eliminate waste* (80%), work toward *continuous, one-piece product flow* (10%), and establish *customer pull* (10%). The percentages indicate relative importance (e.g. 80% of the benefits for implementing lean will come from 'waste elimination' efforts - your benefits may vary).

Waste elimination efforts begin on the shop floor. This is not a management responsibility, it is everyone's responsibility. It is incumbent on the management to provide the tools and training so that all personnel can become engaged in this activity. Some of the lean tools used in waste elimination efforts are:

- Teian systems
- Quick changeover (using single minute exchange of die techniques)
- 5S programs
- Establishment of Standard Work
- JIT supply and production
- Total productive maintenance
- Visual management tools and techniques
- Inventory reduction techniques
- Workspace layout techniques
- Mistake-proofing techniques

Depending on the specific challenges and opportunities that exist within your facility, some of these tools will have more 'bottom-line' impact than others. Careful planning and consideration are required in order to ensure that 'real' benefits are being achieved, and that further waste isn't being introduced into our systems.

Continuous, one-piece flow-Redesign of the work process is the responsibility of the management, and redesigning the work process to allow significant reduction in batch sizes is a considerable undertaking. But why would you want to convert your operations to one-piece flow?

Consider how much (in terms of space, dollars, labour to move, labour to account for, spoilage/obsolescence) exists in your factory. It exists in the forms of Raw Materials (RM), Work-in-Process (WIP), and Finished Goods (FG). In many factories, this amounts to millions of dollars in labour and materials. Also, we have developed (at further, considerable cost) extensive systems to manage this inventory. All of this represents considerable recurring costs, and as we will see, has a negative impact on both quality and delivery as well.

The reasons we have so much inventory (RM, WIP, FG) are two-fold. First, our systems are designed to process *Batches*. Second, inventory provides buffers throughout our processes, reducing the risks associated with unreliable supply lines, poor quality, poor equipment maintenance, and poor production scheduling. By solving these problems, the inventory becomes a hindrance to further improvements. Reducing batch sizes incrementally will expose problems in our production systems. Elimination of the problems will allow further reduction in batch sizes. Some of the tools used to achieve *continuous, one-piece flow* are:

1. Incremental inventory reduction
2. Root-cause analysis/5 Why's
3. Standard work
4. Mistake-proofing
5. Kaizen blitz/Improvement workshops
6. Just-in-Time supply of materials
7. Supply chain management techniques
8. Jidoka/Autonomation

While true 'one-piece flow' may not be achievable in all circumstances, significant reductions are possible, with significant improvements in costs, quality, and delivery.

Customer pull-The concept of 'pulling' value through the production system is contrasted by most systems, which either Make-To-Stock (MTS) or 'PUSH' batches through a *Master Production Schedule* or Material Requirements Planning (MRP/MRPII) software. 'Push' systems produce without regard to demand or the requirements of downstream processes. The difference may seem insignificant, but in reality, pull systems can further (and significantly) reduce costs and improve throughput/delivery.

Implementing 'pull' means redesigning the entire system of production from the point of view of your customers. The system doesn't begin production until an order is received from the customer. This order triggers the entire production system, from the supply chain through each production process. Some of the tools used to implement *customer pull* are:

1. Kanban production scheduling
2. System-wide inventory reduction planning
3. Heijunka scheduling
4. Just-in-Time production
5. Supply chain management
6. Kaizen blitz/improvement workshops
7. Decoupling of MRP/MRPII systems from production scheduling

True 'customer pull' may not be realistic in all production environments. Portions of the system (based on operational requirements or significant cost factors) may require the development of a *hybrid* system (a pull system which contains certain push elements). Such hybrid systems are still very effective in reducing costs and improving responsiveness to customers.

7.4.16 Operations Support

As opposed to *Production Operations*, the rest of the organization exists to support those profit-generating operations. Operations support is separated in many companies from typical management functions (such as Human Resources/Payroll, Accounting/Finance, Sales/Marketing, and Executive/Staff), but they share a common characteristic that distinguishes them from revenue generating departments: they are all considered Non-Value-Added but Essential (NVAE), and, therefore, require special, focused consideration in applying the lean principles.

The aim of leaning support operations is necessarily to:

1. increase service levels to production operations with increased quality and similar cost,
2. provide a similar level of support with significantly less resources, and
3. give reduced levels of support, transferring certain functions to entities (either internal or external to the organization) which can provide those functions at significantly reduced costs.

Some of the tools used to implement lean in *operations support* environments are:

1. Strategic (Hoshin) planning techniques
2. Supply chain management techniques
3. System-wide inventory reduction planning
4. Kanban production scheduling
5. Kaizen blitz/Improvement Workshops
6. Decoupling of MRP/MRPII systems from production scheduling
7. 5S programs

Lean efforts in operations support require a long-term view of where the organization is with respect to *mission, goals,* and *milestones,* as it will entail system-wide modifications that could profoundly impact production operations. Great care must be taken to ensure that service levels are maintained during this (typically 2-5 year) transition, so that cost, quality, and service/delivery levels aren't compromised to our customers.

7.4.17 The Supply Chain

As an organization proceeds with 'leaning' it' internal systems (both on the operations side and on the support side), it will soon run into constraints which will slow the lean effort. And while some of these constraints are internal (and can be overcome), some will be external, in the Supply Chain.

These constraints normally take the following forms:

1. Poor quality products (products which have significant variation)
2. Products don't meet specifications (substitute products don't conform to requirements)
3. Poor service/delivery levels (resulting in critical product shortages)
4. Delivery of product in unnecessarily large batch quantities
5. Poor marking of products/packaging (resulting in poor internal delivery of product to production lines)
6. Internal company processes contain significant overhead in the ordering/purchasing cycle

Supply chain management techniques offer the opportunity to not only considerably reduce the costs associated with raw material: purchases, but can significantly increase both delivered product quality and delivery/services levels. Some of the tools used to apply lean to the *supply chain* are:

1. Strategic (Hoshin) planning techniques
2. Outsourcing of functions in which the company doesn't have a competitive advantage
3. Supplier consolidation, partnering, and replenishment
4. System-wide inventory reduction planning
5. Kanban production scheduling
6. Kaizen blitz/Improvement workshops

From the perspective of the Customer, the Supply Chain is broken down into the following components:

- *First-tier supplier*-a supplier that invoices the customer for goods and services rendered directly by that supplier.
- *Second-tier supplier*-a supplier that invoices the first-tier supplier for goods and services rendered.
- *Third-tier supplier*-a supplier more than one step removed from the first-tier supplier, providing goods and services to second-tier suppliers.

Thus, if we consider 'ourselves' (or our company) the 'customer' (not exactly true in the lean sense, but common usage), our vendors are the 'first-tier' suppliers, their vendors are 'second-tier', and all other vendors are 'third-tier' vendors. Typical *supply chain* lean efforts focus on the *first-tier* suppliers.

7.4.18 The Lean Enterprise

The Lean Enterprise considers that every portion of the system of producing goods and services must continue to eliminate waste, provide a flow of value to the customer, based on the customer pulling this value through the system. This 'Enterprise' exists from the point where our raw materials come from, through the supply chain, through our production system, to our customers, and even including the disposal cycle of our products.

One of the ramifications of this concept is that we can't reduce waste in any portion of the system simply by pushing this waste to another portion of the system. We can't make our suppliers bear the cost of holding inventory we don't want. We should work with them to improve their portion of the system so that their reliability is improved to the point where the inventory (which reduce the risk of poor reliability) is no longer necessary to maintain expected service/delivery levels. We can't expect the customer to bear extra costs, poorer quality, or decreased service/delivery levels due to our system improvements. True waste elimination removes waste from the system, rather than transferring it within the system.

7.4.19 Leaning Your Customers

Believe it or not, as we lean our operational systems, our support system, and our supply chain, we will find that additional lean constraints exist in an unexpected place - with our customers. But how can this be? In lean companies, customers drive everything we do. Big or small orders, the customer is the reason we exist. But customers can exhibit non-lean behaviours, which will drive portions of our system to contain waste, which will add to costs, diminish quality, and reduce our service/delivery levels.

Consider the following scenarios, and how they affect the production systems:

1. If we are a second-tier supplier (or provide goods to a government agency), our customers may be required to obtain competitive bids in order to contain costs. This process forces each competing organization to respond to bids that may not convert to orders. Accordingly, each competing organization must consider the cost of this bid process as overhead in the product pricing process. Since this process does not add value to the products, it is considered waste.
2. Because Economic Order Quantity (EOQ) purchasing is still prevalent in many organizations, the customers may tend to order the products in large batches, periodically throughout the year. These orders hinder one's efforts to create a smooth production flow, and may require one to actually process large batches within the facility. At a minimum, the company will have to maintain large batches until they ship them, which carries both cost and risk, as well as disruption within the systems.
3. Customers often ask for a great many features in the products, without regard to the necessity of these features. In order to meet these inflated specifications, we must design the systems to produce these features (often in conjunction with features requested by other customers as well). The resulting 'feature bloat' has an enormous impact on cost, quality/reliability, and service/delivery.

By working more closely with the customers, a company can gather a better idea of what products they need, what features are required, and how often they should be ordering. If the company is able to prove to the customers that it is the *best* supplier of these goods and services, the company should be able to be their 'sole source' for these items, giving it a huge advantage among its competitors. Thus, the establishment of the *Lean Enterprise* should be the company's ultimate goal: a system which is able to provide the *exact* goods and services required by the company's customers, exactly where and in the required quantities, at competitive cost, and with superior quality.

7.4.20 Keys to Lean Success

The following are some considerations to successful lean implementation:

1. Prepare and motivate people
   - Widespread orientation to continuous improvement, quality, training and recruiting workers with appropriate skills
   - Create common understanding of need to change to lean
2. Employee involvement
   - Push decision-making and system development down to the "lowest levels"
   - Trained and truly empowered people
3. Share information and manage expectations
4. Identify and empower champions, particularly operations managers
   - Remove roadblocks (i.e. people, layout, systems)
   - Make it both directive yet empowering
5. Atmosphere of experimentation
   - Tolerating mistakes, patience, etc.
   - Willingness to take risks
6. Installing *enlightened* and realistic performance measures, evaluation, and reward systems
   - Do away with rigid performance goals during implementation
   - Measure results and not number of activities/events
   - Tie improvements, long term, to key macro-level performance targets (i.e. inventory turns, quality, delivery, overall cost reductions)
7. The need to execute pilot projects prior to rolling out culture across the organization
   - After early wins in operations, extend across ENTIRE organization

7.4.21 Characteristics of Lean Manufacturing

- Integrated single piece continuous workflow.
- Close integration of the whole value chain from raw material to finished product through partnership-oriented relations with suppliers and distributors.
- Just-in-Time processing: a part moves to a production operation, is processed immediately, and moves immediately to the next operation.
- Short order-to-ship cycles times; small batch production capability which is synchronized to shipping schedules.
- Production is based on orders rather than forecasts; production planning is driven by customer demand or 'pull' and not to suit machine loading or inflexible work flows on the shop floor.

- Minimal inventories at each stage of the production process.
- Quick changeovers of machines and equipment allow different products to be produced with one-piece flow in small batches.
- Layout is based on product flow.
- Total quality control. Active involvement by workers in troubleshooting and problem solving to improve quality and eliminate wastes.
- Defect prevention rather than inspection and rework by building quality in the process and implementing real-time quality feedback procedures.
- Team based work organizations with multiskilled operators empowered to make decisions and improve operations with few indirect staff.

7.4.22 Benefits of Lean Production

Establishment and mastering of a lean production system would allow you to achieve the following benefits:

1. Waste reduction by 80%
2. Production cost reduction by 50%
3. Manufacturing cycle times decreased by 50%
4. Labour reduction by 50% while maintaining or increasing throughput
5. Inventory reduction by 80% while increasing customer service levels
6. Capacity in current facilities increase by 50%
7. Higher quality
8. Higher profits
9. Higher system flexibility in reacting to changes in requirements improved
10. More strategic focus
11. Improved cash flow through increasing shipping and billing frequencies

However, by continually focussing on waste reduction, there are truly no end to the benefits that can be achieved.

All types of manufacturers are discovering the advantages of doing a Lean analysis and applying the principles of Lean Manufacturing to their own company. Perhaps you are faced with one or many, of the following challenges:

- Missed order dates
- High product cost relative to the competition
- Declining market share due to delivery time or cost problems
- Limited capacity

If so, Lean Manufacturing can have an immediate and positive impact on your company. Through the process of implementing Lean Manufacturing, you will be able to find ways to achieve a number of benefits. Results will vary, but here are some ways to achieve savings and improvements:

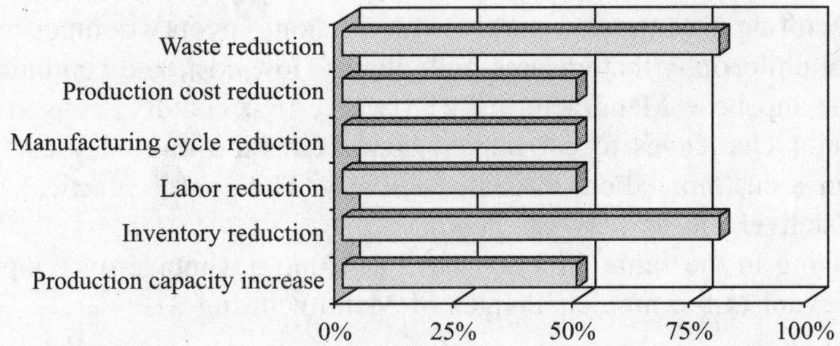

Figure 7.5 Implementation of Lean Manufacturing.

Reduce:

Manufacturing Lead Time	: 50 – 90%
Floor Space Requirements	: 5 – 30 %
Work-in-Process	: 60 – 80%

Increase:

First-Pass Yields	: 50 – 100%
Throughput	: 40 – 80%
Productivity	: 75 – 125%

7.4.23 Applications

Lean techniques are applicable not only in manufacturing but also in the service-oriented industry and the service environment. Every system contains waste, i.e. something that does not provide value to your customer. Whether you are producing a product, processing a material, or providing a service, there are elements which are considered 'waste'. The techniques for analyzing systems, identifying and reducing waste, and focussing on the customer are applicable in any system, and in any industry.

Lean thinking may also be applied for getting rid of bureaucracy in your home office. To run your home office more effectively and faster, you may need just as little as 10% of its current staff. Only executives who have a direct involvement with finding, keeping, or growing customers as well as key support staff—accountants, tax, legal and human resources people - should stay. Sending to an operating unit can rehabilitate others.

7.5 AGILE MANUFACTURING

7.5.1 Introduction

Agile manufacturing is a recent movement viewed by the auto industry, which shares with the consumer electronics industry the distinction of being the pacesetter in manufacturing process innovation, as the next step in its development. It represents the demise of the century-long tradition of manufacturing driven by scale. It aspires to total flexibility without sacrificing quality or incurring added costs.

Agile manufacturing is contrasted with lean production, Toyota's composite of tools, culture, and organizational philosophy that ensures high quality, low cost, and continuous and sustained improvement. The Japanese Manufacturing 21 (twenty-first century) consortium defines it in terms of nine major challenges to car makers, one being the three-day car, three days from customer order for a customized car to dealer delivery. The goal is practical; leading Japanese auto makers can deliver the ten-day car now.

US firms moving in the same direction have a strong advantage over Japanese companies in some areas relevant to the nine challenges of Manufacturing 21.

1. Break dependency on scale and economies of scale (reducing set-up costs is key).
2. Produce vehicles in low volumes at reasonable cost (Nissan's Intelligent Body System, a Lego-block approach that favours existing over newly designed body components, leaves tooling as the only major expense for a new model).
3. Guarantee the three-day car.
4. Replace large centralized plants with distributed clusters of mini-assembly plants located near customers (as much as five days' time is required to ship cars to dealers; Japan's horrendous traffic congestion has become the weak link in JIT inventory management, with suppliers unable to deliver on time).
5. Be able to reconfigure components in many different ways.
6. Make work stimulating (those who carry out Lego-block production should not be treated as Lego blocks).
7. Turn the customer into a "prosumer," an ugly neologism that means proactive something; the idea is that the customer will take an active role in the product design by, for example, configuring options at a computer in a dealer showroom.
8. Streamline ordering systems and establish close relationships with suppliers.
9. Manage the massive volumes of data generated by the production system so as to be able to analyze that data quickly and in an agile fashion.

Agile production would appear to be the blueprint for future manufacturing. Managers in every industry would do well to incorporate the essence of the Manufacturing 21 challenges into their agendas. Publishing, retailing and banking are but a few of the industries likely to rally around agility.

Why Agility In Engineering?

- Customers are expecting:
  - Low volume products
  - High quality products
  - Custom products
- Very short product life cycles, development time, and production lead times are required.
- Customers want to be treated as individuals.

How does Agile manufacturing differ from World Class manufacturing?

Lean or *World Class Manufacturing* is being very good at doing the things you can control. *Agile manufacturing* deals with the things we cannot control. Agility is the ability to thrive and prosper in an environment of constant and unpredictable change. Agility is not only to

Traditional Manufacturing	Gaining Control	World Class Manufacturing	Agile Manufacturing
➢ Complex systems	➢ ERP or MRP II	➢ Lean manufacturing	➢ Enriching customer
➢ Departmentalism	➢ Better customer service	➢ Just-In-Time	➢ Competitiveness Through cooperation
➢ EOQ	➢ Reduced inventory (10–25%)	➢ Total Quality Mgt.	➢ Organizing for change & uncertainty
➢ No employee involve	➢ Lower production costs	➢ Much less costly	➢ People and information
➢ Financial secrecy	➢ Greater flexibility	➢ More responsive	➢ High customer flexibility
➢ High inventory	➢ Better control	➢ Long term profitability	➢ Technology
➢ Inspection	➢ Planned operations	➢ LT improvement	➢ Highly educated & trained work-force
➢ Lack of strategy	➢ Better communication	➢ Productivity improvement	➢ Flexible manufacturing structures
➢ Late delivery		➢ Time to market	➢ Virtual corporations
➢ Long cycle times		➢ Zero defects	
➢ Politics		➢ Inventory turns	
➢ Record inaccuracy			
➢ Shortages/Expedite			
➢ Month-end push			

Figure 7.6 Four steps to agile manufacturing.

accommodate change but to relish the opportunities inherent in a turbulent environment. Here are some of the axioms of agile manufacturing:

- Everything is changing very fast and unpredictably.
- The market requires low volume, high quality, custom and specific products.
- These products have very short life cycles and very short development and production lead times are required.
- Mass production is moribund. Customers want to be treated as individuals. (This leads to a people intensive, relationship driven operation.)
- Perfect quality and very high levels of service are expected and required.
- Products and services become information-rich.

The swift trend towards a multiplicity of finished products with small improvements development and production lead times has led many companies into problems such as inventories, overheads, and lack of efficiency. They are trying to apply the traditional mass production approach without realizing that the whole environment has changed. Mass production does not apply to products where the customers require small quantities of highly customized, design-to-order products, and where additional services and value-added benefits like product upgrades and future reconfigurations are as important as the product itself. While it is difficult for many people to accept that mass production must be replaced with new approaches, there are many telling signs. The proliferation of finished products; the rapidly increasing introduction of new products; short product life cycles; the customer demand for products that specifically address their needs—all these trends negate the fundamental ideas of mass production which have served the western industry so well for decades.

Even World Class Manufacturing and best practice approaches are based on the time-honoured concepts of mass production of standard products. The famed Toyota Production System (TPS) has two kinds of products; type A and type B. Type A is a standard product and type B is custom product. Practitioners of TPS strive hard to eliminate type B parts and products because they do not fit the concepts of one piece part flow and rate-based schedules (*heijunka*).

An agile approach to manufacturing faces the reality that a company must serve customers with small quantities of custom designed parts with perfect quality, 100% on-time delivery, and at very low cost. To adopt agile manufacturing requires that the company already be world class and use lean manufacturing methods. This is a starting point. You can only build agility on a firm foundation. These changes are more apparent in some industries than others. In the consumer electronics and automotive industries, the need for agility is most apparent. New products are coming very fast. The range of products is increasing at an astonishing level. The customers and the markets are becoming continuously more fragmented and specialized. Only the innovative and agile companies will survive these changes.

REVIEW QUESTIONS

1. Define JIT.
2. Discuss more about Toyota production system.
3. Compare Just-In-Time (JIT) and Lean production.
4. What is meant by Kanban system?
5. What are the salient features of automation?
6. What is lean manufacturing?
7. Enumerate the basic elements of lean manufacturing.
8. What are the functions of lean production?
9. What are the benefits of lean production?
10. Mention the advantages of Agile manufacturing.

Chapter 8

OPEN SYSTEM/CIM DATA MODEL

8.1 INTRODUCTION

For the past several years, manufacturing enterprises have increasingly become information integration in line with the introduction of CIM (Computer Integrated Manufacturing) and FA (Factory Automation) systems. The tendency to integrate information has rapidly increased and the use of information integration systems has become widespread. In the field of information processing technology, on the other hand, computer down sizing and using the communications technologies such as the Internet have remarkably evolved. Since enhanced capabilities are now brought about by such technological advances, it seems only natural that they begin to be used in the factories. As a result, a lot of control systems having a host of information functions are introduced into the shop in the factories, making information integration factories real.

In addition to the technical aspects, the manufacturing enterprises should realize the necessity for overseas expansion of factories, lengthening of facilities life cycle, and introduction of management information systems. Therefore, manufacturing enterprises want to make a distributed type of CIM system. In order to make a distributed type of CIM system, the use of open manufacturing systems and packaged factory software is now being highlighted.

8.2 OPEN MANUFACTURING SYSTEMS

The manufacturing system structure is now beginning to switch from centralized CIM type, which is centred on a large number of computers or large-scale database to a distributed type CIM, in which a large number of personal computers and other small-size computers are joined to form a network. Manufacturing system decentralization is made feasible because various small-size computers have improved in terms of capabilities to enhance their reliabilities. In the shop, centralized manufacturing systems are not favoured because they entail heavy investments at the building time for the manufacturing system. Instead, a distributed type is preferred as it is convenient for situations where partial system constructions are called for from time to time.

It is now possible that NC machine tools and PCs (programmable controllers), which have been controlled by dedicated controllers, will be controlled by personal computers. Various boards and software for use in such situations are now commercially available. The resulting environment is suitable for situations where personal computers are used to perform information processes which have so far been executed by large-size computers, and control processes which have previously been conducted by dedicated controllers.

When personal computers are used to construct a distributed manufacturing system as described above, individual applications can be designed while taking into account the operating characteristics of distinctive subsystems or other elements of a manufacturing system unlike the earlier method by which a manufacturing system was constructed while giving priority to overall manufacturing system optimization. Therefore, the introduction of such a distributed manufacturing system is expected to establish a system suitable for workers and managers.

Further, if the product and manufacturing management system are partly changed while a distributed manufacturing system is employed, it is expected that proper follow-up can be promptly taken simply by making changes corresponding to the changed portions without altering the whole system. In other words, when a distributed manufacturing system is constructed, it is expected that an original manufacturing system can be formulated to suit the characteristics of individual enterprises and workplaces and that a created distributed manufacturing system will be able to promptly cope with changes in the market and environment for long periods of time.

To realize such a distributed system, however, it is necessary to have a large number and a wide variety of software programs. It is also necessary to furnish software that matches various characteristics of jobs and operations. If such a software is developed originally, larger investments might be required than in the case of a conventional centralized type. It is therefore necessary for a distributed manufacturing system to employ package software for the purpose of reducing the software development cost and maintenance cost.

To realize a system which makes effective use of the advantages of a distributed system and eliminate the above-mentioned worries, it is essential that the processes of all system configuration elements be completely independent of each other and clearly defined, and that the relationships between such independent elements be clearly established further. This will enable exchange of proper information while all the independent elements are operable in a common operating environment.

To fulfill the above requests from the shop in the factory, it is necessary to ensure that (i) individual system configuration elements clarify their interfaces, (ii) the subsystem functional structure is explicitly defined, and (iii) that compatibility is establishable via a common platform. More specifically, it is necessary to model and clarify all job functions, classify them into appropriate independent units, and freely (openly) combine job functions to establish a proper system functionality.

Establishing an open manufacturing system means to classify manufacturing system component devices and software programs into appropriate units, clearly define their relationships, and provides means of freely combining them on a common platform.

8.3 MODELLING A MANUFACTURING SYSTEM

For manufacturing system overall opening, modelling is frequently done according to an object-oriented modelling method, which regards all manufacturing system elements as objects in order to handle them as independent elements and combines them to form a large-scale system (Fig. 8.1). Examples are CIM Application Framework proposed by SEMATEC, which is a research and development association for US semiconductor manufacturing industry, (i) Object-Oriented Manufacturing Execution System (proposed in 1995) by Manufacturing Science Center (NCMS) in USA, and (ii) OPC (OLE for Process Control) which aims at exercising process control using Microsoft's OLE (Object Linking and Embedding).

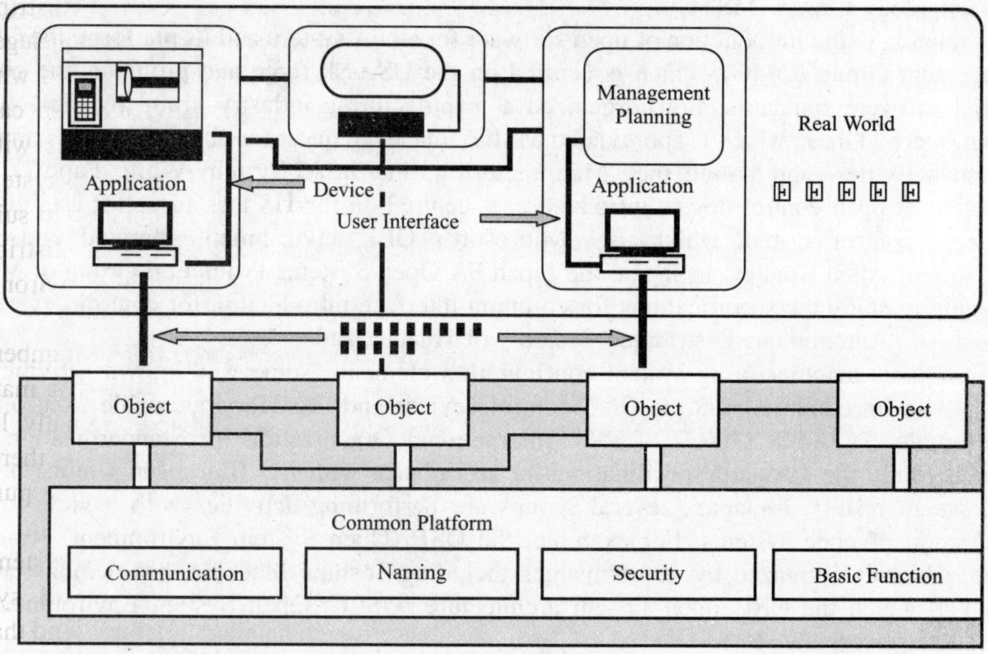

Figure 8.1 Distributed System Concept Structure and Open System.

The basic structure of a recent distributed manufacturing system is such that manufacturing functions are regarded as objects which are independent elements, and combined to create a standard model, and that the objects are combined on a common platform such as CORBA and OLE. The structure of an open system is such that there is no need to be aware of hardware or software differences as long as a common platform is used.

8.3.1 Open Manufacturing Systems

The introduction of open manufacturing systems entails, by its nature, standardization which has to be completed in a combined fashion by several enterprises. It is not to be accomplished by a single device manufacturer or software manufacturer. It is also well to remember that standardization for open system establishment is meaningless if the relevant technical specifications are not released. Operations for standardization must be openly carried out.

These activities are conducted on various levels such as the introduction of open manufacturing system overall structure and open devices. Some typical activities are now described.

As regards the manufacturing system overall structure, there are activities related to the aforementioned "CIM Framework" of SEMATIC in the US and activities related to "Object-Oriented Manufacturing Execution System" (proposed in 1995, which is proposed by Manufacturing Science Center (NCMS). In Europe, there are standardization activities such as "CIM Open System Architecture" which is promoted by CEN, a European standardization organization. In Japan, various activities are conducted, including the object modelling activities which are performed by Japan FA Open Systems Promotion Group of Manufacturing Science and Technology Center (MSTC).

In relation to the introduction of open software for an FA system and its platform, the Object Management Group (OMG), which is centred on the US task force and promotes the object-oriented software standardization, organized a manufacturing industry group (Manufacturing Special Interest Group which is abbreviated MSIG), made a request for objectification, compiled relevant activities, and issued the "Manufacturing Enterprise System White Paper"). The promotion of open control device introduction is centred on the US task force for establishing a process control system which uses Microsoft's OLE environment operated under the leadership of Allen Bradley. In Japan, the Japan FA Open Systems Promotion Group of MSTC has begun to study the specifications for common interface introduction for control devices on the basis of "Autonomous Distributed System" of Hitachi, Ltd.

As regards manufacturing system configuration elements, some well-known activities for open system introduction relate to CNC control devices and PCs (Programmable Controllers). For example, TC184/SC1/WG7 of ISO (International Organization for Standardization) has begun to study the standard specifications in accordance with EU (European Union) project development results. In Japan, several groups are performing activities with a view toward introduction of open systems. For example, the OSE (Open System Environment) Research Group, which is organized by 19 companies including Toshiba Machine, has demonstrated a system in which the CNC open system architecture (OSEC: Open System Environment for Control)) is incorporated.

8.4 DEVELOPMENT OF MANUFACTURING EXECUTION SYSTEM FOR MACHINING

In order to realize the open manufacturing system, JOP is developing the Open MES system. The developed MES is shown in Fig. 8.2.

The MES has an object oriented Model which is shown in Fig. 8.3. (This model is based on the SEMATEC model.)

The MES system is developed by IBM, Japan, Toyota Machinery Ltd., and some software houses.

8.5 FUTURE OF OPEN MANUFACTURING SYSTEM USE

Manufacturing systems have become increasingly information integration. There is now an increased demand that manufacturing system functions be modelled to offer open systems,

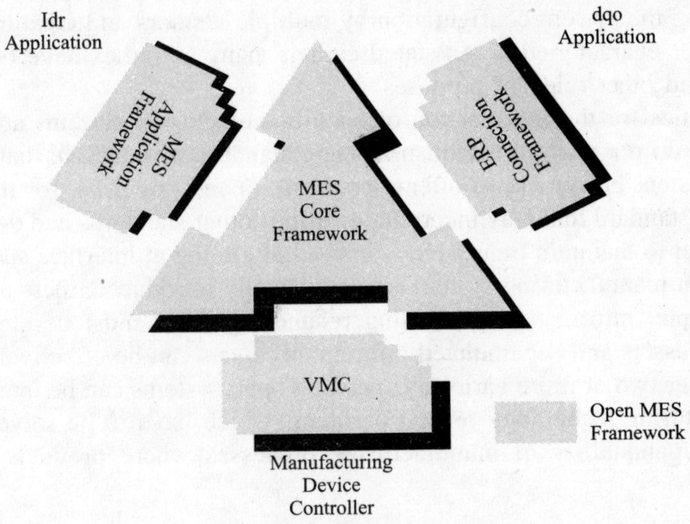

Figure 8.2 MES Framework.

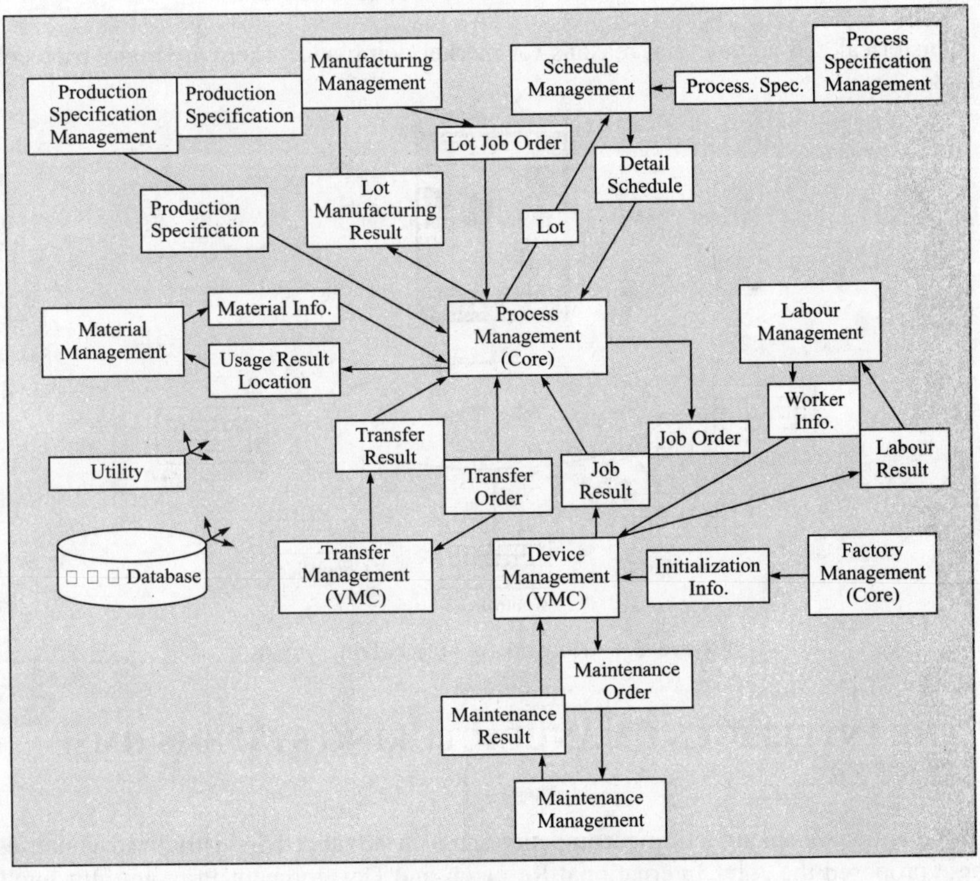

Figure 8.3 Object Model in MES.

thereby facilitating the system configuration by multiple vendors and creating systems suitable for individual user characteristics. As stated earlier, many activities have been performed for standardization and other relevant purposes.

Recent activities for the introduction of open manufacturing systems are characterized by the fact that they do not wish to establish a single standard or interface that covers the whole manufacturing system but intend to offer a practical, open system by, for instance, setting up a loosely defined standard that may incorporate various other standards and describing the target process as a model to maintain transparency instead of aiming at interface standardization only.

When an open manufacturing system is to be actually introduced, there are many problems to fix. For example, information processing related problems must be cleared by checking whether invested assets and accumulated information assets can be effectively used in an open system and whether two or more variously specified open systems can be interconnected. There are also manufacturing technology related problems which have to be solved by assuring the reliability and dependability of manufacturing processes where products are continuously manufactured.

8.6 NETWORKS, PROTOCOLS AND STANDARDS

Due to historical and commercial reasons for market dominance, there are many protocols and standards.

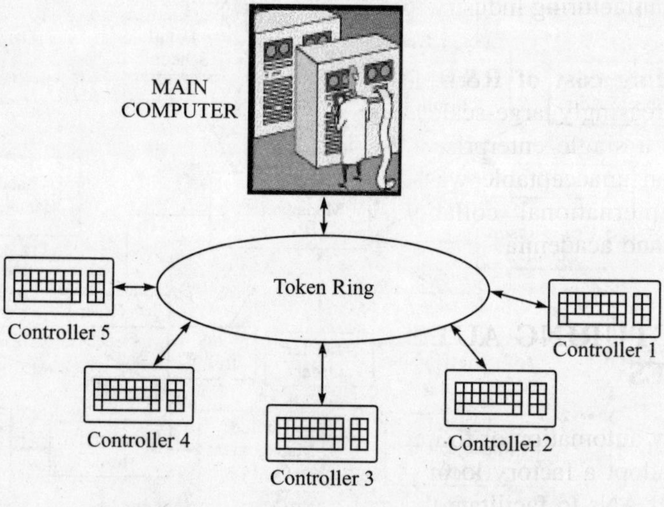

Figure 8.4 Networked Automation system.

8.7 THE INTELLIGENT MANUFACTURING SYSTEMS (IMS) CENTER

In order to improve manufacturing competitiveness in advanced industrialized countries, the Japanese proposed the Joint International Research and Development Program into Intelligent

Manufacturing Systems (IMS) Program. Its objective was to develop next-generation manufacturing systems jointly among partners from Japan, Australia, Canada, the European Union (EU), Norway, Switzerland and the United States. It encompasses the systematization of all the knowledge accumulated in manufacturing industries and the establishment of advanced manufacturing systems capable of flexibly integrating the whole range of corporate activities, from order-booking through design, production and marketing. At the same time, the program has the objective of maintaining a balanced relationship between people and machines, preserving and improving the vitality of manufacturing industry.

IMS addresses a common set of issues due to the rapid changes that the manufacturing industries are facing:

1. **Globalization**. Globalization of manufacturing industries is progressing as action is taken to locate production in or adjacent to markets, reduce risks arising from fluctuations in exchange rates, and rectify trade imbalances.
2. **The changing labour environment**. Manufacturing industry is afflicted by an increasing shortage of technicians, and a growing preference of young engineers to pursue other career paths.
3. **The changing market environment**. The proliferation of lifestyles in economically advanced societies is inimical to broad, uniform markets. Consumer demands are becoming increasingly diversified and sophisticated.
4. **Environmental issues**. The mounting concern about environmental degradation requires manufacturing industry to place environmental considerations at the heart of its activities.
5. **The spiralling cost of R&D**. The sound development of manufacturing industry requires increasingly large-scale and complex R&D projects. In many instances, it is not feasible for a single enterprise to tackle such projects. Duplication of such projects represents an unacceptable waste of resources. The only solution to this impasse is increased international collaboration among companies, public sector research institutions and academia.

8.8 MANUFACTURING AUTOMATION PROTOCOL (MAP) ACTIVITIES

To encourage factory automation or to implement computer integrated manufacturing (CIM), it is indispensable to adopt a factory local area network (LAN) system. MAP is an international standard for factory LANs to facilitate the interconnection between various factory automation equipment and devices from various vendors. It is becoming increasingly important in factory automation.

8.9 RESEARCH AND DEVELOPMENT IN MANUFACTURING

Throughout Japan, research and development in production engineering focuses on computer integrated manufacturing (CIM), factory automation (FA), electronic packaging, precision machining, and inspection/recognition technologies. In 1985, Japan's Ministry of International Trade and Industry (MITI) produced a road map for the development of Computer Integrated Manufacturing in Japan. The overall road map is shown in Fig. 8.5.

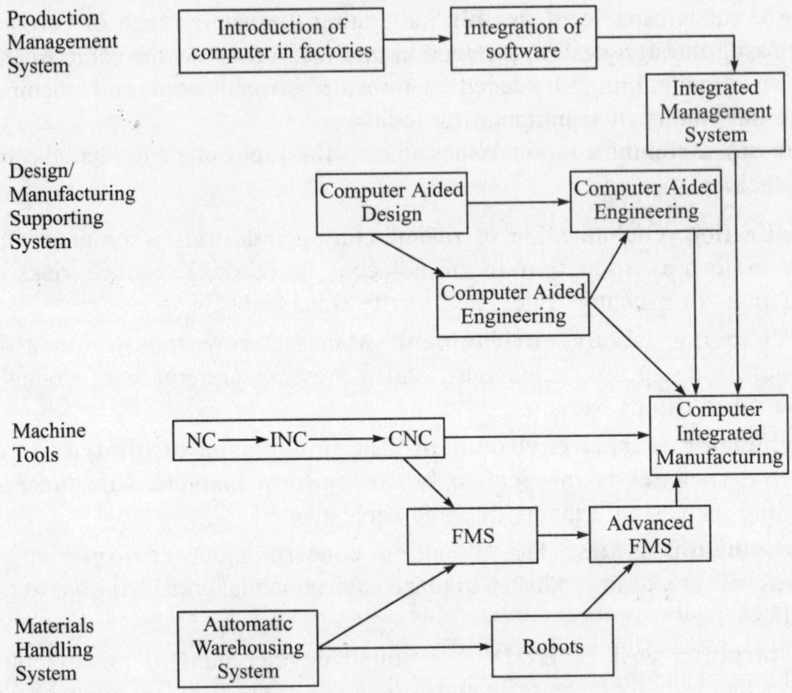

Figure 8.5 Japan's Development of Computer Integrated Manufacturing (MITI).

8.10 COMPUTER INTEGRATED MANUFACTURING/MAP

In 1985, the Japanese initiated a project to develop a standard protocol for factory automation. The International Robotics and Factory Automation Association was established that year for the purpose of setting the standard for communication between equipment used on the factory floor. The approach, called mini MAP, was based on the lower four levels of the MAP manufacturing automation protocol introduced by General Motors in 1980s. The first prototype demonstrations were made in 1992 and 1993 at the Japan Exhibition Center at Makuhari Messe, where over 20 equipment makers demonstrated factory integration of their equipment using the mini MAP standard protocol. To quote Hitachi, "CIM is the only technology capable of handling the explosive diversification of consumer goods in today's fast-paced market environment driven by increasingly shorter concept to finished product."

The development of Japan's factory automation capabilities began with Toyota's prototype integrated factory in 1972. In 1978, MITI initiated a national R&D project to develop a flexible manufacturing factory, which was demonstrated in 1985 at MITI's mechanical engineering laboratory in Tsukuba. The problem of standardizing the communication interface between equipment vendors was then addressed by the International Robotics and Factory Automation

Association. By 1990, Japan was the world leader in installed flexible manufacturing system (FMS) lines and industrial robots and was working to develop a full CAD/CAM (computer-assisted design/manufacturing) capabilities for computer integrated manufacturing. FANUC, a major supplier of industrial robots and control systems for General Motors, began implementing full CAD/CAM capabilities for its own factory automation in the late1980s.

Leading Japanese electronic manufacturing and assembly companies have developed their own advanced production and computer integrated manufacturing capabilities. In-house capability protects intellectual property, assures timely delivery, reduces costs, and cuts down on response times required for new product introduction. Such capabilities provide the Japanese with self-reliance and the ability to implement real "concurrent engineering" methodologies. Most companies, however, are selling their proprietary manufacturing technology in the open market in order to increase the return on their equipment investment. The availability of the equipment in the open market is usually preceded by long periods of utilization in Japanese facilities. Sony demonstrated to the JTEC panel a full range of production and assembly equipment developed for component manufacturing and final assembly applications. Sony's equipment sales now represent 2% of the company's sales. In the United States, Polaroid Corporation has installed a complete assembly line that uses over 100 Sony robots.

8.10.1 Production Engineering Capabilities

Production technology is a key strength of Japanese firms. Most leaders in the electronics industry have established separate organizations to manage developments in production technology. Their goals are to maximize manufacturing efficiency and high quality output. On a corporate level, production engineering research laboratories provide the critical link between product development and the development and application of advanced production equipment and processes. Every major business unit within a firm has its own production engineering staff and production engineering laboratory to develop process design and process technology. Production technology development drives concurrent equipment, component, and product developments— the final product design and development are synchronized with production technology development. The importance of production engineering in Japanese firms is demonstrated by the fact that in companies like Hitachi and NEC, the production engineering departments report directly to the company presidents, alongside the other company research laboratories.

Most leading Japanese electronics companies also have materials characterization laboratories. Expertise in both materials and manufacturing processes guarantees high-quality outputs and improved reliability. It was evident to JTEC panelists that in Japan both are the focus of attention when quality or reliability problems occur; likewise, improvements in materials, package architectures, and manufacturing processes eliminate failure causes.

8.10.2 Equipment Development Capabilities

Without exception, every company visited by the JTEC panel has its own equipment development teams for designing and building production equipment. Sony's factory automation and precision products group includes a thousand members. The FA systems division makes advanced equipment to manufacture new products developed within Sony. The robotics and FA components division developed Sony's Multi Assembly Robot Technology (SMART) systems.

In addition, Sony has another 300 people working within the machine tool, production technology, and production engineering divisions to develop production systems for customers. In 1993, Nippondenso, Japan's leading auto electronics firm, has over 200 people at its Kota site to develop production equipment and processes. This commitment has led to advanced production systems. Nippondenso's main four-storey building at Kota handles chip assembly, hybrid manufacturing, SMT board, and also final assembly, burn-in, and testing for more than 120 models of engine controllers on one 100-metre manufacturing line. The production line was replaced or upgraded every three to four years. Most of the Kota plant's equipment has been produced in-house. JTEC panel members have never seen such levels of commitment to process development in US electronics firms.

Japanese industry as a whole has pumped in massive resources on the design and development of complex automated equipment, and the range of equipment development expertise within individual companies is tremendous. That expertise covers all the technological areas required to be self-sufficient and dominant in electronics manufacturing. The consensus in Japan is that equipment provides a major competitive advantage and that equipment development technology is mandatory in order to lead to the introduction of new products. Japanese drivers for equipment development include cost reduction, machine accuracy, reliability, speed, flexibility, peripheral integration, ease of operation, and links to CAD (CAD-CAM integration). Numerous Japanese companies are aggressively producing and marketing electronics manufacturing equipment worldwide.

Equipment development teams

The Japanese approach to equipment development is built on the team concept. Numerous individuals with varied technical backgrounds are teamed together during the development process. Members of the development team are ultimately transferred to the manufacturing facility for implementation and to continue equipment enhancements to maximize production efficiency. This ensures highly skilled technical personnel to operate, maintain, and continuously implement changes required to enhance the equipment.

Modular systems

The concept of modular equipment, both in hardware and software, has become a prime objective for the Japanese equipment developers. Their goal is to design independent subsystems that can be quickly reconfigured to meet the demands of rapid product change with minimal cost. Sony's Multi Assembly Robot Technology (SMART) has a modular architecture and is presently being marketed worldwide.

8.11 DATABASE IN CIM

This concept document provides an overview of object models and relational data models, and provides a summary description of a persistence framework.

8.11.1 Relational Databases and Object Orientation

Relational databases and object orientation are not entirely compatible. They represent two different views of the world: in an RDBMS, all you see is data; in an Object-Oriented system, all you see is behaviour. It is not that one perspective is better than the other: the Object-Oriented model tends to work well for systems with complex behaviour and state-specific behaviour in which data is secondary, or systems in which data is accessed navigationally in a natural hierarchy (for example, bills of materials). The RDBMS model is well suited to reporting applications and systems in which the relationships are dynamic or ad hoc.

The fact, however, remains that a lot of information is stored in relational databases, and if Object-Oriented applications want access to that data, they should be able to read and write to an RDBMS. In addition, Object-Oriented systems often need to share data with non-Object-Oriented systems. It is natural, therefore, to use an RDBMS as the sharing mechanism.

While object-oriented and relational design share some common characteristics (an object attribute is conceptually similar to an entity's column), fundamental differences make seamless integration a challenge. The fundamental difference is that data models expose data (through column values) while object models hide data (encapsulating it behind its public interfaces).

8.11.2 The Relational Data Model

The relational model is composed of entities and relations. An entity may be a physical table or a logical projection of several tables also known as a view. Fig. 8.6 illustrates ORDER, LINEITEM, and PRODUCT tables and the various relationships between them. A relational model has the following elements:

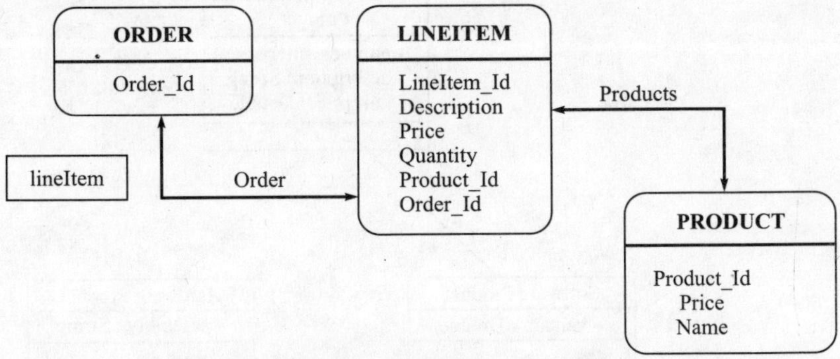

Figure 8.6 A Relational Model.

An entity has columns. Each column is identified by a name and a type. In Fig. 8.6, the LINEITEM entity has the columns LineItem_Id (the primary key), Description, Price, Quantity, Product_Id and Order_Id. (The latter two are foreign keys that link the LINEITEM entity to the ORDER and PRODUCT entities.)

An entity has records or rows. Each row represents a unique set of information which typically represents an object's persistent data.

Each entity has one or more primary keys. The primary keys uniquely identify each record (for example, Id is the primary key for LINEITEM table).

Support for relations is vendor specific. The example illustrates the logical model and the relation between the PRODUCT and LINEITEM tables. In the physical model, relations are typically implemented using foreign key/primary key references. If one entity relates to another, it will contain columns which are foreign keys. Foreign key columns contain data which can relate specific records in the entity to the related entity.

Relations have multiplicity (also known as cardinality). Common cardinalities are one to one (1:1), one to many (1:m), many to one (m:1), and many to many (m:n). In the example, LINEITEM has a 1:1 relationship with PRODUCT and PRODUCT has a 0:m relationship with LINEITEM.

8.11.3 The Object Model

An object model contains, among other things, classes (see [UML01] for a complete definition of an object model). Classes define the structure and behaviour of a set of objects, sometimes called object **instances**. The structure is represented as attributes (data values) and associations (relationships between classes). Fig. 8.7 illustrates a simple class diagram, showing only attributes (data) of the classes.

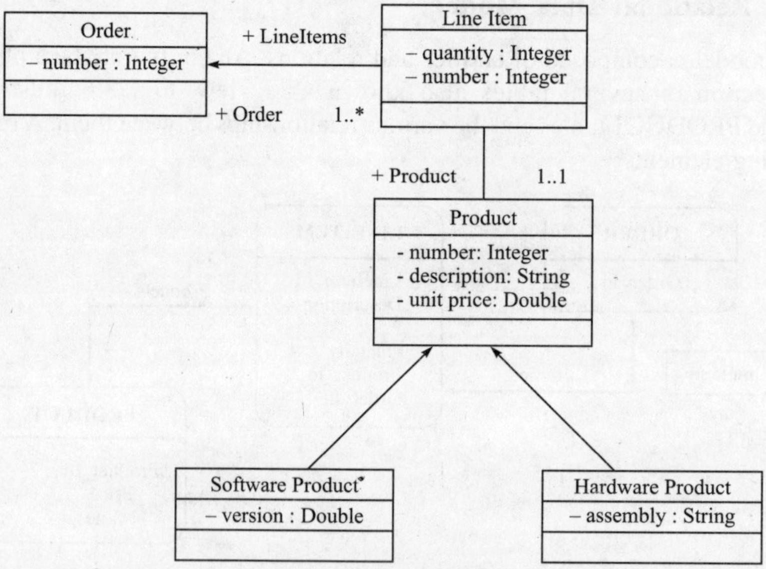

Figure 8.7 An Object Model (Class Diagram).

An Order has a number (the Order number), and an association to 1 or more (1..*) Line Items. Each Line Item has a quantity (the quantity ordered).

The object model supports inheritance. A class can inherit data and behaviour from another class (for example, software products and hardware products inherit attributes and methods from Product class).

8.11.4 Persistence Frameworks

The majority of business applications utilize relational technology as a physical data store. The challenge facing object-oriented applications developers is to sufficiently separate and encapsulate the relational database so that changes in the data model do not "break" the object model, and vice versa. Many solutions exist which let applications directly access relational data; the challenge is in achieving a seamless integration between the object model and the data model.

Database application programming interfaces (APIs) come in standard flavours (for example, Microsoft's Open Database Connectivity API, or ODBC) and are proprietary (native bindings to specific databases). The APIs provide data manipulation language (DML) pass through services which allow applications to access raw relational data. In object-oriented applications, the data must undergo object-relational translation prior to being used by the application. This requires considerable amount of application code to translate raw database API results into application objects. The purpose of the object-relational framework is to generically encapsulate the physical data store and to provide appropriate object translation services.

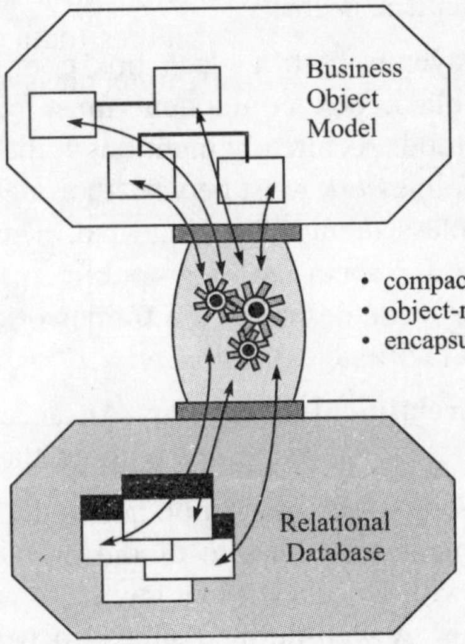

Figure 8.8 The Purpose of a Persistence Framework.

Application developers spend over 30 per cent of their time implementing relational database access in object-oriented applications. If the object-relational interface is not correctly implemented, the investment is lost. Implementing an object-relational framework captures this investment. The object-relational framework can be reused in subsequent applications, reducing the object-relational implementation cost to less than 10 per cent of the total implementation costs. The most important cost to consider when implementing any system is maintenance. Over 60 per cent of the total costs of a system over its entire life cycle can be attributed to maintenance. A poorly implemented object relational system is both a technical and a financial maintenance nightmare.

8.11.5 Essential Characteristics of an Object-relational Framework

(i) **Performance**. Consideration must be given decomposing objects into data and composing objects from data. In systems where data throughput is high and critical, this is often an Achilles heel of an inadequately designed access layer.

(ii) **Minimize design compromises**. A familiar pattern to object technologists who have built systems which utilize relational databases is to adjust the object model to facilitate storage into relational systems, and to alter the relational model for easier storage of objects. While minor adjustments are often needed, a well-designed access layer minimizes both object and relational model design degradation.

(iii) **Extensibility**. The access layer is a white box framework which allows application developers to extend the framework if certain functionality is desired in the framework. Typically, an access layer will support, without extension, 65-85 per cent of an application's data storage requirements. If the access layer is not designed as an extensible framework, achieving the last 15–35% of an application's data storage requirements can be very difficult and costly.

(iv) **Documentation**. The access layer is both a black box component, and a white box framework. The API of the black box component must be clearly defined, well-documented, and easily understood. As already mentioned, the access layer is designed to be extended. An extensible framework must be very thoroughly documented. Classes which are intended to be subclassed must be identified. The characteristics of each relevant class's protocol must be specified (e.g. public, private, protected, final). Moreover, a substantial portion of the design of the framework of an access layer must be exposed and documented to facilitate extensibility.

(v) **Support for common object-relational mappings**. An access layer should provide support for some basic object-relational mappings without the need for extension.

(vi) **Persistence interfaces**: In an object-oriented application, the business model for an object application captures semantic knowledge of the problem domain. Developers should manipulate and interact with objects without having to worry too much about the data storage and retrieval details. A well-defined subset of persistent interfaces (save, delete, find) should be provided to application developers.

8.12 COMMON OBJECT-RELATIONAL SERVICES

Today, common patterns are emerging for object-relational applications. IT professionals who have repeatedly crossed the chasm are beginning to understand and recognize certain structures and behaviours which successfully exhibit object-relational applications. These structures and behaviours have been formalized by the high-level CORBA Services specifications (which apply equally well to COM/DCOM-based systems). The CORBA service specifications which are applicable and useful for object-relational mapping are:

- Persistence
- Query
- Transactions
- Concurrency
- Relationships

The reader is encouraged to refer the appropriate CORBA specifications for further details.

8.12.1 Persistence

'Persistence' is a term used to describe how objects utilize a secondary storage medium to maintain their state across discrete sessions. Persistence provides the ability for a user to save objects in one session and access them in a later session. When they are subsequently accessed, their state (for example, attributes) will be exactly the same as it was in the previous session. In multi-user systems, this may not be the case since other users may access and modify the same objects. Persistence is interrelated with other services discussed in this section. The considerations such as relationship, concurrency and others is intentional (and consistent with CORBA's decomposition of services).

Following are examples of specific services provided by persistence.

(i) **Data source connection management**: Object-relational applications must initiate connection to the physical data source. Relational database systems typically require identification of the server and the database. The specifics of connection management tend to be database vendor specific and the framework must accordingly be designed in a flexible accommodating manner.

(ii) **Object retrieval**: When objects are restored from the database, data is retrieved from the database and translated into objects. This process involves extracting data from database specific structures retrieved from the data source, marshalling the data from database types into the appropriate object types and/or classes, creation of the appropriate object, and setting the specific object attributes.

(iii) **Object storage**: The process of object storage mirrors object retrieval. The values of the appropriate attributes are extracted from the object, a database specific structure is created with the attribute values (this may be an SQL string, stored procedure, or special remote procedure call), and the structure is submitted to the database.

(iv) **Object deletion**: Objects that are deleted from within a system must have their associated data deleted from the relational database. Object deletion requires that appropriate information be extracted from the object, a deletion request be constructed (this may be a SQL string, stored procedure, or special remote procedure call), and the request submitted to the database. Note that in some languages (e.g. Smalltalk and Java), explicit deletion is not supported; instead, a strategy called **garbage collection** is supported. Persistence frameworks supporting these languages must provide an alternative way to remove data from the database once applications no longer reference the data. One common way is for the database to maintain **reference-counts** of the number of times an object is referenced by other objects. When the reference count for an object drops to zero, no other objects reference it, and it may be possible to delete it. It may **be acceptable** to delete objects with a reference count of zero since, even when an object is no longer referenced, it may still be queried. A database-wide policy on when object deletion is allowed is still needed.

8.12.2 Query

Persistent object storage is of little use without a mechanism to search for and retrieve specific objects. Query facilities allow applications to interrogate and retrieve objects based on a variety of criteria. The basic query operations provided by an object-relational mapping framework are through to find unique operation. This operation will retrieve a specific object, find and return a collection of objects based on a query criteria.

Data store query facilities vary significantly. Simple file-based data stores may implement rigid home-grown query operations, while relational systems provide a flexible data manipulation language. Object-relational mapping frameworks extend the relational query model to make it object-centric rather than data centric. Pass-through mechanisms are also implemented to leverage relational query flexibility and vendor-specific extensions (e.g. stored-procedures).

Note that there is some potential conflict between data-based query mechanisms and the object paradigm: database query mechanisms are driven by **values** of attributes (columns) in a table. In the corresponding objects, the principle of encapsulation prevents us from seeing the values of attributes; they are **encapsulated** by the operations of the class. The reason for encapsulation is that it makes applications easier to change: we can alter the internal structure of a class without breaking about dependent classes as long as the publicly visible operations of the class do not change. A query mechanism based on the database is dependent on the internal representation of a class, effectively **breaking** encapsulation. The challenge for the framework is to prevent queries from making applications so fragile that they cannot be changed.

8.12.3 Transactions

Transactional support enables the application developer to define an atomic unit of work. In database terminology, it means that the system must be able to apply a set of changes to the database, or it must ensure that none of the changes are applied to the operations within a transaction either execute successfully or the transaction fails as a whole. Object-relational frameworks at a minimum should provide a relational database-like commit/rollback transaction facility. Designing object-relational frameworks in a multi-user environment can present many challenges, which requires a lot of thought.

In addition to the facilities provided by the persistence framework, the user understand how to handle errors. When a transaction fails or is aborted, the system must be able to restore its state to a stable prior state, usually by reading the prior state information from the database. Thus, there is a close interaction between the persistence framework and the error handling framework.

8.12.4 Concurrency

Multi-user object-oriented systems must control concurrent access to objects. When an object is accessed simultaneously by many users, the system must provide a mechanism to ensure modifications to the object and store in a predictable and controlled manner. Object-relational frameworks may implement pessimistic and/or optimistic concurrency controls.

Pessimistic concurrency control requires that the application developer specify their intent when the object is retrieved from the data store (for example, read only, write lock, …). If objects are locked, other users may block when accessing the object and wait for the lock to be relinquished. Pessimistic concurrency should be used and implemented with caution as it is possible to create deadlock situations.

Optimistic concurrency control assumes that it is unlikely that the same object will be simultaneously accessed. Concurrency conflicts are detected when the modifications are saved to the database. Typically, if the object has been modified by another user since its retrieval, an error will be returned to the application indicating failure of the modify operation. It is the application's responsibility to detect and handle the error. This calls for the framework to cache the concurrent values of objects and compare them against the database. Optimistic concurrency is less costly if there are only a few concurrency conflicts, but more expensive if the number of conflicts is fairly large (because of the need to redo work when conflicts occur).

All applications using shared data must use the same concurrency strategy; you cannot mix optimistic and pessimistic concurrency control in the same shared data or corruption may occur. The need for a consistent concurrency strategy is best handled through a persistence framework.

REVIEW QUESTIONS

1. Prepare a chart showing the data required and data generated in the various segments of a manufacturing industry.
2. What is database?
3. What are the objectives of a database?
4. What are the disadvantages of a database?
5. How does the conventional file system differ from a database?
6. What are the four classes of data in manufacturing?
7. What are the desirable features of a database management system?
8. Describe the three database models.
9. Study typical data in a factory and analyze how you can classify the data.
10. What is a relational database? Give an example.
11. Describe the architecture of a database management system.
12. What are the essential requirements of a relational database management system?
13. What is a query language? What are the major types of query language?
14. What is SQL?
15. Give an example of the use of SQL to access a database.
16. What is the role of PDM in drawing office management?
17. What are the features of PDM which help speedy product development?
18. What are the benefits of PDM?

REFERENCES

1. International Robot FA Center: Prospectus for Japan FA Open Systems Promotion Group (1996).
2. NCMS: Object-Oriented Manufacturing Execution Systems: A New Approach to Agile Manufacturing Information and Control Systems, Final Report, Vol. 1 (1996).

3. OMGMSIG: Manufacturing Enterprise Systems, A White Paper by OMGMSIG, Version 1.0 (1996).
4. OSE Research Group: OSE Architecture Version 2.0 (1996).
5. SEMATECH: Computer Integrated Manufacturing (CIM) Application Framework, Specification 1.2 (1995).
6. SEMATECH: Computer Integrated Manufacturing (CIM) Application Framework, Specification 1.3 (1996).

Chapter 9

NC AND CNC MACHINES

9.1 INTRODUCTION

Earlier, design and manufacturing activities were done independently and in a sequential manner without any relation among them as shown in Figure 9.1.

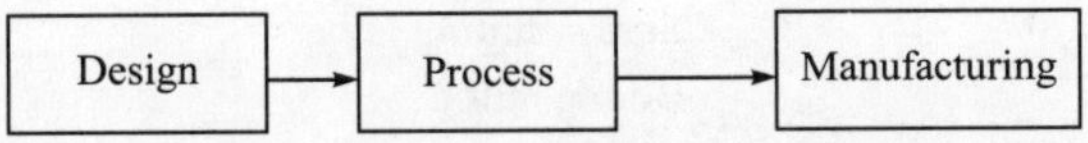

Figure 9.1 The Design Process.

The figure explains the product flow. In this process plan form, the intermediate stage links design and manufacturing phases. Therefore, the designer has to take more care on the manufacturing activities before the product enters the actual manufacturing, and hence the time involved in it is more when compared with actual manufacturing. The process plan is the stage which translates process design requirement into a set of manufacturing activities.

Today, market trends provide unstable demand for the product, therefore, products are redesigned continuously and introduced into the market to cope with the need and competition. Hence, firms continually strive to reduce time taken to put the product into the market. This is achieved ultimately by reducing the product life cycle time by reducing manufacturing time and hence firms are ready to invest in flexible production systems rather than dedicated ones. These changes put an end to the segregation of design and manufacturing phases. In order to ensure a place in the market over the competitors by reducing the product life cycle time which inturn reduced the manufacturing time made the firms to change from the conventional to automatic machine. The basic automation is provided by **numerical control machines.**

Definition

Numerical control machine is defined as a form of programmable automation in which the process is controlled by numbers, letters and symbols.

9.2 HISTORICAL BACKGROUND

Many of the achievements in CAD/CAM have a common origin in Numerical Control. In late 1940s John T. Parson conceived a method of using a punched card containing coordinate position data to control a machine tool. The machine was directed to move in small increments, thus generating a desired surface. Initial work involved the development of a prototype NC milling machine, by retrofitting a conventional tracer mill with position servomechanism demonstrated during 1952. Shortly thereafter, the machine tool builders began their work, which resulted in the APT (Automatically Programmed Tool) language.

The US Air Force commissioned MIT to develop the first *numerically controlled* machine in 1949. It was demonstrated in 1952.

Motivation: To manufacture complex curved geometries in 2D or 3D was extremely expensive by mechanical means (which usually would require complex jigs to control the cutter motions).

Inventions in Manufacturing and its Developments

Year	Inventions
1750	Screw-driven lathe
1751	Slide lathe- 1st metal lathe
1770	Screw-cutting lathe
1775	Boring mill
1813	Interchangeability of parts
	Simon North horse pistols
1817	Planning machine
1845	Turret lathe
1847	Milling machine-Brown and Sharpe
	Making twist-drill helical grooves
1946	ENIAC Computer [Electronically Numerically Integrated and Computer]
1950	1946 Hardwired relay control
	1947 APT language for NC
1960s	❖ Industrial robot
	❖ Interactive computer graphics
1970s	❖ CNC computer
	❖ DNC/FMS
	❖ CAD/CAM
	❖ PLC device and cell control computer vision
	❖ 3D CAD
1980s	❖ Solid modelling
	❖ Factory networking
	❖ MAP/TOP
	❖ CIM
	❖ Concurrent engineering
1990s	Intelligent Manufacturing System

9.3 BASIC COMPONENTS OF AN NC

- Program of Instruction
- Machine Control Unit
- Machine Tool

9.3.1 Program of Instruction

The program of instruction is a detailed step-by-step set of directions which tell the machine tool what to do. There are two methods of input to the NC system: Manual Data Input (MDI) and direct link with computer.

9.3.2 Machine Control Unit

This consists of electronics and hardware that read and interpret the program of instruction and convert it into mechanical action.

The **tape reader** is an electromechanical device for winding and reading the punched tape containing the program. Data on the tape are read into data buffer. The signal output channels are connected to the servomotors and other controls in machine tool. Through these channels the instructions are sent to the machine tool from the controller. To make certain that the instructions are properly executed by the machine, feedback device/channel should be present.

Sequence controls coordinate the activities of the other elements in the controller unit. Use of the digital computer has permitted substantial improvements to be made in the controls for NC.

1. Computer numerical control
2. Direct numerical control
3. Adaptive control.

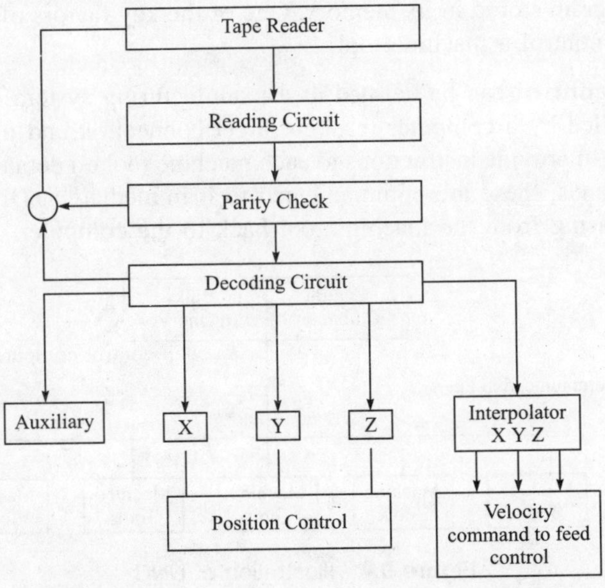

Figure 9.2 The Data Processing Unit of MCU.

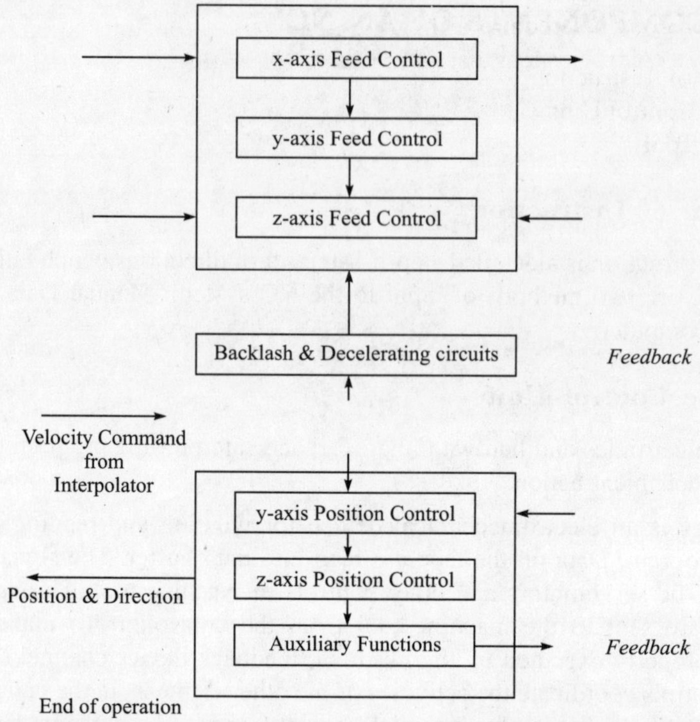

Figure 9.3 The Control Loop Unit of an NC Machine.

Computer Numerical Control (CNC) involves the replacement of the conventional hard-wired NC controller unit by a small computer which is used to perform some or all the basic NC functions by program stored in its memory. One of the key factors of CNC is that only one computer is used to control a machine tool.

Direct numerical control can be defined as a manufacturing system in which a number of machines are controlled by a computer through direct connection and in real time. The DNC computer is designed to provide instructions to each machine tool on demand. When the machine needs control commands, these are communicated to it immediately. DNC also involves data collection and processing from the machine tool back to the computer.

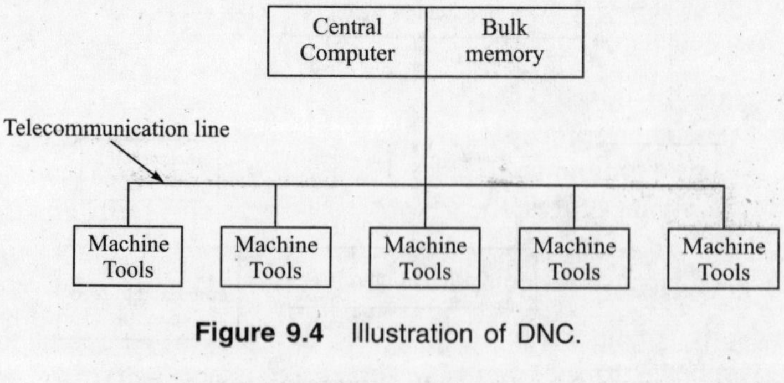

Figure 9.4 Illustration of DNC.

Adaptive control does not require a digital computer for implementation. Adaptive control machining denotes a control system that measures one or more process variables (such as cutting force, speed, feed, etc.) and manipulates feed and or feed in order to optimize the machining process.

One of the important developments of DNC is flexible manufacturing system. A flexible manufacturing system is a group of NC machines which are interconnected by a material handling system. All the machines and the work handling system are controlled by computer.

Machine Tool. The last component of an NC machine is machine tools and related controllers for its actuation. NC machines start from simple drilling machine to versatile machining centres. Automatic tool changers are integrated parts of an NC machine which is indexed by a proper indexing mechanism.

9.4 STEPS IN NC

The following stages are required for using NC machines:

1. Process plan
2. Part programming
3. Preparation of punched tape
4. Verification of punched tape
5. Actual production.

9.4.1 Process Plan

Once the design of the part is over, it flows into the various manufacturing departments in the form of drawing which is to be interpreted in terms of manufacturing.

This process plan which is prepared with the help of "Route Sheet". Specify the sequence of work flow.

9.4.2 Part Programming

Part program is the sequence of steps in which machining process takes place to complete the product. There are three ways of preparing a part program. They are :

- Manual part program
- Computer assisted part program.

Manual part program In this type of machining feed, speeds depend on programmer's knowledge about the machining process. The person who prepares the part program is called part programmer. The part programmer prepares the sequence of steps for the machining operation has to follow. While programming logic should be followed. For example, turning should be done before threading or drilling should be complete before tapping. The part programmer prepares the manuscript containing cutter and workpiece relative positions.

Computer assisted part program In computer assisted part programming method part, the job of the part programmer is to specify the workpiece geometry and operation sequence and tool path. Once these data are fed into the system, the computer does the computational tasks and generates the part program automatically.

9.4.3 Preparation of Punched Tape

Punched tape is prepared by using a typewriter like device which punches holes wherever necessary or the tape is punched directly on the computer using a device called a *tape punch*. The punched tape may follow either ANSI or EIA (Electronics Industries Association).

9.4.4 Verification of Punched Tape

The punched tape is verified by feeding the program into the machine using the proper arrangement plotting the table movements or by running the machine directly to produce the product (Acid test).

9.4.5 Production

After the changes are verified in the program, the actual production of the component takes place.

9.5 VERIFICATION OF NUMERICAL CONTROL MACHINE TOOL PROGRAMS

A manufacturing process of great significance, with a long history, is **milling**. A Numerical Control (NC) milling machine cuts a part out of a block of metal or other material (the "stock") by moving a cutting tool through space. For three-axis machining, a rotating cutting tool, with a vertical axis of rotation, does the cutting. An "NC program" specifies the movement. Typically, the program consists of a "cutter location file," which is a long list of coordinates for successive tool locations. The tool is moved to the first spatial location, then moves along a straight line (or, say, a circular arc) to the next, then along a straight line from the second location to the third, and so on, until the list of locations is completed. (Actually, the file also contains other information, such as tool sizes, and instructions for when tools must be changed.) For five-axis machines, two additional coordinates giving the angular orientation of the tool axis are also specified and the tool linearly interpolates both spatial and angular coordinates.

Typically, these NC programs are created by tool programmers. The programmers begin with a geometric representation of the part expressed in Constructive Solid Geometry (CSG), or some sort of boundary representation using splines, and generate series of tool movements that are supposed to mill the part from the stock.

As with computer programs, NC programs are error-prone. Bugs can produce cuts that are too deep or leave too much material. Thus, an important practical problem is: "Given a surface of known equation and a file of NC tool movements, does the shape that the tool cuts match the mathematical shape to within a given tolerance?" It has been claimed that "current methods of verifying NC part programs result in one of the highest non-recurring cost factors in producing NC machined parts within the aerospace industry." It is also a significant cost within the aircraft and automotive industries, among others.

The early approach to finding bugs was to use "proofing runs" on wood or foam; this is still a popular technique. Later systems would move a graphical display of a cutter over a display of the part. The user could visually check for errors as the tool moved on the graphics display. However, only gross errors could be detected in this way. In the 70s, work began on programs that actually detected errors in NC programs or verified that none existed.

Some of these methods use "complete" representations of the current state of the partially cut workpiece. Researchers investigated the feasibility of using CSG systems or solid modellers for the simulation of NC programs. Others developed systems based on octree or boundary representation of the workpiece. For each method, every tool movement would update the workpiece, and at the end the workpiece was compared to the desired piece. Unfortunately, all of these approaches were slow. Each tool movement subtracts a fairly complex shape from the current workpiece representation and requires significant updating. For example, for a simple three-axis movement with a ball-end cutter, the swept volume is bounded by parts of two spheres, three cylinders, and two planes; the number of movements is typically in the order of 10,000.

To avoid this problem, a number of people took advantage of the fact that it is not necessary to represent the entire state of the workpiece to verify the correctness of an NC program. The desired shape is known in advance, and this information can be used to simplify the process. Several researchers used an image-space approach, where an extended Z-buffer is provided for each pixel in a given graphical view of the part, and each tool movement updates the pixel information for each pixel that it passes over CGTech's Vericut simulator uses such an approach. These methods are faster than the complete methods but they are view-dependent. Because of this, errors might not be visible from a certain view direction and would therefore be missed. (Consider a nearly flat surface viewed edge-on. A large surface would be mapped into a thin line represented by only a few pixels, and errors that occurred where there were no pixels would be missed.)

Another approach is to grow vectors normal to the surface of the part. These vectors are "mowed down" by the passage of the cutting tool, and their lengths are checked at the end of the run. Vectors that stick out above the surface indicate excess material and vectors cut below the surface point to gouges. One might use pixels to select these vectors (one vector for each pixel). Or one might use properties of the tool size and the surface curvature to pick fewer vectors while still guaranteeing that they will find all errors greater than a user-set tolerance.

We have proposed a recent modification of the extended Z-buffer, called ray representations (ray-reps). A ray-rep is the intersection of a grid of parallel rays in some direction with the part. (Note that the intersection of a ray and a part consists of zero, one, or several line segments.) Given such a representation, all the operations of CSG can be computed quickly by using a specialized parallel computer for processing ray-reps, called a ray casting engine. All of the approaches using modified Z-buffers or vectors introduce error. Even if the intersection with the tool path movement is calculated exactly (more on this below), in effect the part surface is being replaced by a set of discrete points. What can we say about the correctness of the surface at points where there are no vectors? All of these approaches use enough vectors to be fairly accurate in practice, but with a few exceptions none of them attempt to analyze errors introduced by the approach to check whether they fall within tolerance.

The ray-rep representation seems an ideal candidate for such analysis. Even with the special-purpose parallel computer, one wishes not to use more rays than are needed. But how to analyze the errors introduced remains an open question, as is the determination of the optimal number of rays to use. It is also interesting to look at other possible applications of this representation and to analyze the relationship between the spacing of the rays and the error introduced.

A second area that needs more study is the one of intersecting rays with surfaces. Intersecting rays with part surfaces is reasonably well understood. However, when the surfaces are represented parametrically using B-splines or similar methods, computing intersections generally require a slow interactive technique. One must also compute the intersection of rays with the envelopes sweep out by tool movements. This is not too difficult for three-axis movements of ball-end cutters. But the same cannot be said of the intersection of a ray with a three-axis toroidal-end cutter movement. Most existing codes use a polygonal approximation to the envelope or a large number of discrete tool locations. Unfortunately, no error analysis is available.

For five-axis cutter movement, an efficient method has been given for computing the intersection within a given error bound by using three-axis tool movement as an approximation and relying on tree-based localization technique to quickly decide which of the three-axis movements is the one intersected first that remains open is to find a method for generating a polygonal approximation for the five-axis envelope with a provable error bound that generates a reasonable number of triangles. Finding such a provably good approximation technique is an interesting and important open problem.

9.6 CLASSIFICATIONS OF NC MACHINE TOOLS

Based on motion type

- Point-to-point
- Continuous path

Based on control loops

- Open loop
- Closed loop

Based on power supply

- Electric
- Hydraulic
- Pneumatic

Based on positioning system

- Incremental
- Absolute

9.6.1 Based on Motion Type

In this **point-to-point** type the motion of the tool which performs the actual work is far more important than the path it travelled, i.e. it moves from one point to another to do the specific task, e.g. the drilling machine.

Figure 9.5 shows the continuous path motion in an NC machine. This combine both the linear and circular motions. The path of the cutter is continuously controlled to generate the profile of the workpiece. One of the distinguishing features of continuous motion is that simultaneous control of more than one axis of tool movement can be achieved, e.g. lathe and milling machine.

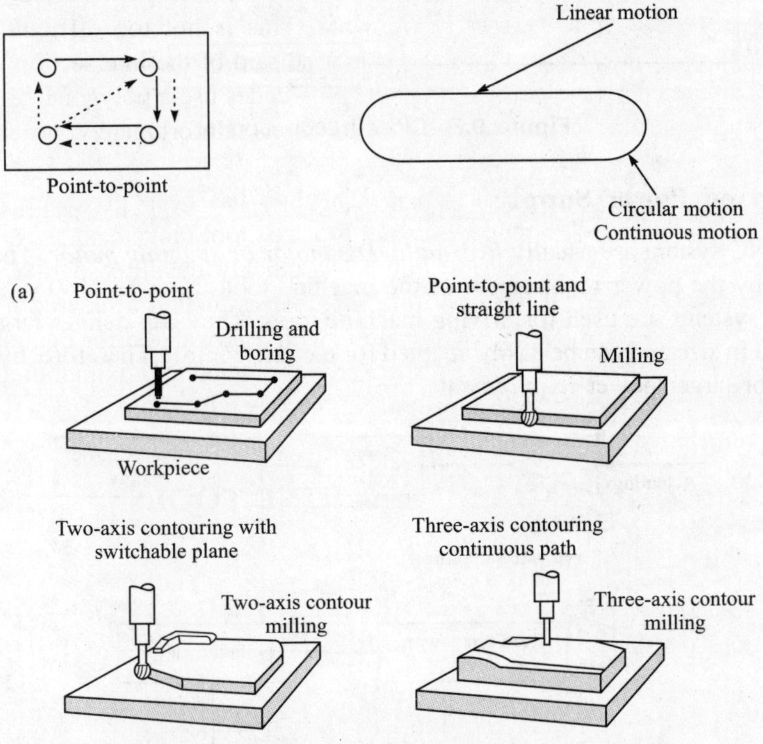

Figure 9.5 Motion System.

Straight cut motion is also available where the tool movement is parallel to either the *x*-axis or the *y*-axis.

9.6.2 Based on Control Loops

The term 'open loop' means there is no feedback of the process and the action of the controller has no reference to the result it produces.

In closed loop control system, the controller measures the actual position the lead screw travelled with the needed, and adjustment will be made by the comparator. Extreme care must be taken care in designing the closed loop control system. By increasing the amount of feedback signal the loop is made more sensitive.

The design of control and choice of the loop type, to meet critical performance and cost specification, require knowledge of the nature of the controlled machine and load torque.

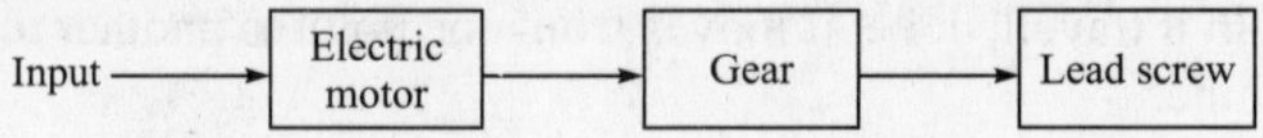

Figure 9.6 Open Loop Control.

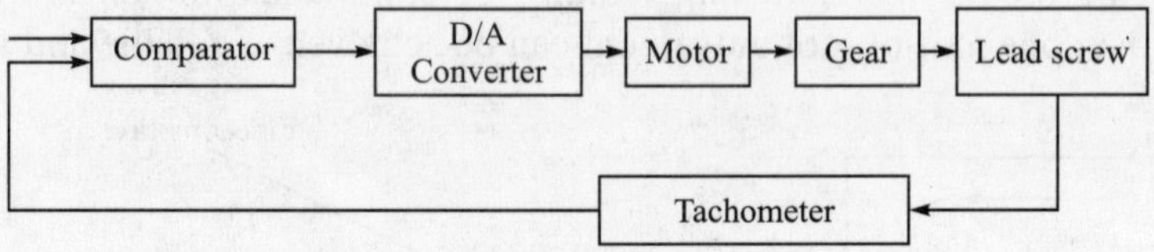

Figure 9.7 Closed Loop Control.

9.6.3 Based on Power Supply

Drives for an NC system are usually *hydraulic, DC motor or stepping motor*. The type selected is determined by the power requirement of the machine tool.

Hydraulic systems are used for driving machine tools. They can deliver large power being relatively small in size and can be easily adapted for electric control. Therefore, hydraulic system will be used for larger power requirements.

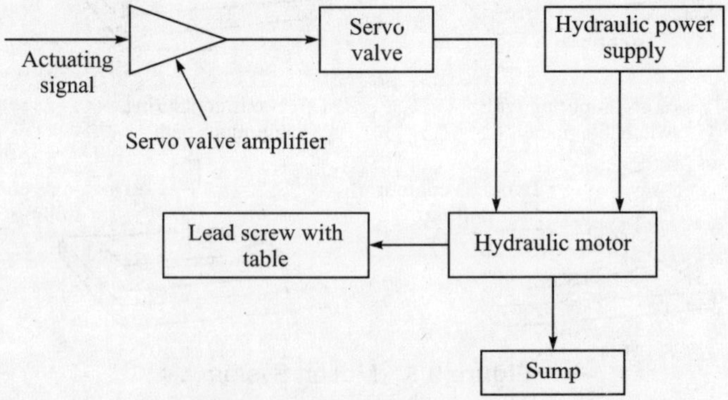

Figure 9.8 A Hydraulic System.

The main components of a hydraulic system are:

1. Hydraulic power supply
2. Servo valve
3. Sump
4. Hydraulic motor

Hydraulic systems develop more angular acceleration than an electric motor of same peak power. But the disadvantage is that the care to be taken in checking the oil leakage from transmission line and line must be kept clean and protected against contamination.

The DC motor is the most used drive for NC machine tool because its speed and voltage can be varied smoothly and easily.

The stepping motor is an incremental digital control device. It translates an input pulse sequence into a proportional angular movement, rotating one angular increment for each input pulse. The position of shaft is found out by the pulse frequency.

9.6.4 Based on Positioning System

The position control system will be classified as absolute and incremental systems. An absolute system is one in which all positions are calculated with respect to the origin.

In incremental system, the positions are calculated with respect to the previous position.

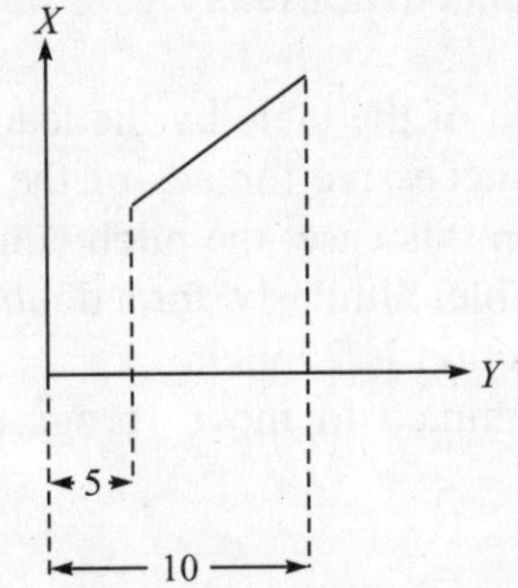

Figure 9.9 Absolute System.

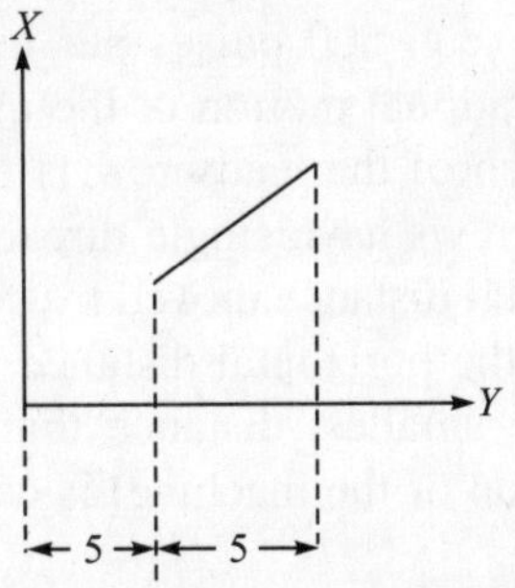

Figure 9.10 Incremental System.

Most of the point-to-point systems use absolute positioning method. If we compare absolute with incremental:

1. An incremental positioning control loop needs simpler circuit.
2. In incremental system, inspection of punched tape is easier because the sum of all positioning commands are zero (i.e. the end point of a machining process becomes the starting point of the other).
3. Mirror imaging program in incremental type is easy. The reason is that we have to change the command from positive (+) to negative (–) using mirror axis as the original coordinate point.
4. In case of tool breakage or power supply failure in absolute system, the machinist moves the tool holder to the home point, changes the tool, and restarts the operation, but this is difficult in case of incremental programming method.

9.7 BASICS OF MOTION CONTROL AND FEEDBACK FOR NC MACHINES

In most modern NC machines, the actuator is a servo-motor, and the control system is a closed loop feedback system, with an encoder (see Figure 9.11) to provide feedback on the actual position of the table. This feedback is used to control the speed of the motion.

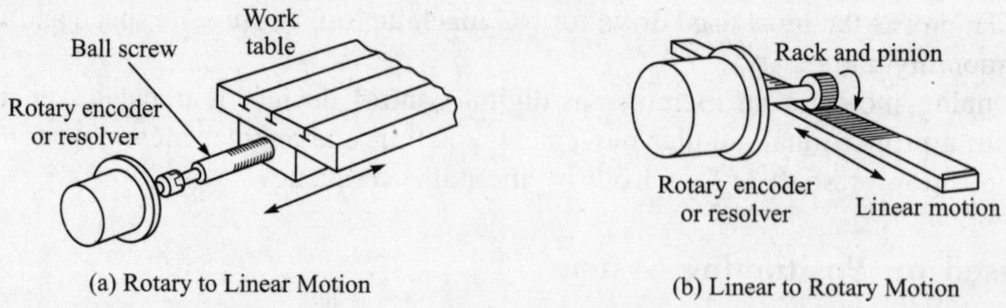

Figure 9.11 Two Types of Encoder Configurations

Machines with stepper motors move in discrete length units, since each pulse to the stepper motor makes it rotate through a fixed, finite angle. Likewise, encoders usually give out pulse outputs (e.g. 500 pulses per revolution).

Rotational motion of the motor is converted to linear motion of the table by the leadscrew. The pitch of the leadscrew is the horizontal distance between successive threads of the screw. Most screws have single threads (called single start screw) — in this case, the pitch equals the horizontal distance moved by the table in one revolution of the table. Similarly, for a double start thread, the horizontal distance moved by the table in one revolution is 2*pitch.

The smallest distance that a machine table can be programmed to move (programming resolution of the machine) is called a Basic Length Unit (BLU).

Advantages of NC Machine

1. Easier to program; easy storage of existing programs;
2. Easy to change a program
3. Avoids human errors
4. NC machines are safer to operate
5. Complex geometry is produced as cheaply as simple ones
6. Usually generates closer tolerances than manually operated machines

Disadvantages

1. High initial investment.
2. Require properly trained personnel.
3. Cost of maintenance is high

9.7.1 Control of NC Machines

The most common motor types for larger machines are servo-controlled motors. Depending on the hardware, the smallest distance unit that the machine tool can be moved by is referred to as a BLU (Basic Length Unit). In practice, the BLU is equivalent to the accuracy of the machine tool.

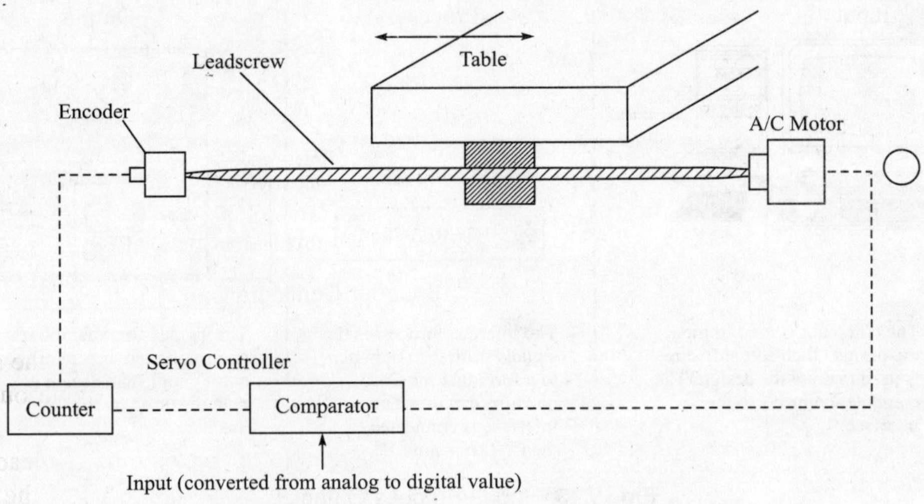

Figure 9.12 A Schematic of the Servo Control.

Structurally, every NC machine tool has two components:

- The conventional machine tool, with servo-motors to drive the lead screws.
- A Machine Control Unit (MCU) or the controller.

The MCU is made up of a Data Processing Unit (DPU) and a Control Loops Unit (CLU).

Data processing unit

- Input device [RS-232 port/ Tape reader/ Punched tape reader].
- Data reading circuits and parity checking circuits.
- Decoders to distribute data to the axes controllers.

Control loops unit

- Interpolator to supply machine-motion commands between data points.
- Position control loop hardware for each axis of motion.

9.7.2 What Does 'CNC' Mean

'CNC' means "Computer Numerical Control". To put it precisely, a *computer* converts the design into *numbers* which the computer uses to *control* the cutting and shaping of the material.

A CNC production facility needs three pieces of equipment:

A computer

The computer is used to draw the design. However, the design is only a picture and the CNC machine cannot use this to manufacture the product. The computer software must also convert the drawing into numbers (coordinates) that the CNC machine can use when it starts to cut and shape the material.

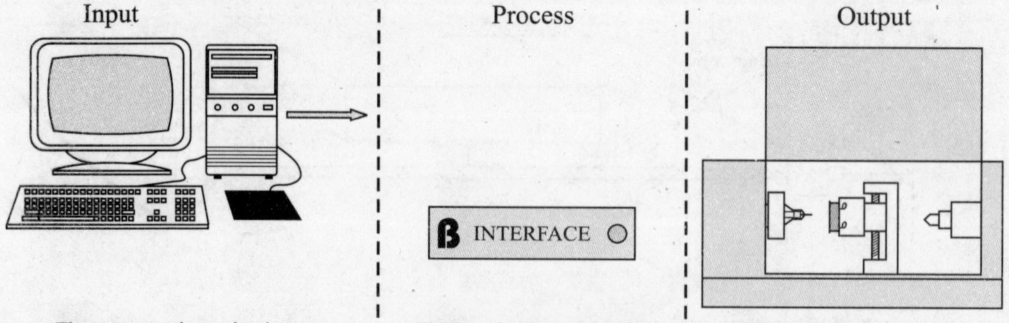

The computer is used to input the design. Tech soft software is used to draw the design. The computer connects to the interface.

The interface processes the signals from the computer to a form that the CNC machine can use. The interface is connected to the CNC machine.

The signals from the interface control the movement of the cutting tool. The design is manufactured on the CNC machine.

Fig. 9.13 Input–Process-Output.

An interface

A computer cannot be directly connected to a CNC machine. The computer is connected to an interface. This converts the signals from the computer to a form that the CNC machine understands. The signals are in the form of digital signals when they are sent to the CNC machine.

CNC (Computer Numerical Control) machine

The signals from the interface control the motors on the CNC machine. The signals determine the way the vice moves. The vice moves in three directions X, Y and Z (horizontally, vertically and depth). The signals also control the speed of the cutting tool.

CNC axes movement

The coordinate system used for the tool path must be identical to the coordinate system used by the CNC machine. The standards for machine axes are established according to the industry standard report EIA RS-267A.

CNC milling machines can perform simultaneous linear motion along the three axis and are called **three-axes machines**.

More complex CNC machines have capability of executing additional rotary motions.

CNC machine tool positioning modes

Within a given machine axes coordinate system, CNC can be programmed to locate tool positions in the following modes; incremental, absolute, or mixed.

CNC/DNC application

In the Industrial Automation (IA) market space, one of the most misunderstood segments is that of CNC/DNC. CNC (Computer Numerical Control) stands for machinery or machine tools that utilize proprietary or PC-based controls that interpret NC or numerical control programs to make parts. Examples include CNC milling machines, CNC turning centres or lathes, CNC punch

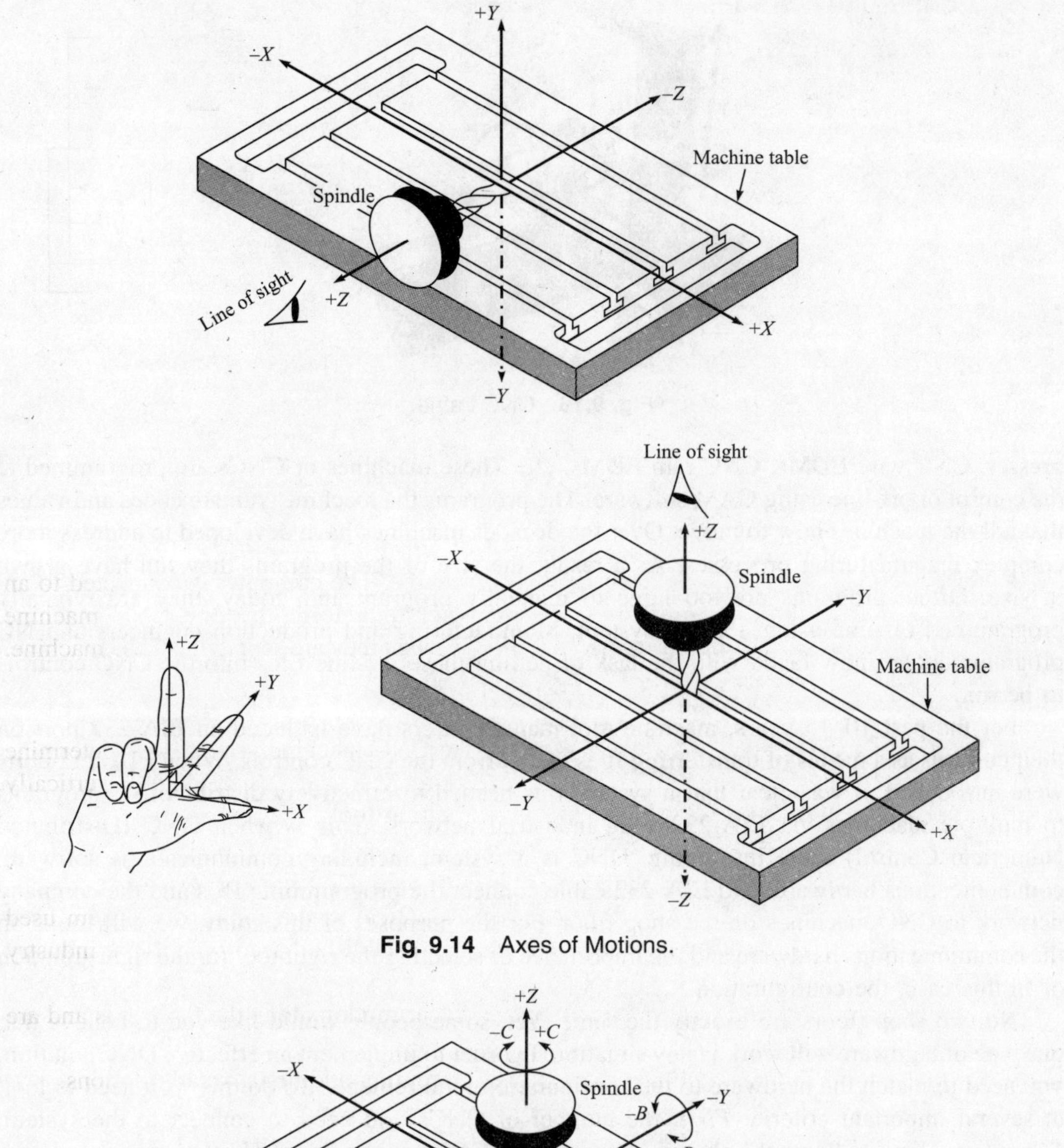

Fig. 9.14 Axes of Motions.

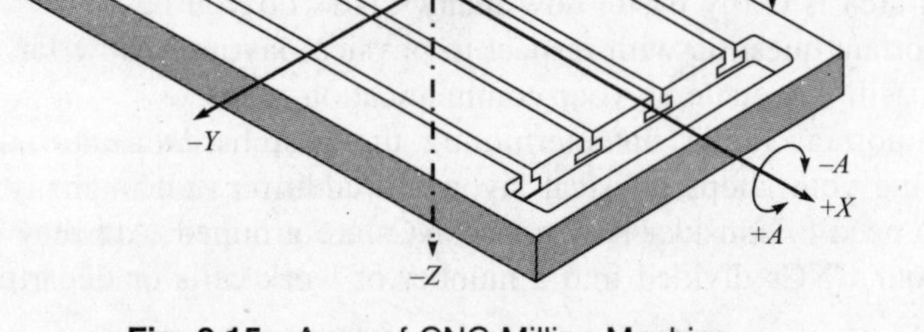

Fig. 9.15 Axes of CNC Milling Machine.

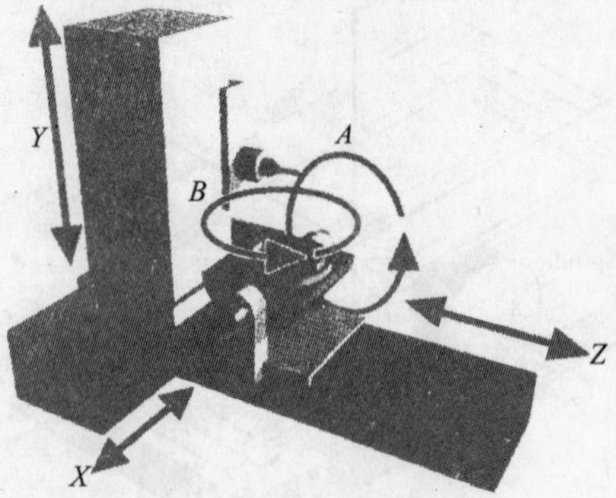

Fig. 9.16 CNC Lathe.

presses, CNC wire EDMs, CNC ram EDMs, etc. These machines or CNCs are programmed at the control or off-line using CAM software. The programs the machines run are codes and values that tell the machines how to move. Over the decades machines have developed to address more complex manufacturing processes. As a result, the size of the programs they run have grown in size. Often, programs are too large to manually program and, today, they are typically programmed off-line using a CAM system. Manufacturing and production engineers and NC programmers are now faced with the task of getting these off-line files into the CNC controls to be run.

For the past 10–15 years, machine tool manufacturers have included an EIA-232 port on their controls as a means of transferring files to and from the CNC controls. When EIA-232 ports were introduced, it was clear that a system was needed to effectively distribute NC programs to multiple machines via EIA-232 – an industrial network. This is when DNC (Distributed Numerical Control) came into being. DNC is a system, including communications software, communications hardware, and EIA-232 cable connect the programming PCs and the company network to CNC machines on the shop floor. For the purposes of this study, we will focus on the communications hardware and the importance of selecting the right tool for the right situation or in this case, the configuration.

No two shop floors are exactly the same. Yet, some people would like you to believe that one type of hardware will work in any situation. In order to implement an effective DNC solution, you need to match the hardware to the configuration requirements. To do this, you need to look at several important criteria. *First*, the number of CNCs you need to connect to the system. *Second*, you have to know the physical layout of the CNCs on the floor? *Third*, you have to know the performance specifications need to be met such as the maximum sustained baud rate and flow control. The first area is really basic: how many CNCs do you have?

Another important question with respect to physical layout is how far away are the CNCs from the PC you will designate as your communications server.

The most important factor in determining the best hardware for the DNC system is accurately assessing your shops physical layout. In addition to how many CNCs you need to connect, you also need to consider how your CNCs are grouped. Are they in one large area or group? Or, are your CNCs divided into a number of work cells or departments?

The last main consideration in selecting the right serial communications hardware for your DNC system is the specific hardware requirements of the CNCs you need to network. There are many details you have to consider when selecting the right hardware for any DNC system. The number of CNCs you need to connect, how those CNCs are physically laid out, and the actual communications parameters or requirements of each machine are all items that need to be considered in the planning process.

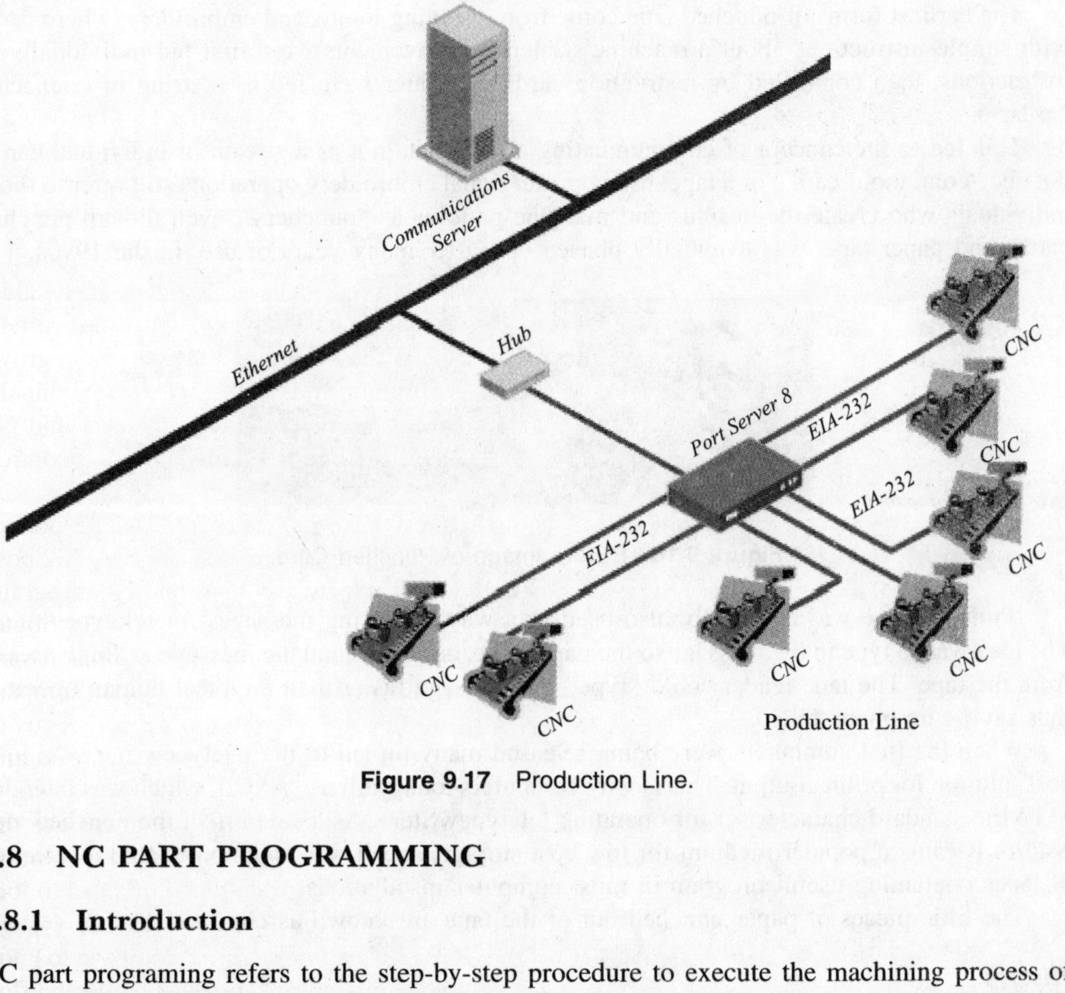

Figure 9.17 Production Line.

9.8 NC PART PROGRAMMING

9.8.1 Introduction

NC part programing refers to the step-by-step procedure to execute the machining process of any part. Using the instruction provided in the program, the CNC machine tool will be operated. Earlier, NC part program involved the preparation of punched tape which is used as an input medium. There are four different ways by which part program can be prepared :

1. Manual part programming
2. Computer assisted part programming
3. CAD/CAM based programming
4. Part program based on solid model.

As the trend towards integrating the CAD/CAM increases, the third approach of programming gets more popular than was case earlier.

9.8.2 Punched Tape

Punched tape is an old-fashioned form of data storage, consisting of a long strip of paper in which holes are punched to store data.

The earliest forms of punched tape come from weaving looms and embroidery, where cards with simple instructions about a machine's intended movements were first fed individually as instructions, then controlled by instruction cards, and later were fed as a string of connected cards.

This led to the concept of communicating analog data not as a stream of individual cards, but one "continuous card", or a tape. Many professional embroidery operations still refer to those individuals who create the designs and machine patterns as "punchers", even though punched cards, and paper tape, was eventually phased out, after many years of use, in the 1990s.

Figure 9.18 Picture Image of Punched Card.

Punched tape was eventually also used as a way of storing messages for teletypewriters. The idea was to type in the message to the paper tape, and then send the message at "high speed" from the tape. The tape reader could "type" the message faster than a typical human operator, thus saving on phone bills.

When the first computers were being released many turned to the teletypewriter as a low-cost solution for printer output. This is why computers today still use ASCII, which was intended to be the standard character set for operating teletypewriters. As a side effect the punched tape readers became a popular medium for low cost storage, and it was common to find a selection of tapes containing useful program in most computer installations.

The little pieces of paper punched out of the tape are known as **chad.**

History

The punched card actually predates computers considerably, originated by Joseph Jacquard in 1801 as a control device for the Jacquard looms. Such cards were also used as an input method for the primitive calculating machines of the late 19th century.

The version by Herman Hollerith, patented on June 8, 1887 and used with mechanical tabulating machines in the 1890 US Census, was a piece of cardboard about 90 mm by 215 mm, with round holes. This was the same size as the dollar bill of the time, so that storage cabinets designed for money could be used for his cards. The early applications of punched cards all used specifically-designed card layouts. It wasn't until around 1928 that punched cards and machines were made "general-purpose". In that year, punched cards were made a standard size, corresponding to the US currency of the day.

To compensate for the cyclical nature of the Census Bureau's demand for his machines, Hollerith founded the *Tabulating Machine Company* (1896) which was bought by Thomas J. Watson SR., founder of IBM in 1914. IBM manufactured a wide variety of business machines and eventually married the punched card to its early computers, encoding binary information as patterns of small rectangular holes.

The IBM punching format with rectangular holes eventually won out over the Univac 90-character format, which used 45 columns (2 characters in each) of 12 round holes. IBM (Hollerith) punched cards are made of smooth stock, .007 of an inch thick. There are about 143 cards to the inch thickness; a group of such cards is called a *deck*.

Functional details

The method is quite simple: On a piece of light-weight cardboard, successive positions either have a hole punched through them or are left intact. The rectangular bits of paper punched out are called chads. Thus, each punch location on the card represents a single binary digit (or "bit"). Each column on the card contained several punch positions (multiple bits).

The IBM card format, which became standard, held 80 columns of 12 punch locations each, representing 80 characters (since 12 bits is more than enough for representing a character, not all combinations were used.) originally coded: 1 punch (digit[0-9]) was a digit, 2 punches (zone[12,11,0] + digit[1-9]) was a letter, 3 punches (zone[12,11,0] + digit[1-7] + 8) was a special character, later the introduction of EBCDIC allowed columns with as many as 6 punches (zones[12,11,0,8,9] + digit[1-7]).

The IBM cards could be used in two different modes. In the "text mode", a column represented a digit or other character. As these cards had 12 bits in a column, while computers only used 6-bit or 8-bit characters internally, not every combination of holes was legal. On the other hand, in the "binary mode" every column was treated as a simple bitfield, and every combination of holes was permitted (one column was then read into 2 bytes). In this "binary mode", cards could be made in which every possible punch position had a hole: these were called "lace cards." Some other computers (e.g. the IBM 700/7000 series scientific computers) "binary mode" treated every row as two 36-bit words, in columns 1-72, ignoring the last 8 columns.

Other coding schemes, sizes of card, and hole shapes were tried at various times. Often the text was also printed at the top of the card, allowing humans to read the text as well, if the cards were produced by a card-punch machine (called a "key-punch"), which was like a large, very noisy typewriter. There were also cards with all the punch positions perforated so programming or data could be punched out manually, one hole at a time, with a device like a blunt pin with its wire bent into a finger-ring on the other end.

In applications requiring storage of multi-character data, such as words or large numbers, the card columns were used in groups known as *fields*.

Electromechanical equipment (called unit record equipment) for punching, sorting and printing the cards was manufactured. These operations did not require the use of a computer. For applications in which accuracy was critical, the punching might be checked by use of a verifier.

The card readers used an electrical (metal "brush") or, later, optical sensor to detect which positions on the card contained a hole. They had high-speed mechanical feeders to feed hundreds of cards through in a very short time.

Advantages

One of the key advantages of this system is that a computer was not required to encode information onto the cards — the typewriter-like card-punch machine was all that was needed — and "key-punch operators" (who did nothing but punch cards full-time on such machines) were in great demand.

Quality control was often having two different operators key the same data, with the 2nd using a card-verifier instead of a card-punch. If a card failed verification, the card-verifier would stop, letting the operator replace the card with a corrected one.

When the time came to transfer the information thus encoded into the computer, the process could occur at very high speed (either by the computer itself or by a separate device that "read" the cards and "wrote" the data onto magnetic tapes (or, later, on removable hard disks) that could then be mounted on the computer), thus making best use of expensive computer time.

Obsolescence

Punched-card systems fell out of favor in the mid to late 1970s, as disk and tape storage became cost-effective, and interactive terminals meant that users could edit their work with the computer directly rather than requiring the intermediate step of the punched cards.

However, their influence lives on through many standard conventions and file formats. The terminals that replaced the punched cards displayed 80 columns of text, for compatibility with existing software. Many programs still operate on the convention of 80 text columns, although strict adherence to that is fading as newer systems employ graphical user interfaces with variable-width type fonts.

Hanging chads

The term for the punched card area which is removed during a punch is *chad*. One notorious problem with a punched card system of tabulation is the incomplete punch; this can lead to a smaller hole than expected, or to a mere slit on the card, or to a mere dimple on the card. Thus, a chad which is still attached to the card is a *hanging chad*. This technical problem actually influenced the 2000 US presidential election; in the state of Florida, voting machines which used punched cards to tabulate votes generated improperly rendered records of several hundred votes, spread out over an entire state, which tipped the vote in favour of George W. Bush over Albert Gore, thus influencing history.

Various methods of punching punched card

Here are eight basic techniques now in use to read the digital information from punched paper or plastic tapes. Three different mediums are used to sense a hole in the tape: air, mechanical fingers and light. Only the method of producing an output signal that represents the presence of a hole is shown—design details of complete reading stations or additional electronics that may be necessary have been omitted.

Air sensing

1. An old method of reading punched tape pneumatically, used widely in the player piano. Tapes are wide, of coarse pitch. The technique is simple, rugged, not very fast. When a hole in the tape opens the port in the metal bar, atmospheric pressure allows the spring to expand the bellows to give a direct mechanical push.

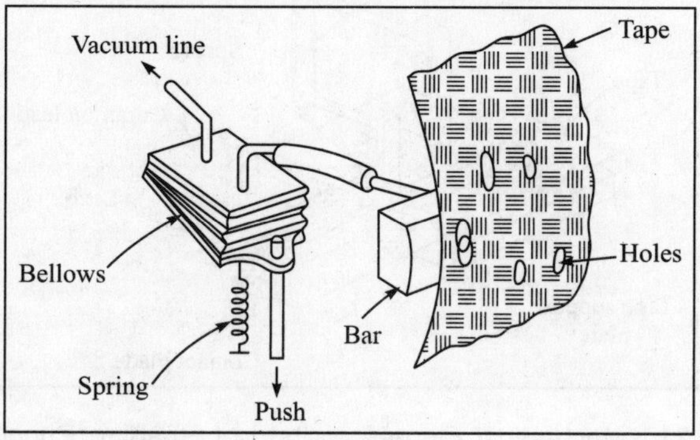

Figure 9.19 Air Sensing Method of Reading Punched Card (Pneumatic).

2. The hot wire anemometer technique is the newest way being used to read punched tape. A small wire, about 0.0005 inch diameter by about 1/16th inch long, is supported in a duct and heated by current. When hole in tape opens, airflow changes wire temperature, hence resistance, providing a low-level electrical signal. Capable of very high reading rates.

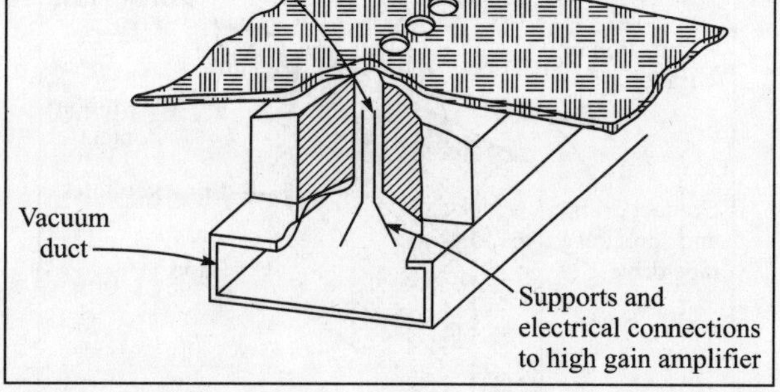

Figure 9.20 Air Sensing Method of Reading Punched Card (hot wire anemometer technique).

Mechanical fingers

3. A conventional method of reading stepped tape. The tape is supported by a block with a hole for each channel. While the tape is in motion, bail holds contact levels up. When tape stops, fingers rest on tape. Where there is a hole, finger enters it, making electrical contact. This design has many variations, such as offset fingers for more contact room,

double-contact blades for isolated channels, etc. Contact blade spring period can be critical. Cost moderate. Speed about 5 pitches per second. Tape about 1 inch wide, 8 channels plus sprocket on 1/10th inch centers; other sizes can be used.

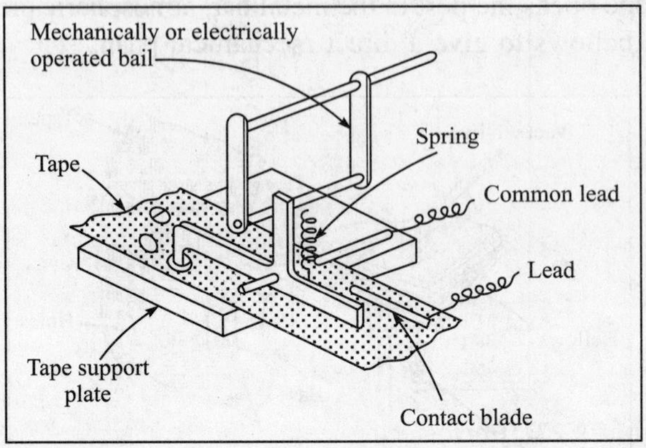

Figure 9.21 Mechanical Fingers Method of Reading Punched Card.

4. A conventional method of reading program tape. Tape can travel one direction only - reversal will damage contacts. A wide variety of contacts readily available from stock. Maximum speeds with stepped tape motion about 40 steps per second; with continuous tape motion, as high as 60 inches per second. Contact life several million cycles. As many as 50 channels can be placed on a 3-inch wide tape, each channel controlling one circuit. Sprocket pitch about 1/10 inch. Although contacts can operate small current relays, electronic equipment may be better. Mechanical cost moderate, electronic cost can vary.

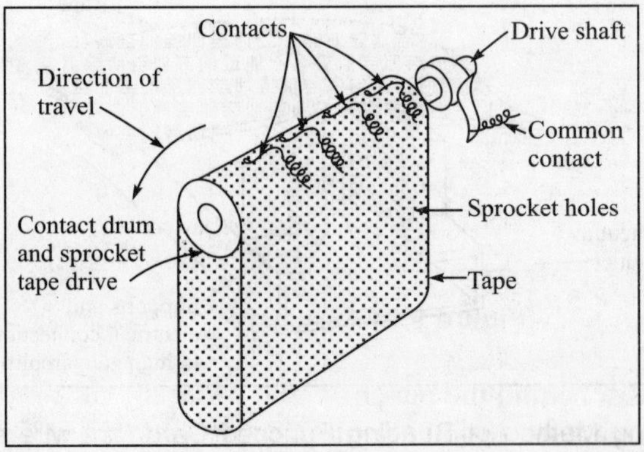

Figure 9.22 Mechanical Method of Reading Punched Card.

Light sensing

5. The basic technique for reading punched tape photoelectrically. Fast, reliable, quiet, with no moving parts. May be used with stepped or continuously moving tape. Uses one photoelectric cell or photomultiplier tube for each channel. Prisms, mirrors, light pipes, lenses, etc. used to spread light beams from channels to obtain room for photocells. Cost intermediate.

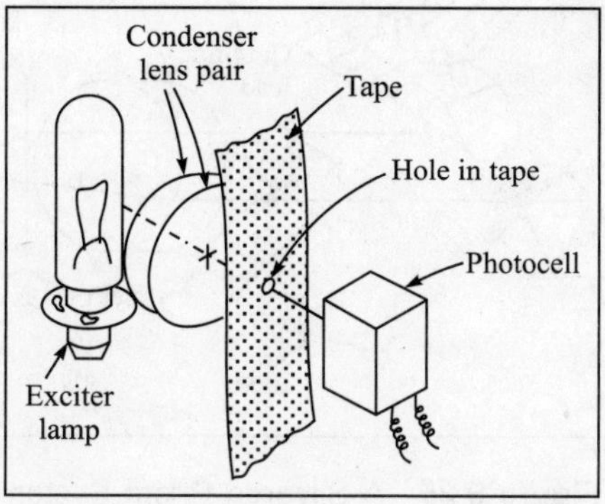

Figure 9.23 Photoelectric Method of Reading Punched Tape.

6. Photoelectric system using one photocell and a mechanically driven scanning disc to give a serial output from the tape. The scanning disc replaces electronic equipment otherwise required to obtain serial output for single line transmission of the tape information.

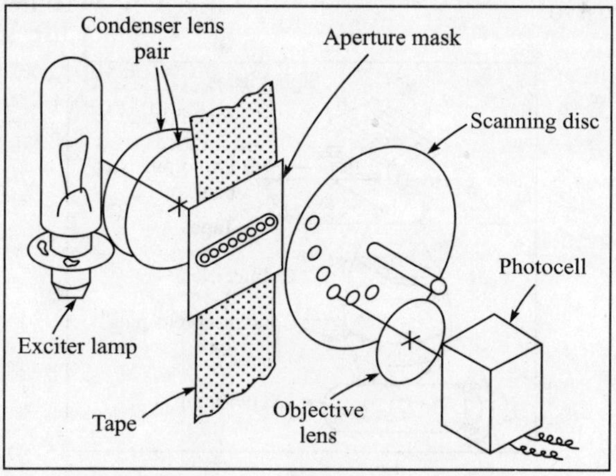

Figure 9.24 Photoelectric System.

7. A silvered prism behind the punched tape reflects light from the exciter lamp back out through the same hole, to fall normally on the phototransistors. This same arrangement will also read opaque tape with printed dots by reflected light (the prism being inoperative). Approximately the same ranges and limitations on reading speeds and channel spacing as in the more common arrangement.

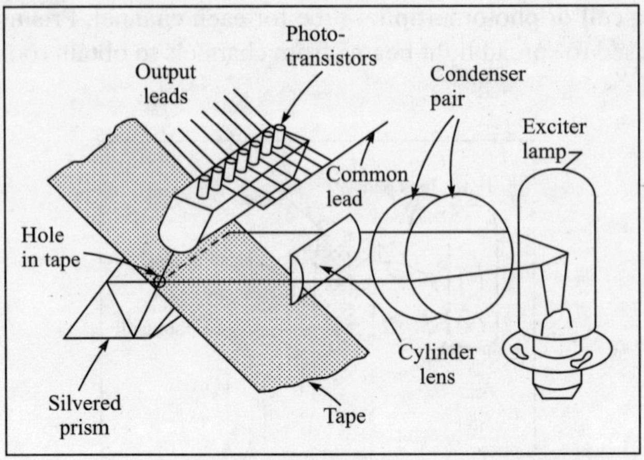

Figure 9.25 A Silvered Prism System.

8. An optional method for high-speed reading of 6, 7 or 8 channels (plus sprocket) using small phototransistors (about 0.080 inch diameter by 9/16 inch long) mounted in phenolic block directly over the centre of the 1/10th inch spaced information channels. Reading speeds from about 7 to 1,000 characters per second, depending on tape drive mechanism. (Present phototransistors have a maximum limit of about 10 kc). Cost relatively expensive.

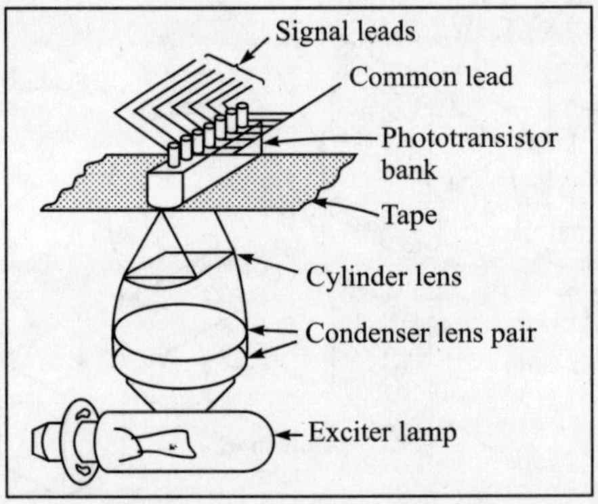

Figure 9.26 Phototransistors.

Preparation of punched tape

Punched Tape can be prepared either following *EIA (Electronic Industry Association) Standard or ISO/ASCII (American Standards Code for Information Interchange) Standard.*

EIA method

In this method has eight track organized with seven bits of information codes. The number 1 to 8 represent the number of track.

1. Tracks 1 to 4 are used for numerical data such as dimension, speed, feed, etc.
2. The sprocket holes at 3 and 4 are offset and equally spaced holes for forwarding or rewinding the tape through reader.
3. Track 5 is for parity check.
4. Track 6 is used to represent “zero”.
5. Track 7 marked X, is used to code alphabets to identify various machine operation(a- z).
6. Track 8 is used to punch End of Block.

In this system for each number there must be a odd number holes to be punched. If the number of holes in row is not odd, then additional hole is to be punched in the fifth track. This odd parity check looks for any errors or mechanical failures in the tape preparation.

ISO/ASCII code

This is also a seven bit code having 8 tracks. Here the number of holes corresponding to the NC word is even. This method becomes popular these days because it permits the information to be readily exchanged with computers.

1. Numbers are coded in Binary Coded Decimal (BCD) form on first four tracks.
2. Tracks 5 and 6 for numerical values.
3. Sprocket holes is between tracks 3 and 4.
4. Track 7 is for coding alphabets.
5. Track 8 is for parity check to produce even number of holes.

9.9 PART PROGRAMMING METHODS

As seen earlier there are four ways of part program can be done. They are:

1. Manual part programming
2. Computer assisted part programming
3. CAD/CAM based programming
4. Part program based on solid model

9.9.1 Manual Part Programming

It is one of the conventional ways of NC programming used for generation of parts of simple geometry.

In the control of NC machines, the programmer (usually called the part programmer) writes the instructions for the machine tool controller to specify the following:

1. Which tool should be loaded on the machine spindle
2. The cutting conditions (speed, feed, coolant ON/OFF, etc.)
3. The starting and end points of a motion segment
4. How to move the tool with respect to the machine.

The program is directly fed into the machine either using punched card or transferred from a computer memory.

Representation in Punched Tape									Character	Name of the Character
8	7	6	5	4	F	3	2	1		
	b_7	b_6	b_5	b_4		b_3	b_2	b_1	Symbol	
					●				NULL	NULL
●				●	●				BS	BACKSPACE
				●	●			●	TAB	TABULATION
				●	●		●		EB	END OF BLOCK
●				●	●	●		●	CR	CAR RETURN
●		●			●				SP	SPACE
●		●			●	●		●	%	PROGRAM START
		●		●	●				(	CONTROL OUT
●		●		●	●			●	)	CONTROL IN
		●		●	●		●	●	+	PLUS SIGN
		●		●	●	●		●	−	MINUS SIGN
●		●		●	●	●	●	●	/	OPTIONAL BLOCK
		●			●				0	
●		●		●	●			●	1	
●		●		●	●		●		2	
		●			●		●	●	3	
		●			●	●			4	
●		●		●	●	●		●	5	
		●			●	●			6	
●		●		●	●	●		●	7	
●		●		●	●				8	
		●		●	●			●	9	
	●	●		●	●		●		:	ALIGNMENT FUNCTION
	●				●			●	A	
	●				●		●		B	
	●				●		●	●	C	
	●				●	●			D	
	●				●	●		●	E	
	●				●	●	●		F	
	●				●	●	●	●	G	
	●				●				H	
●	●				●			●	I	
●	●				●		●		J	
	●				●		●	●	K	
●	●				●	●			L	
	●				●	●		●	M	
	●				●	●	●		N	
●	●				●	●	●	●	O	
	●				●				P	
●	●				●			●	Q	
●	●				●		●		R	
	●				●		●	●	S	
●	●				●	●			T	
	●				●	●		●	U	
	●				●	●	●		V	
●	●				●	●	●	●	W	
●	●				●				X	
	●				●			●	Y	
					●		●		Z	
●	●				●	●	●	●	DEL	DELETE

Fig. 9.27 EIA Method of Coding Punched Tape.

Representation in Punched Tape									Character	Name of the Character
8	7	6	P	4	F	3	2	1		
b_7	b_6	b_5		b_4		b_3	b_2	b_1	Symbol	
					●				SP	SPACE
					●			●	1	
					●		●		2	
			●		●		●	●	3	
					●	●			4	
			●		●			●	5	
			●		●	●			6	
					●				7	
				●	●				8	
			●	●	●			●	9	
				●	●		●	●	EOB	END OF BLOCK
		●			●				0	
		●	●		●			●	/	OPTIONAL STOP
		●	●		●		●		S	
		●			●		●	●	T	
		●	●		●	●			U	
		●			●	●		●	V	
		●			●	●			W	
		●	●		●	●		●	X	
		●	●	●	●				Y	
		●		●	●			●	Z	
		●	●	●	●		●		BS	BACK SPACE
		●	●	●	●		●	●	,	
		●		●	●	●		●	TAB	TABULATE
	●				●				–	MINUS SIGN
	●		●		●			●	J	
	●		●		●		●		K	
	●				●		●	●	L	
	●		●		●	●			M	
	●				●	●		●	N	
	●				●	●	●		O	
	●		●		●	●	●	●	P	
	●		●	●	●				Q	
	●			●	●			●	R	
	●		●	●	●		●	●	%	PROGRAM START
	●	●	●		●				+	PLUS SIGN
	●	●			●			●	A	
	●	●			●		●		B	
	●	●	●		●		●	●	C	
	●	●			●				D	
	●	●	●		●	●		●	E	
	●	●	●		●	●	●		F	
	●	●			●	●	●	●	G	
	●	●		●	●				H	
	●	●	●	●	●			●	I	
	●	●	●	●	●		●			LOWER CASE ALPHABETS
●	●	●	●	●	●	●				UPPER CASE ALPHABETS
	●	●	●	●	●	●	●	●	DEL	DELETE
					●				CR	CARRIAGE RETURN

Fig. 9.28 ISO/ ASCII Method of Coding Punched Tape.

The motion of the machine tool with respect to the part depends on the machine and the controller capabilities. Many drilling machines use PTP control for machining. Most milling machines can perform contouring, that is, they can move the tool along specified geometric paths. These paths are typically made up of segments that are linear or circular arcs. The part programmer writes the step-by-step sequence of machine tool movement with respect to workpiece.

9.9.2 Computer Assisted Part Programming

Due to development in the CAD/CAM integration, today CNC part programs are prepared by drafting the component final shape and raw material on the offline programming software, which generates the part program corresponding to the controller selected. In this method, the programming job has been divided into two categories:

- Part Programmer Job
- Computer Job.

Part programmer job

The part programmer job is to provide the following information to the computer: Defining the geometry of workpiece and operation sequence with tool path.

Computer job

The computer job in computer assisted part program is to:

1. translate the input data to machine understandable form,
2. perform necessary arithmetic calculation,
3. compute cutter offset compensation, and
4. generate NC part program.

Numerical control words

Different types of words used in NC machine forms the block of instruction; these are:

Sequence number (N-words)	:	Identifies the block.
Preparatory word (G-words)	:	Prepares the controller for instructions to follow.
Feed rate(F-words)	:	Specifies the feed in machining.
Cutting speed(S-word)	:	Specifies the cutting speed of the process.
Coordinates word	:	Specifies the *x*-, *y*-, *z*-axis value.
Tool selection (T-word)	:	Specifies the tool used in operation.
Miscellaneous function (m-word)	:	Specifies the miscellaneous or auxiliary functions available in the machine tool.

9.9.3 Punched Tape Programming Format

In both the EIA and ISO/ASCII standard method, there are three ways of format for compiling of NC data into suitable block of instructions. They are:

- Fixed sequential format
- TAB sequential format
- Word address format

(i) Fixed sequential format

In this, the information is given in a particular sequence and the block must contain a fixed number of characters. Even if the values of the function remain constant in the next stage, the value will be specified again. Each block will be terminated by the EOB (End of Block) statement.

N001 G00 X 20 Z0 F35 S1000 EOB
N002 G01 X 30 Z0 F35 S1000 EOB
N003 G01 X 30 Z-10 F35 S1000 EOB

The advantage of this format is that the whole block can be read at the same instant, instead of reading character by character.

(ii) TAB sequential format

The NC words for each block of instruction will be provided in the same sequence, but each word is preceded by the character TAB. One of the advantages of this format when compared with earlier format is that if the instructions remain unchanged in the following block, the instructions need not be specified again. But the character TAB has to be punched.

N001 TAB G00 TAB X 20 TAB Z0 TAB F35 TAB S1000 EOB
N002 TAB G01 TAB X 30 TAB Z0 TAB F35 TAB S1000 EOB
N003 TAB G01 TAB X 30 TAB Z-10 TAB F35 TAB S1000 EOB

(iii) Word address format

In this format, NC words are used to identify the function which is followed by the numbers corresponding to the function. It doesn't follow any sequence as the two earlier formats. The block will be terminated by the EOB statement.

N001 G00 X 20 Z0 F35 S1000 EOB
N002 G01 X 20 EOB
N003 Z-10 EOB

Various M-Codes

M-code	Description
M00	Program Stop
M02	Optional Stop
M03	Spindle Rotate in clockwise
M04	Spindle Rotate in anticlockwise
M05	Spindle Stop
M06	Tool Change
M08	Coolant On
M09	Coolant Off
M10	Vice Open
M11	Vice Close
M98	Subprogram Call
M99	Subprogram Exit
M30	Program Stop and Rewind

Various G-Code

G-code	Description
G00	Rapid Traverse
G01	Linear Interpolation
G02	Circular Interpolation in Clockwise
G03	Circular Interpolation in Anticlockwise
G04	Dwell
G20	Input Inch Data
G21	Input Metric Data
G28	Move the Tool to Reference Point
G32	Threading Cycle
G70	Finishing Cycle
G71	Turning Cycle
G72	Facing Cycle
G73	Pattern Repeating
G76	Thread Cutting Cycle
G98	Feed per Minute
G99	Feed per Revolution

9.9.4 NC Part Programming

***Facing operation: for* 3 mm**

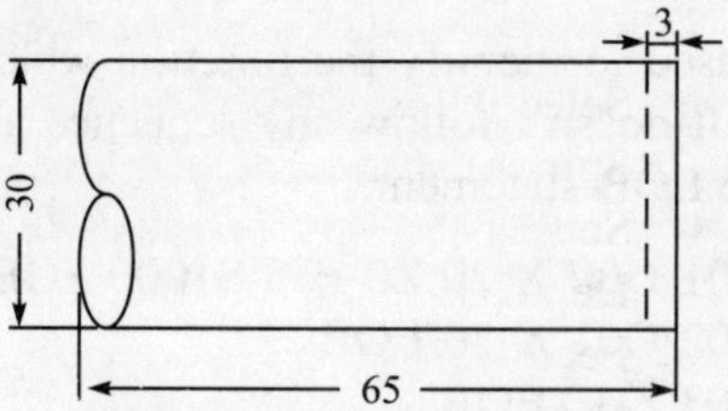

Program	Detailed explanation of codes used
[Billet X30 Z65	Specifying the raw material size
G21 G98	Metric data input and feed per minute
G28 U0 W0	Moving tool to reference point
M06 T0101	Selecting tool No. 1
M03 S1000	Spindle speed is set to 1000 rpm
G00 X30 Z1	Tool will be moved to initial position which the above the workpiece.
G01 Z-1.0 F45	Giving depth of 0.5 mm
G01 X0	Moving tool to spindle centre
G01 Z1	Moving tool up the level of workpiece
G00 X30	Moving tool to X30
G01 Z-2.0	
G01 X0	
G01 Z1	

G00 X30	
G01 Z-3.0	
G01 X0	
G01 Z1	Moving tool up the level of workpiece
G00 X30	Moving tool to X30
G28 U0 W0	Moving tool back to home position
M05	Spindle stop rotation
M30	Program stop and rewind

Simple turning operation

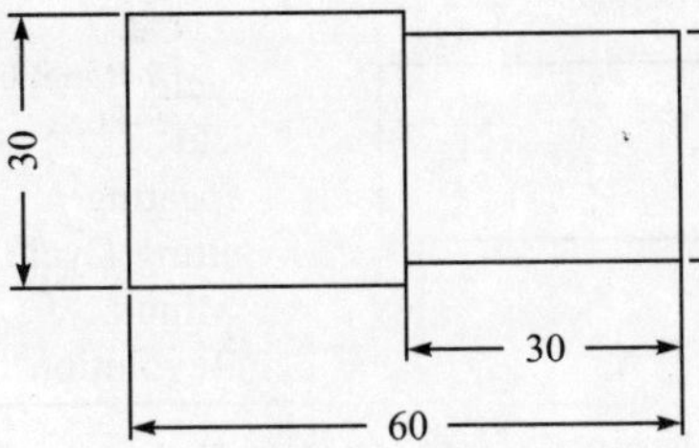

Program	**Detailed explanation of codes used**
[Billet X30 Z60	Specifying the raw material size
G21 G98	Metric data input and feed per minute
G28 U0 W0	Moving tool to reference point
M06 T0101	Selecting tool No. 1
M03 S1000	Spindle speed is set to 1000 rpm
G00 X30 Z1	Setting tool just above the workpiece level
G00 X29	Depth of cut is given (1 mm)
G01 Z –30 F35	Feed is given
G01 X30	Moving the tool out in the *x*-direction
G00 Z1	Moving the tool above the workpiece rapidly
G00 X28	Depth of cut is given (1 mm)
G01 Z-30	
G01 X30	
G00 Z1	
G00 X27	
G01 Z-30	
G01 X30	
G00 Z1	
G00 X26	
G01 Z-30	
G01 X30	
G00 Z1	
G00 X25	
G01 Z-30	
G01 X30	

G00 Z1	
G28 U0 W0	Moving tool to reference point
M05	Spindle stop rotation
M30	Program stop and rewind

G90 single turning cycle (Box turning cycle)

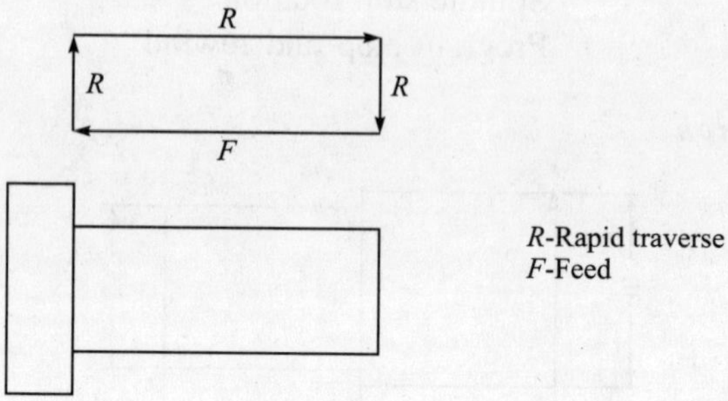

Syntax:

G90 *X* (*U*) *Z* (*W*) *F*

where,

X – Diameter of workpiece
U – Incremental distance from current tool position to final in *x*-axis
Z – Length of workpiece
W – Incremental distance from current tool position to final in *z*-axis
F – Feed.

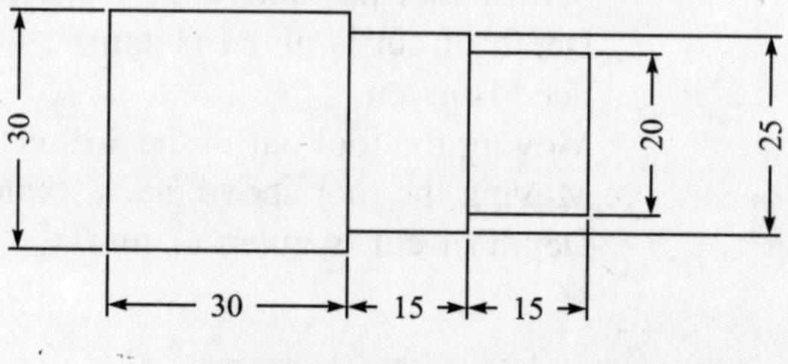

Program	**Detailed explanation of codes used**
[Billet X30 Z60	Specifying the raw material size
G21 G98	Metric data input and feed per minute
G28 U0 W0	Moving tool to reference point
M06 T0101	Selecting tool No. 1
M03 S1000	Spindle speed is set to 1000 rpm
G90 X29 Z-30 F35	Box turning cycle
X28	
X27	
X26	
X25	

G90 X24 Z-15 F35
X23
X22
X20
G28 U0 W0
M05
M30

G90 taper turning cycle :

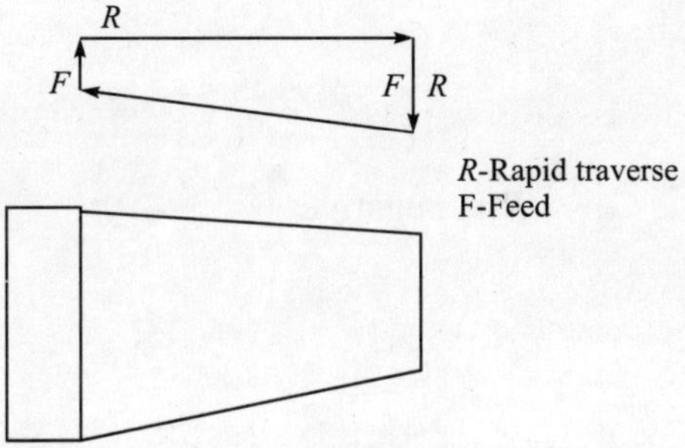

Syntax:

G90 *X* (*U*) *Z* (*W*) *R F*

X – Diameter of the workpiece in which the movement is to beign
U – Increment value of tool from current position in the *x*-axis
Z – Feed of the tool (length in which tool to move)
W – Increment value of tool from current position in the *z*-axis
R – Difference between the initial radius to final radius of workpiece

Note

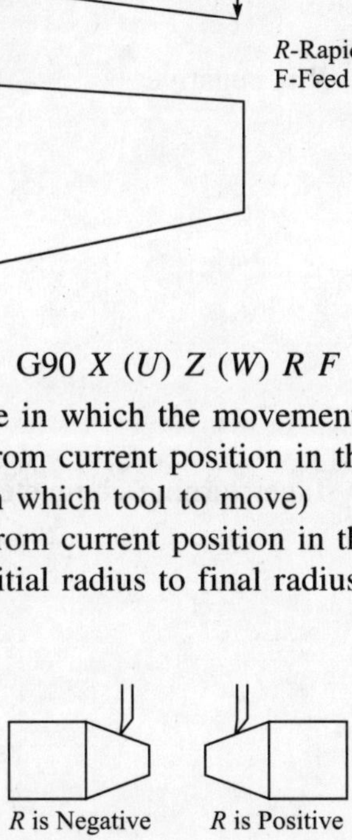

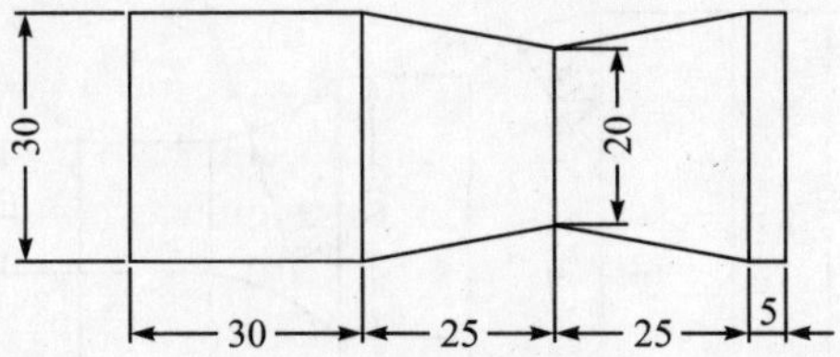

Program	Detailed explanation of codes used
[Billet X30 Z110	Specifying the raw material size
G21 G98	Metric data input and feed per minute
G28 U0 W0	Moving tool to reference point
M06 T0101	Selecting tool No. 1

M03 S1000	Spindle speed is set to 1000 rpm
G00 X30 Z1	Placing tool just above the workpiece rapidly
G90 X30 Z-30 R0.0 F30	$R = (D - d)/2$
X29 R0.5	
X28 R1.0	
X27 R1.5	
X26 R2.0	
X25 R2.5	
X24 R3.0	
X23 R3.5	
X22 R4.0	
X21 R4.5	
X20 R5.0	
G90 X30 Z-55 R0.0 F30	*R* is negative
X30 R-0.5	
X30 R-1.0	
X30 R-1.5	
X30 R-2.0	
X30 R-2.5	
X30 R-3.0	
X30 R-3.5	
X30 R-4.0	
X30 R-4.5	
X30 R-5.0	Taper turning completed
G28 U0 W0	
M05	
M30	

Circular interpolation

G02 clockwise direction

G03 anticlockwise direction

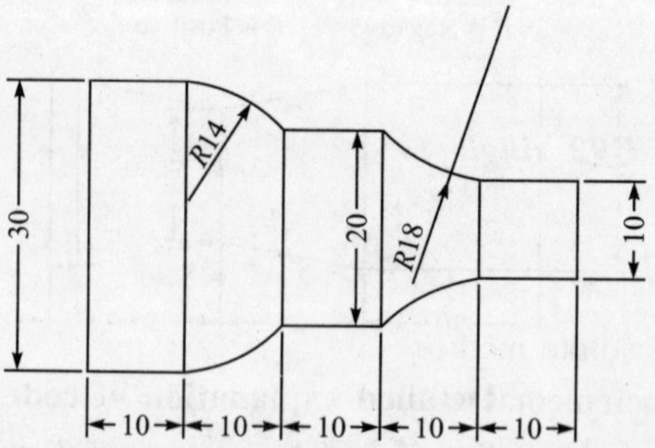

The operations involved are :

1. Turn to φ10.
2. G02 clockwise circular interpolation R 18

3. Turn to φ20.
4. G03 anticlockwise circular interpolation R14

Program	Detailed explanation of codes used
[Billet X30 Z60	Specifying the raw material size
G21 G98	Metric data input and feed per minute
G28 U0 W0	Moving tool to reference point
M06 T0101	Selecting tool no. 1
M03 S1000	Spindle speed is set to 1000 rpm
G00 X30 Z1	Placing tool just above the workpiece rapidly
G90 X30 Z-10 F45	Turn to φ10.
X29	
X28	
X27	
X26	
X25	
X24	
..	
..	
X10	
G02 X20 Z-20 R 18 F30	Clockwise circular interpolation
G90 X30 Z-30 F40	Turn to φ20.
X28	
X26	
X24	
X22	
X20	
G03 X30 Z-40 R 14 F30	Anticlockwise circular interpolation
G28 U0 W0	
M05	
M30	

Threading operation : G92 single threading cycle

Syntax :

$$G92\ X(u1)\ Z(w1)\ F$$

X – Depth of cut in absolute method
u_1 – Depth of cut in incremental method
Z – Length of thread in absolute method
w_1 – Length of thread in incremental method
F – Lead/pitch of thread.

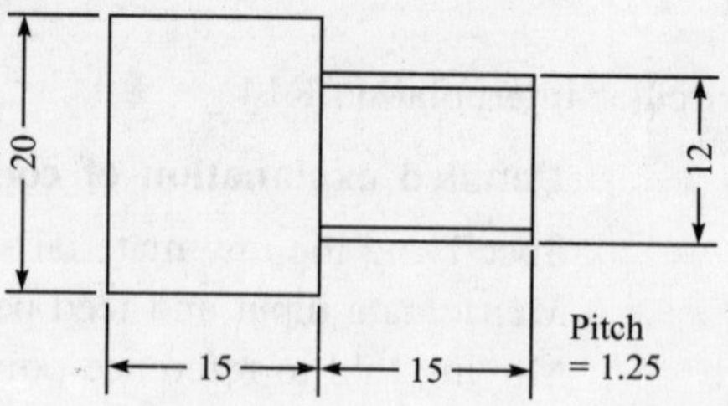

The operation involved are :

- Turn to φ12
- Threading for 12 × 1.25 pitch.

Program	**Detailed explanation of codes used**
[Billet X30 Z60	Specifying the raw material size
G21 G98	Metric data input and feed per minute
G28 U0 W0	Moving tool to reference point
M06 T0101	Selecting tool No. 1 (TURNING TOOL)
M03 S1000	Spindle speed is set to 1000 rpm
G00 X20 Z1	Placing tool just above the workpiece rapidly
G90 X20 Z-15 F45	
X19	
X18	
X17	
X16	
X15	
X14	
X13	
X12	
G28 U0 W0	Moving to reference point to change tool
M06 T0202	Selecting Tool 2 as threading tool.
G00 X12 Z1	Position tool above workpiece
G92 X12 Z-15 F1.25	Threading operation starts
X11.5	
X11.0	
X10.5	
X10.0	
X9.83	
G28 U0 W0	
M05	
M30	

9.10 MODERN MACHINING SYSTEMS: INTERPOLATION AND DNC

Many parts that need to be machined have very complex shapes (for example, dies for casting or plastic injection moulding of common parts). Such shapes require complex mathematical functions to describe them. As we have seen, we can only program the machine tool to move along straight lines or circular arcs. *How can we machine a part with geometry that is more complex?*

(a) Hardware solution

Pantograph machines. It is very common in Hong Kong and China.

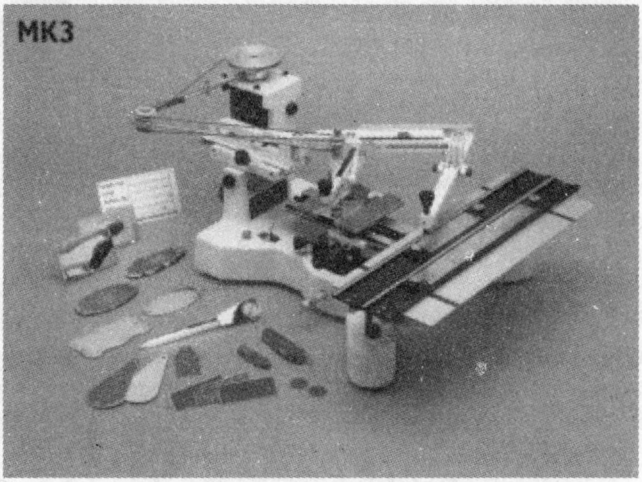

Figure 9.29 A Commercial Pantograph Machine

Typically, pantographs can be used for cutting quite complex 2D as well as 3D shapes. This is a dominant method for cutting moulding dies in the SAR. The method usually involves first making a plastic/wood model of the required mould in 2:1 (double size) or 1.5: 1 ratio. This model is used as a template to move a stylus around it which guides a cutting tool (usually a flat end mill, a ball-end mill, or sometimes a grinding tool) across the die block till the die is machined. By using the pantograph, the scaled model can be directly converted to the correct sized die. The models are usually larger than the actual part (die) since more details can be carved into the wood/ plastic by the craftsmen preparing the model.

(b) Software solution: Interpolation, and the use of DNC

In many modern industries, complex geometric shapes can be directly machined by using three-axis or five-axis machines, and CAM software. This software typically converts the CAD data of a part into the required NC part program. Since most parts have very complex geometry, the mathematical functions used to describe the shape of the part are not simple — they usually partition the entire surface into a set of smaller areas, called patches, each of which is described by means of a 3- or 4-degree polynomial function in x- and y- variables. On three-axis machines, such shapes are machined by first generating the profile of the part along many subsequent layers at different values of the z-coordinate. The part is machined by cutting the profile at each layer, proceeding to the next deeper layer, and so on. Of course, the geometry of each layer is described by a sequence of complex mathematical functions (not straight lines or arcs). Hence, we must first convert this data into a format that the NC controller can understand, i.e. straight lines and arcs. This process is called linear approximation (circular approximation). The approximation can result in thousands of small linear segments along which the tool must move for each layer. Thus, the NC program for such linear approximations (circular approximations) can be quite large.

Many NC controllers cannot handle such large part program files. Therefore, several machine tool companies allow the user to drip-feed the part program into the NC machine (that is, the part program is transferred to the controller in several small segments). As the controller finishes a fixed percentage (e.g. 90%) of the previous segment, the next segment is transferred to the controller from a computer. This is more popular in CNC (Computer Numerically Controlled) machines.

It is even possible that for a complex machine shop with several machine tools, the central computer can decide, depending on the status of each machine tool, which part must be machined on which machine tool. In such cases, the central computer is linked to each CNC machine, and uploads the required part program to the selected machine at the correct time. Such control is called DNC (Direct Numerical Control), and is popular in some advanced manufacturing systems.

Part programming contains geometric data about the part and motion information to move the cutting tool with respect to the workpiece. Basically, the machine receives instructions as a sequence of blocks containing commands to set machine parameters, speed, feed and other relevant information. A block is equivalent to a line of codes in a part program.

N135 G01 X1.0 Y1.0 Z0.125 T01 F5.0

9.11 AUTOMATICALLY PROGRAMMED TOOLS

APT uses English-like language statements to define part shape and tool motion as well as machine tool dependent data, such as feed rates and spindle speeds. This data is contained in an APT part program. An APT processor program is used to read these statements, interpret the meanings, and perform all the necessary calculations in order to generate a series of cutter location points that define the cutter path. The APT processor is a computer program which runs on a mainframe computer, possibly at a central site with time-sharing facilities. The generalized APT output is converted to the particular format G-code required by the CNC machine using a postprocessor program. Although G-codes are fairly well standardized, differences do exist between machine suppliers and CNC machines can be supplied with a variety of subsets of available codes. A description of postprocessor programs is given in this Chapter.

APT is a three-dimensional system which can be used to define complex geometrical shapes and to control up to five axes CNC machines. A major advantage of APT is that it has developed into an accepted standard for machine tools. There are many versions of the APT language available, each having particular benefits and characteristics. The prime disadvantage of APT is that it uses English-like commands to define geometry instead of the much more convenient graphical methods. Other disadvantages are that it requires extensive computing capability and can have a slow response (particularly with programmers who use the system intermittently and require multiple overnight runs on time-shared facilities).

The four types of statements in the APT language are:

- *Geometry statements* which define primitive elements such as points, lines, circles, planes, cones and spheres.
- *Motion statements* which describe the tool path in relation to the part geometry.
- *Postprocessor statements* which give specific machine tool code information as well as feeds and speeds.
- *Auxiliary statements* which give part and tool tolerances.

APT part programs usually list the part and the post processor reference number followed by the program statements as follows:

PARTNO
MACHIN
Geometry statements
Motion statements and machine tool commands
FINI

Geometry statements

The general form of geometry statements is:

symbol = geometry type/descriptive data

For example, a point, named P1, can be defined at coordinates X = 100 mm, Y = 200 mm and Z = 300 mm by the statement:

P1 = POINT/100.0, 200.0, 300.0

A line, named L1, can be defined between two points, P1 and P2, by the statement:

L1 = LINE/P1, P2

or through the point, P1, and parallel to another line, L2, by the statement:

L1 = LINE/P1, PAREL, L2

A plane, named PL1, can be defined between three points, P1, P2 and P3, by the statement:

PL1 = PLANE/P1, P2, P3

or through the point, P1, and parallel to another plane, PL2, by the statement:

PL1 = PLANE/P1, PAREL, PL2

A circle, named C1, can be specified by a centre position, P1, and a radius of 100 mm by the statement:

C1 = CIRCLE/CENTRE, P1, RADIUS, 100.0

or by a centre position, P1, and tangent to line, L1, by the statement:

C1 = CIRCLE/CENTRE, P1, TANTO, L1.

Motion statements

Motion statements use the defined geometry primitives in order to define tool movement. The general form of motion statements is:

motion statement/descriptive data

The instruction to go from an initial starting point, P1, and from which all other points are referenced, is given by the statement:

FROM/P1

The instruction to go to the point, P2, is given by the statement:

GOTO/P2

or to go to the absolute position X = 100 mm, Y = 200 mm, and Z = 300 mm is given by the statement:

GOTO/100.0, 200.0, 300.0

The instruction to move by an incremental (or delta) amount from the present position is given by the statement:

GODLTA/100.0, 200.0, 300.0

Contouring commands are used to direct the tool along two intersecting surfaces. The drive surface guides the side of the cutter, the part surface defines the position of the bottom of the cutter, and the check surface defines the limit of current tool motion. Modifier words, such as TO, ON, PAST or TANTO, are used to govern the position of the tool in relation to the check surface, as shown in Figure 9.30. Motion statements, GOLFT **(go to the left)**, GOFWD **(go forward)** and GOUP **(go up)**, are also used to control the cutter motion.

The instruction for the tool to move forward, with the drive surface, S1, on the left-hand side, and past the check surface, S2, is given by the statement:

GOLFT/S1, PAST, S2.

Postprocessor and auxiliary statements

Statements which specify machine tool related functions, such as those covered by F-, S-, T-, and M-codes, are defined in the postprocessor statements. For example, the instruction to set the feedrate at 100 mm per minute is given by the statement:

FEDRAT/100, MPM

The instruction to set the spindle at 1500 rpm in a clockwise direction is given by the statement:

SPINDL/1500, RPM, CLW

The instruction to set the coolant on is given by the statement:

COOLNT/ON

Other auxiliary statements are used to define cutter size, part number, and curve tolerance. The instruction to define a 50 mm diameter cutter is given by the statement:

CUTTER/50.0

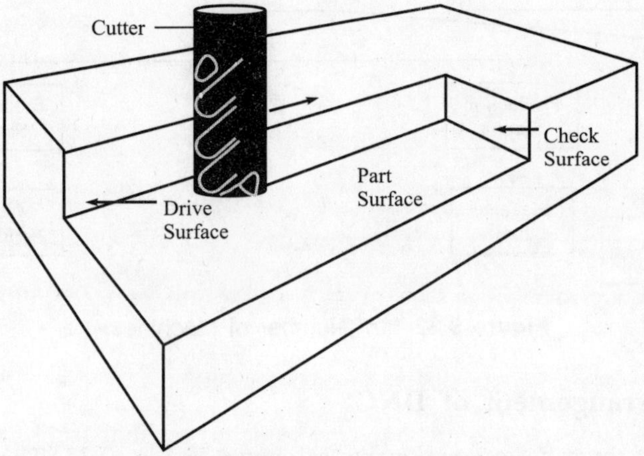

Figure 9.30 APT Contouring Surfaces.

9.12 DIRECT NUMERICAL CONTROL (DNC)

The technology of DNC started in the early 1960s itself as an attempt to feed the numerical data directly into the machine by avoiding punched tape. The attempt has succeeded due to use of the computer for data feeding, storage, and control of the NC machine tool. DNC can be defined as a type of manufacturing system in which several NC and CNC machines are operated/ controlled by a computer.

Direct numerical control is a form of NC machine in which a number of machines are controlled by a single computer directly. The tape reader, the least reliable component of CNC is omitted. Depending on the computing capability, memory, access or transfer rate of data's single computer have "n" number of machines to be controlled. DNC computer provides the instruction to the various machines on demand and on-line through a communication line. DNC also has feedback data collection from the machine tool that it controls.

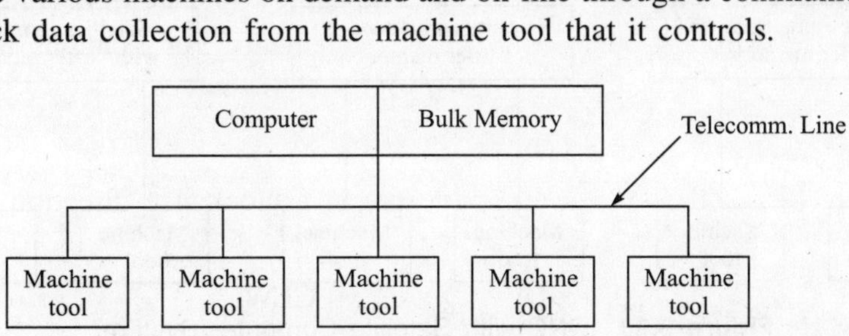

Figure 9.31 General Block Diagram of an DNC System.

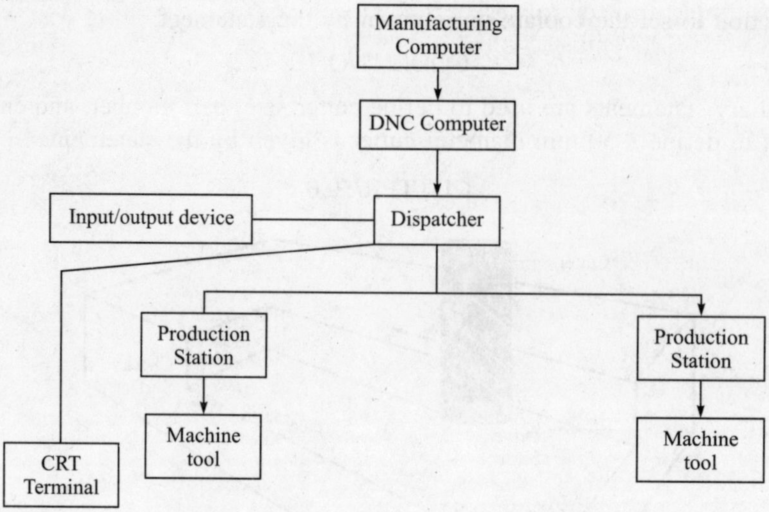

Figure 9.32 "*n*" Number of Machines.

9.12.1 Actual Arrangement of DNC

The DNC consists of three basic components as shown in Fig. 9.33. They are as follows:

1. A computer with memory for storage.
2. Telecommunication lines.
3. Machine tool.

The basic function of the computer is to call the part-program from storage and send the same through telecommunication line when the need arises. It also receives the feedback from the machine tool. This two-way communication occurs on-line or in real time. In addition to the main computer, sometimes additional satellite computer is also used depending on the number of machines, computational capability, etc.

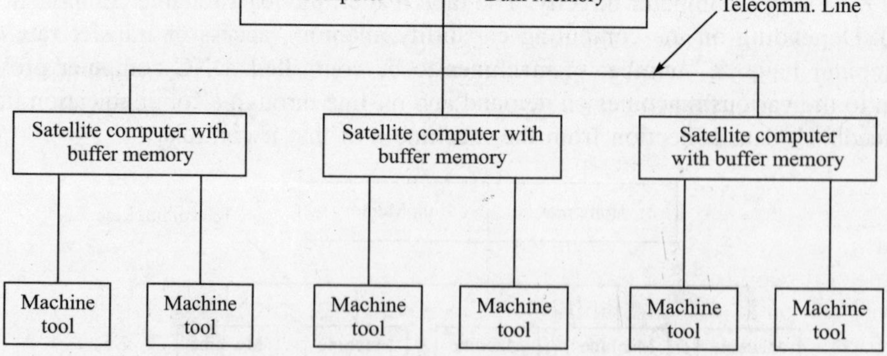

Figure 9.33 DNC with Satellite Computer System.

Besides performing the control task of NC machine, the computer can also be used to prepare bill of materials, inventory control, purchase order, etc. CNC machines are at the lowest hierarchy in the DNC system.

9.12.2 Function of DNC

The main functions of the DNC are :

1. Part program storage
2. Data collection, processing and report generation
3. Communication.

Part program storage

The part program storage must have the following capabilities :

1. Stored program should be accessible whenever the need arises.
2. The system should allow new program entry when the machining is on.
3. Editing of part program to be provided.
4. Part program stored will be in the form CLFILE (Cutter Location File) and this stored program should be converted into program of instruction by the post processor.
5. Storage system should be provided with security options.

Data collection, processing and report generation

Since the telecommunication link has two-way access of data (one, transfer data from the central computer to the machine tool and another collected data or request from the machine tool to the central computer), collected data will be processed according to the request priority and generate the report about the production rate, tool condition, machine usage, etc. These generated data will be used for monitoring the process condition, and if any changes are to be made, they will be carried out.

Communication

It is an important function of a DNC because with the help of the communication system, all the functions can be carried out by the DNC. The main functions are :

1. Communication between the central computer and the machine tool
2. Communication between the central computer and NC part program terminal
3. Communication between the central computer and bulk memory address

Advantages

1. Eliminate punched tape and tape reader.
2. Part program will be stored in the form of CLFILE.
3. Greater flexibility.
4. More computational capability.
5. It forms the basic of the FMS system.
6. Generation of the report about the production rate, tool condition, machine usage, etc.

7. Part program can be stored in any form of storage device.
8. The computer can control any number of NC machine from anywhere.

9.13 ADAPTIVE CONTROL

Adaptive control is the technique of automatically adjusting of cutting parameters such as speeds, feeds and depth of cuts, etc. to an optimum satisfaction level during machining operation, and the maximum metal removal which will result in minimum machining cost. It involves the in-process measurement of cutting parameters, together with an on-line computer fed with information about the cost of running the machine tool.

Adaptive control technique was originated in the early 1960s at Bendix Research Lab, which is sponsored by US Airforce. Earlier versions of AC systems uses analog control system but now it uses microprocessor-based controllers.

In the NC systems, the cutting speed and feeds are prescribed by the part programmer. The determination of these cutting parameters depends upon the experience and knowledge of the part programmer about the workpiece and tool materials coolant conditions, radius of cutter and other related parameters.

Actually, adaptive control is a feedback system as shown in Fig 9.34 in which the cutting speed and feed automatically adopt themselves to the actual conditions of the machining process and these are varied according to changes in the work conditions as cutting progresses.

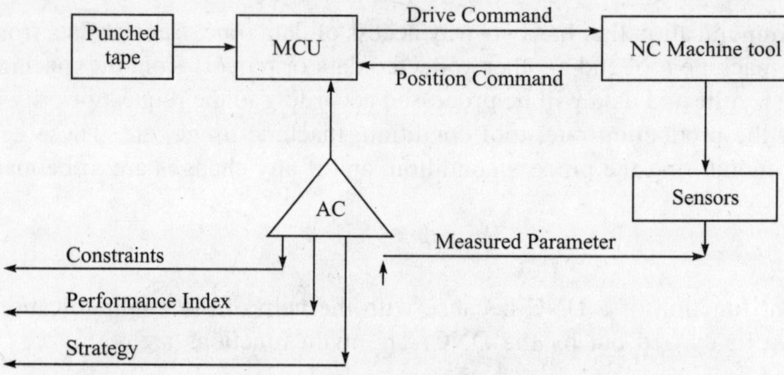

Figure 9.34 Block Diagram of an Adaptive Control System.

After analysis of all the measured machining parameters and given parameters such as machining constraints, appropriate strategy and performance index, etc. the whole system is adjusted towards its optimum conditions to improve the performance level of the system.

9.13.1 Types of Adaptive Control System

The adaptive control systems can be further classified as :

- Adaptive control with optimization (ACO)
- Adaptive control with constraints (ACC).

Adaptive control with optimization system

In this system, performance is optimized according to the given prescribed index of performance. The index of performance is usually an economic function, e.g. maximum production rate or minimum machining cost. It is the most delicate closed loop control system which optimizes the conditions automatically. This type of adaptive system is the early version of AC machine. This attempts to maximize the ratio of material removal rate to tool wear rate, this is called *performance of index*.

Performance index = Material removal rate/Tool wear rate

One of the difficulties associated with the calculation of performance of index is the measurement of on-line tool wear rate.

Figure 9.35 shows that the first developed adaptive control system consists of an NC system, sensor units and an adaptive controller. The sensors are mounted to measure temperature, torque and vibrations, these measured variables are used by the adaptive controller to obtain feed and speed rates for optimum cutting condition.

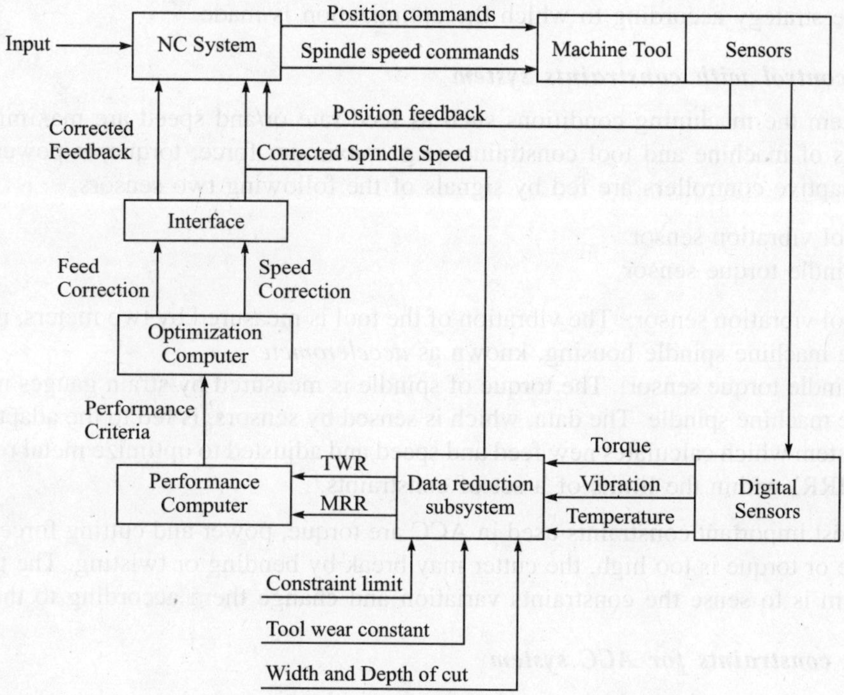

Figure 9.35 Adaptive Control Optimization Controller.

An adaptive controller consists of:

1. Interface circuit
2. Optimization computer
3. Performance computer
4. Data reduction subsystem

The data reduction subsystem is fed with on-line measurement of feed, speed and constraints. This system produces signals corresponding to the material removal rate and tool wear rate each.

$$\text{MRR} = \mu = W \times D \times S$$

where,

S = Spindle speed in rpm.
W = Width of cut.
D = Depth of cut.

$$\text{TWR} = K_1\mu + K_2 T + K_3\, d\theta/dx$$

where,

$d\theta/dx$ = rate of change of torque.
T = tool temperature

and K_1, K_2, K_3 are workpiece and tool constants.

The calculated performance of index is fed into an optimization computer unit, which contains the strategy according to which the optimization is made.

Adaptive control with constraints system

In this system the machining conditions such as feed rate or/and speed are maximized within given limits of machine and tool constraints (e.g. maximum force, torque or power).

The adaptive controllers are fed by signals of the following two sensors:

- Tool vibration sensor
- Spindle torque sensor

(a) Tool vibration sensor: The vibration of the tool is measured by two meters, mounted on the machine spindle housing, known as *accelerometer*.

(b) Spindle torque sensor: The torque of spindle is measured by strain gauges mounted on the machine spindle. The data, which is sensed by sensors, is fed to the adaptive control system which calculates new feed and speed and adjusted to optimize metal removal rate (MRR) within the limits of a set of constraints.

The most important constraints used in ACC are torque, power and cutting force. When the cutter force or torque is too high, the cutter may break by bending or twisting. The principle of ACC system is to sense the constraints variation and change them according to the strategy.

Important constraints for ACC system

Some important constraints which are taken into considerations during setting of parameters at the time of the operation of machine are as follows:

1. Maximum and minimum spindle speed
2. Maximum allowed torque
3. Minimum and maximum allowed chip load
4. Maximum permitted feed rate
5. Impact chip load
6. Maximum allowed vibration

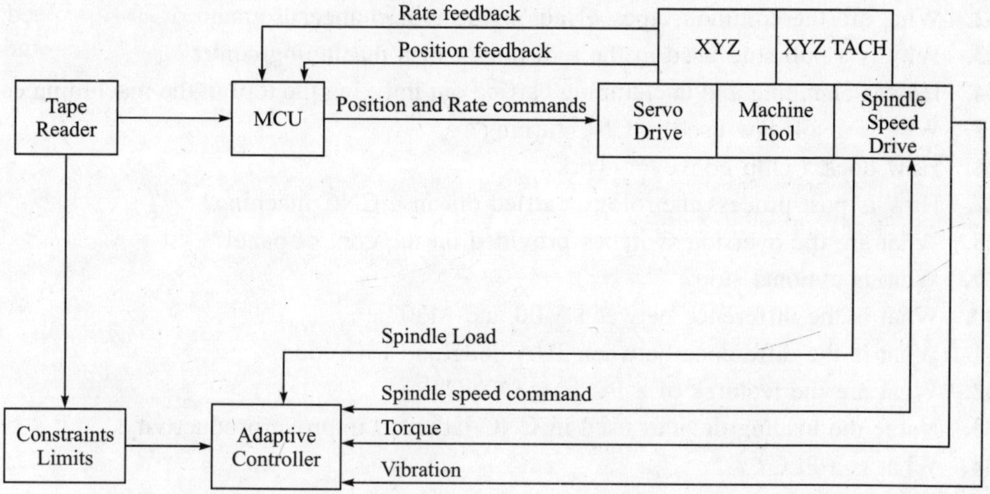

Figure 9.36 Block Diagram of Adaptive Control Constraint System.

REVIEW QUESTIONS

1. Write and explain the standard format of GO J code in a CNC lathe.
2. What is a user macro?
3. What is the purpose of a tool presetter?
4. What are the D and H codes in a machining centre?
5. How is a tapered thread cut in a machining centre?
6. How is a program number designated?
7. What is the difference between absolute and increment programming?
8. Specify nine linear axes as per ISO.
9. Specify the different combinations of axes of a five-axis machining centre.
10. What are the axes of a four-axis machining centre?
11. Name the roughing and finishing tool holders in a CNC lathe for turning.
12. Name the turning inserts for the above.
13. What is the difference between BT taper and ISO taper for machining centres?
14. What is the code for tool nose radius compensation in a lathe?
15. What is the G code for cutter diameter compensation in milling?
16. Describe the types of milling operations available in CAM software.
17. Describe the common tool motion statements available in CAM software.
18. Discuss the stock boundary definition and area clearance in CAM software.
19. What is the speciality of a rapping tool holder in a machining centre? What is the significance of rigid tapping?
20. Why is ceramic tools advantageous than carbide tools in CNC turning?
21. Why is a curvic coupling preferred in indexing devices in a CNC machine?

22. What are the common types of automatic tool changers?
23. Why is a pull stud used in the tool holder of a machining centre?
24. How is clamping and unclamping carried out in fixing the tool in the machining centre?
25. Why is a soft jaw used in CNC turning?
26. How does a chip conveyor work?
27. How is post process metrology carried out in a CNC machine?
28. What are the override switches provided on the control panel?
29. What is optional stop?
30. What is the difference between M00 and M30?
31. What is the difference between JUA code and ISO code?
32. What are the features of a PC based CNC?
33. Name the loading devices used in CNC lathes to improve productivity.
34. What is a PLC?
35. How is tool life monitored in a CNC lathe?
36. How is tool offset measured in the machine?
37. What is the purpose of post processing?
38. What is a CL data file?
39. What is a universal machining centre?
40. Why is a hall nose cutter used in machining of curved surfaces?
41. What is background programming?
42. What is meant by conversational programming?
43. What is a soft key in a CNC panel?
44. What is dry run?
45. What is simulation?
46. What is DNC operation of CNC machines?
47. What is radius programming?
48. In which G function do you incorporate chamfer command?
49. What G code do you use to machine sharp corners?
50. What is a canned cycle?
51. What are the various methods of selecting a hole set for drilling holes in CAM software?
52. What are the advantages of using CAM softwares?
53. What is a word address block format?
54. Differentiate among design model, workpiece model and manufacturing model in CAM software.
55. Define the process, write the operation sheet, sketch the set-ups and write the NC program for turning the components.

Chapter 10

DESIGN CONSIDERATION OF CNC MACHINES

10.1 INTRODUCTION

Today, the machine tools available are in varied designs. The main reason for changing from conventional machines to CNC machine is to increase the productivity by lesser human intervention. Another reason is, the quality and repeatability of the outcome of the product from the CNC machines are so high when compared with the products made in conventional machine.

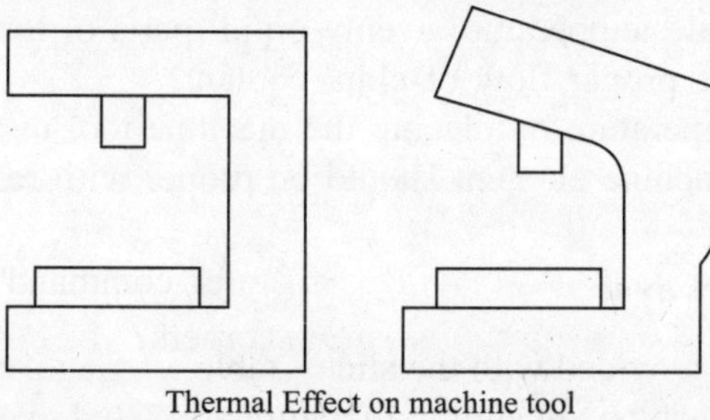
Thermal Effect on machine tool

Figure 10.1 Thermal Effect on CNC Machine.

Since in CNC machines the operations are carried out automatically with high speed, feed, etc., utmost care is to be taken in the designing of CNC machines. If proper designing is not done, it is possible that substantial damage might occur. Unexpected deformation can occur due to:

1. Excessive pressure of tool on workpieces.
2. Deformation of machine part due to excess thermal effects.
3. Vibration due to high speed.
4. Friction on slideway.
5. Overshoot of table and tool on which operation change takes place.

Due to the above factors, defect may arise in the improper designing of the CNC machine. CNC should be of *"robust construction with minimum friction and negligible backlash"*. The main machine elements which are considered to be of more importance in designing the CNC machines are:

1. Main structure of the CNC machine
2. Slide and slide ways.

10.1.1 Main Structure of CNC Machine

The main structure of the CNC machine consists of a bed, a column, head stock, etc. which are rigidly connected to the bed. As the spindle rotating, the cutting force is transmitted to machine bed, frame finally to the foundation. The forces encountered are:

- Static forces due to self weight of machine and pressure of the cutting tool on the workpiece.
- Inertia force due to rapid acceleration and deceleration.

These two forces may cause the table to bend or deform, leading to an error around 40μmm. This micro error will magnify as the operation continues and lead to rejection in the final output of the product. This may also arise due to the thermal effect. The thermal effect arises because of high spindle speed and high depth of cut used in the CNC machine.

The various considerations to avoid these effects are:

1. Provide correctly designed mild steel structure.
2. Use ribs, angle plates, etc. to increase the stiffness of machine structure.
3. Design must distribute temperature evenly on all parts of the machine.
4. Design must provide proper flow of chip, coolant.
5. Reduce ambient temperature by placing the machine tool in an air-conditioned room.
6. Alignment of the machine element should be proper with respect to each other.

10.1.2 Slide and Slideways

Generally, machine tools are provided with the slides, table, carriage, etc. to carry the workpiece or tool. These elements of machine are sliding in nature and fixed over column, bed, etc. which are called *slideways* of machine member. Hence, these slideways should be rigid and accurately designed.

Repeatability and accurate position are important characteristics of CNC machine.

Inaccuracies arise due to the **stick-slip** phenomenon (jerky motion due to sticking of oil) which prevents smooth starting from rest. This can be avoided by using anti-friction bearing arrangements which changes sliding friction to rolling friction.

Different methods available to avoid stick-slip are :

1. Hydrostatic type slideways.
2. Anti-friction type slideways.
3. Wear resistant type slideways.

Hydrostatic type slideways

The machines with horizontal slideways, the cutting force falls on the sliding surfaces. This leads to increase in friction, and wear further leads to decrease in accuracy, so a constant film of lubricants or air must exist to avoid direct metal contact. Hydrostatic slide ways are further classified as:

a. Oil lubricated slideways
b. Air lubricated slideways.

Oil lubricated slideways: In this type, friction is minimized by forcing oil under pressure between mating surfaces, and the pressure is automatically varied according to the load on the surface, resulting from the weight of the moving element and the cutting force.

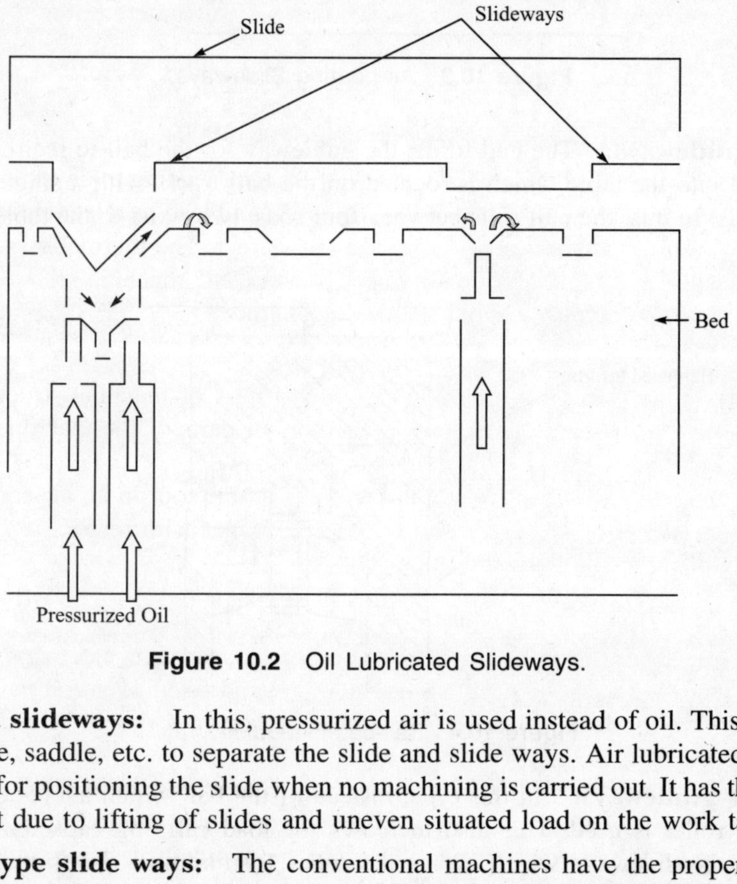

Figure 10.2 Oil Lubricated Slideways.

Air lubricated slideways: In this, pressurized air is used instead of oil. This pressurized air will lift the table, saddle, etc. to separate the slide and slide ways. Air lubricated slide ways are used especially for positioning the slide when no machining is carried out. It has the disadvantage of misalignment due to lifting of slides and uneven situated load on the work table.

Anti-friction type slide ways: The conventional machines have the property of stick-slip due to high sliding friction at low velocity. To avoid this sliding, contact is avoided by making the contact in point or line by converting sliding friction to rolling friction. Anti-friction slideway is of two types:

a. Ball-bearing guideways
b. Roller bearing guideways.

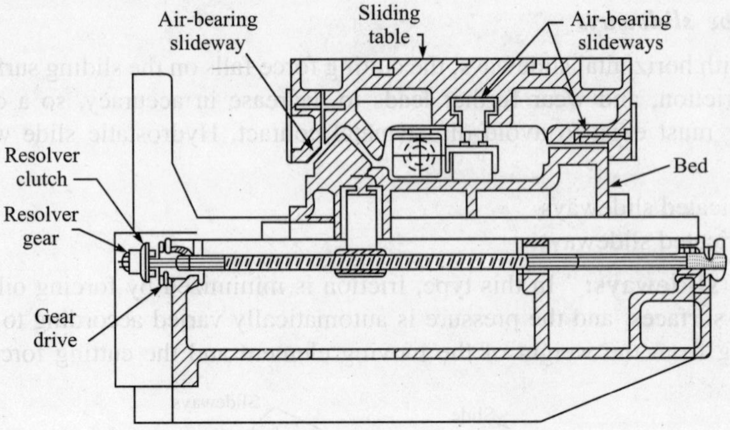

Figure 10.3 Air-bearing Slideways.

Ball-bearing guideways: The bed forms the guideway for the ball to recirculate. Hardened inserts are fitted into the table which is located on the ball track, with a single flat surface in contact with balls. In this, the ball rolls between four rods, two fixed to the table and two to the bed.

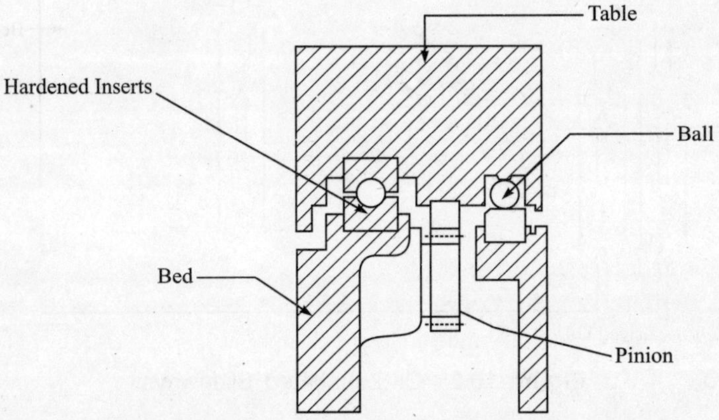

Figure 10.4 Ball-bearing Guide Way.

Roller bearing guideways: In this type, instead of the ball which avoids direct contact of the surface, the roller is used. This also increases the load carrying capacity. Hardened steel inserts are used in slideways which act as the roller to minimize the friction. This kind of slideways are best suited for boring and milling machines.

Wear resistant slideways: The machine tool varies its ability to resist wear and tear of slide ways. Lathe and planning machines operate under heavy load when compared with grinding machine. Hence, different types of bearing surface are required. In this cases, the slideways system in which non-metallic or plastics inserts are used. These inserts stick to the underside of the moving element.

The inserts are made of two or more materials in which one reduces coefficient of friction and the other increases strength, load bearing capability, and also self lubricating property. They are placed in position by scraping the saddle or bed first and then fitting a pre-measured steel strip on it. Recent machines use PTFE (Poly Tetra Fluoro Ethylene) to make positive contact. The various materials used are:

Turcite–B, Ferobestos—CA, SKC—3 and Moglice.

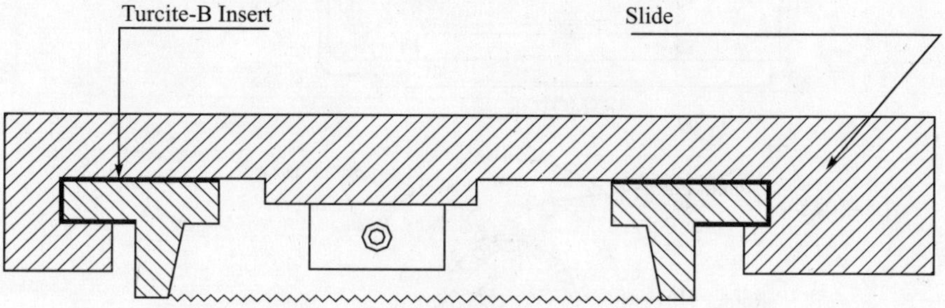

Figure 10.5 Plastic Insert for Slideways.

10.2 SPINDLE DRIVES OF CNC MACHINE TOOL

Two main spindle drives available for a CNC machine are:

1. DC variable drives.
2. AC variable drive

10.2.1 DC Variable Drives

The DC motors are widely used drive for CNC machine due to its easy adjustability of voltage and speed. To satisfy many demands of the machine tool, the drives must be controllable for wide range of power and speed, especially main drive system. Hence, the DC drive should be designed to give constant power over a wide range.

Figure 10.6 shows the DC motor. Its principle is based on the rotation of an armature windings within a magnetic field. Armature (rotor) and the field winding is a stator which is stationery member of a DC motor. Both are wound on iron cores, to make the flux path effective. The armature windings are connected to the commutator which is a cylinder of copper windings insulated by mica and mounted on the rotor shaft. Carbon bush connects the commutator and the armature.

The above figure explains the characteristics of a DC motor. The different types of DC motor available are:

1. Permanent magnet DC motor
2. DC motor with analog thyristor
3. DC motor with digital control.

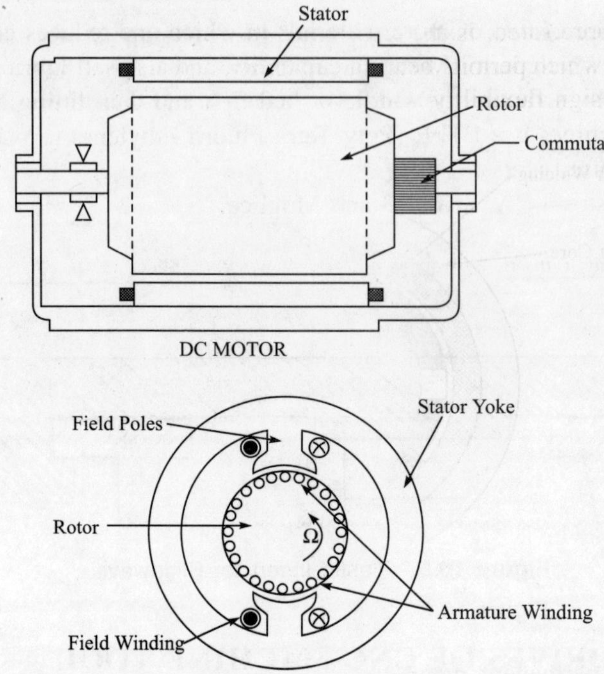

Figure 10.6 DC Motor.

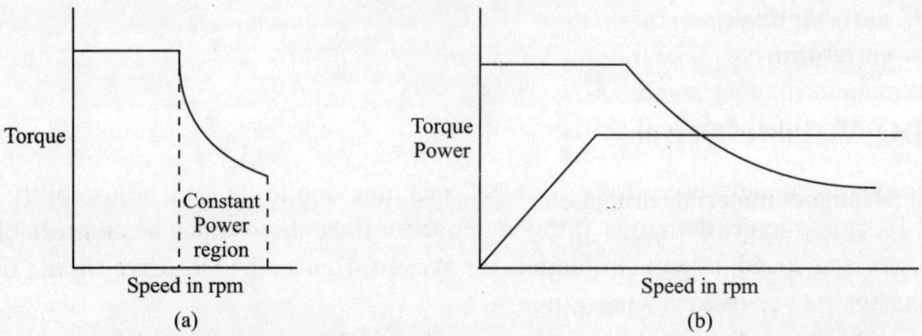

Figure 10.7 (a) Speed–Torque Characteristics of DC Motor. (b) General Characteristics of a DC Motor.

Permanent magnet DC motor

Of all the changes in materials technology over the last several years, advances in permanent magnets have had the largest impact on electric machines. Permanent magnets are often suitable as replacements for the field windings in machines, that is, they can produce the fundamental interaction field. This does three things.

First, since the permanent magnet loss is less, it eliminates the energy required for excitation, usually improving the efficiency of the machine.

Second, since eliminating the excitation loss reduces the heat load, it is often possible to make PM machines more compact.

Finally, and less appreciated, is the fact that modern permanent magnets have very large coercive force densities which permit vastly larger air gaps than conventional field windings, and this, in turn, permits design flexibility which can result in even better electric machines.

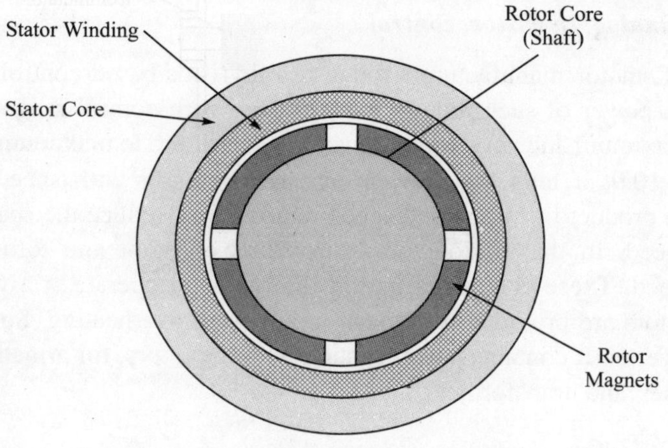

Figure 10.8 Permanent Magnet Motor.

The advantages of permanent magnet DC motor come with a price. Permanent magnet materials have special characteristics which must be taken into account in machine design. The highest performance permanent magnets are brittle ceramics, some have chemical sensitivities, all are sensitive to high temperatures, most have sensitivity to demagnetizing fields, and proper machine design requires understanding the materials well. These notes will not make you to seasoned permanent magnet machine designer. They are, however, an attempt to get started, to develop some of the mathematical skills required and to point out some of the important issues involved.

Permanent magnet materials are, at core, just materials with very wide hysteresis loops. This hysteresis loop is so wide that you can see the effect of the permeability of free space.

It is usual to display only part of the magnetic characteristic of permanent magnet materials the third quadrant of Fig. 10.8, because that is where the material is normally operated. Note a few important characteristics of what is called the "demagnetization curve". The permanent flux density *Br*, is the value of flux density in the material with zero magnetic field *H*. The coercive field *Hc* is the magnetic field at which the flux density falls to zero. Shown also on the curve are loci of constant energy product. This quantity is unfortunately named, for although it has the same units as energy, it represents real energy in only a fairly general sense.

It is the product of flux density and field intensity. As you already know, there are three commonly used systems of units for magnetic field quantities, and these systems are often mixed up to form very confusing units. We will try to stay away from the English system of units in which field intensity *H* is measured in amperes per inch and flux density *B* in lines (actually, usually kilolines) per square inch. In CGS units, flux density is measured in Gauss (or kilogauss) and magnetic field intensity in Oersteds. And in SI, the unit of flux density is the Tesla, which is one Weber per square metre, and the unit of field intensity is the Ampere per metre. Of these, only the last one, $A = m$ is obvious. A Weber is a volt-second. A Gauss is 10 Tesla.

Since permanent magnet motors do not require external power supply for producing magnetic flux, there will not be any thermal loss in stator. Another advantage is the frame size required for a given power output is less due to its high coercive action.

DC motor with analog thyristor control

Al most all the DC motor manufacturers today use thyristor based controller.

The maximum power of such motors is 6000 rpm, with control range of 1:4. By keeping the field voltage maximum and vary armature voltage from zero to maximum during A – B stage. If we see Figure 10.9, it tells that power increases linearly with speed because power is proportional to the product of torque and speed. During B-C reduce the speed current which in turn, increases speed. In this period, power remains constant and torque varies inversely proportional to speed. These conditions permit the motor to operate at a wide range of speed variation. Thermistors are provided to prevent motor from overheating. Speed control is taken care of by the converter. It contains power section with thyristors, for armature and field circuit, semiconductor fuses and transformers.

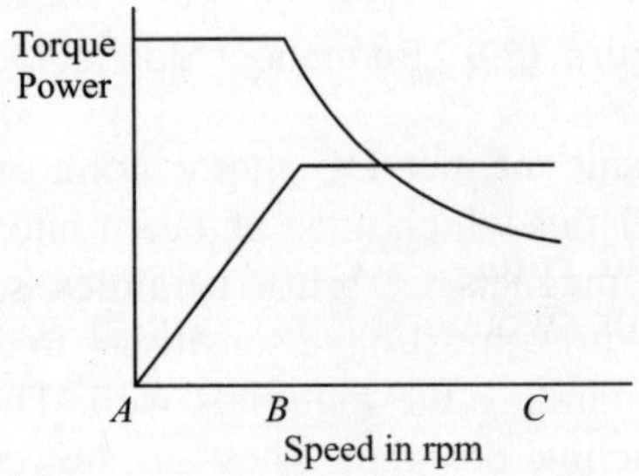

Figure 10.9 Characteristics of DC Motor.

The converter also consists of monitoring devices for motor protection. Faults occurring are indicated by the LED.

AC motor: general principle

As a general rule, the conversion of electrical power into mechanical power takes place in the rotating part of an electric motor. In dc motors, the electric power is conducted directly to the armature (i.e. rotating part) through brushes and commutator. Hence, in this sense, a dc motor can be called a **conduction motor**. However, in ac motors, the rotor does not receive electric power by conduction but by induction in exactly the same way as the secondary of a two-winding transformer receives its power from the primary. That is why, such motors are known as *induction motors*. In fact, an induction motor can be treated as a rotating transformer, i.e. one in which primary winding is stationary but the secondary is free to rotate.

Of all the ac motors, the polyphase induction motor is the one which is extensively used for various kinds of industrial drives.

10.3 CONSTRUCTION

An induction motor consists essentially of two main parts:

1. Stator and
2. Rotor

10.3.1 Stator

The stator of an induction motor is, in principle, the same as that of a synchronous motor or generator. It is made up of a number of stampings, which are slotted to receive the windings. The stator carries a three-phase winding and is fed from a three-phase supply. It is wound for a definite number of poles, the exact number of poles being determined by the requirements of speed. The greater the number of poles, the lesser the speed and vice versa. The stator windings, when supplied with three-phase currents, produce a magnetic flux, which is of constant magnitude but which revolves at synchronous speed (given by Ns = 120 f/P). This revolving magnetic flux induces an e.m.f. in the rotor by mutual induction.

10.3.2 Rotor

A rotating part of an electrical or mechanical device.

(i) **Squirrel cage rotor:** Motors employing this type of rotor are known as squirrel cage induction motors.

(ii) **Phase-wound or wound rotor:** Motors employing this type of rotor are known as 'phase-wound' motors or 'wound' motors or 'slip-ring' motors.

Advantages

1. It has very simple and extremely rugged, almost unbreakable construction (especially squirrel cage type).
2. Its cost is low and it is very reliable.
3. It has sufficiently high efficiency. In normal running condition, no brushes are needed, hence frictional losses are reduced. It has a reasonably good power factor.
4. It requires minimum of maintenance.
5. It starts up from rest and needs no extra starting motor and has not to be synchronized. Its starting arrangement is simple, especially for squirrel cage type motor.

Disadvantages

1. Its speed cannot be varied without sacrificing some of its efficiency.
2. Just like a dc shunt motor, its speed decreases with increase in load.
3. Its starting torque is somewhat inferior to that of a dc shunt motor.

10.4 STEPPER MOTOR

Stepper motor is an incremental digital control device which translates an input pulse sequence into a proportional angular movement, rotating one angular increment for each input pulse. The shaft position is determined by the number of pulses and its velocity by the frequency of pulses. In recent years, these motors have become more popular. The reason for this is, they can be controlled directly by computer, microprocessor and PLC.

The unique feature of the stepper motor is that its output shaft rotates in a series of discrete angular intervals or steps, one step is taken each time a command pulse is sent. When a definite number of pulses are sent, the shaft rotates through a definite angle. This makes the use of stepper motor in open loop systems. Torque ranges from 1μN-m to 40 N-m, the power output ranges from 1 W to 2500 W.

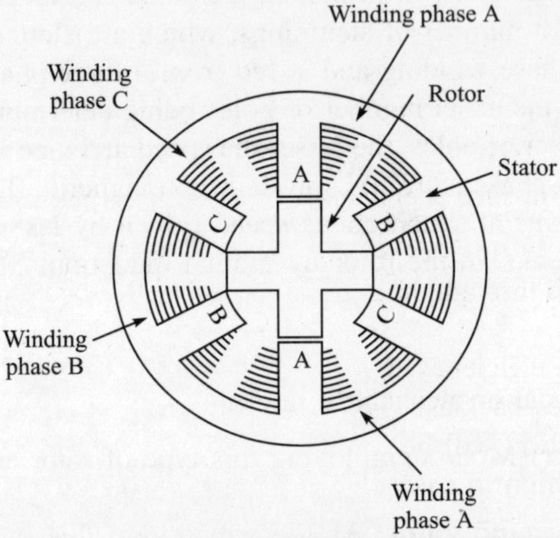

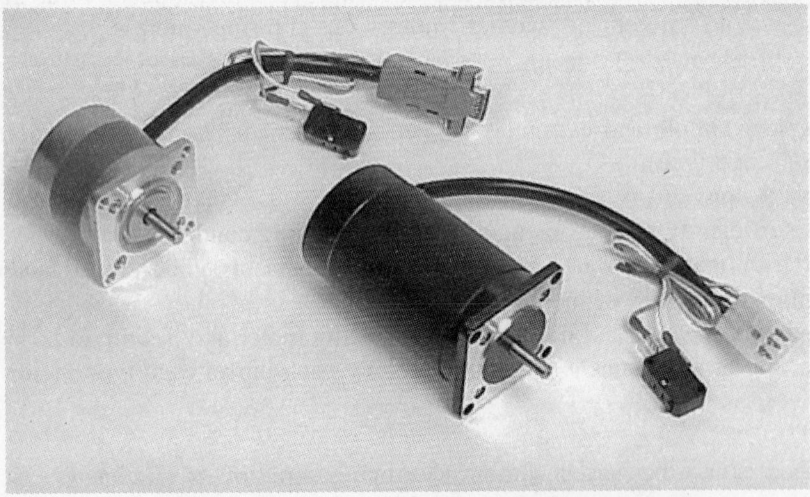

Figure 10.10 The Picture of the Stepper Motor.

The angle through which the motor shaft rotates for each command pulse is called *step angle.* The smaller the step angle, the greater the number of steps per revolution and higher the resolution or accuracy of positioning is obtained. The step angles start from 0.72° to as large as 90°. But the standard or common step angles are 1.8°, 2.5°, 7.5°, 15°.

$$\text{Step angle } \beta = \frac{360°}{\text{No. of stator phases} \times \text{No. of rotor teeth}}$$

$$\text{Resolution} = \text{No. of steps/revolution} = 360°/\beta$$

10.4.1 Types of Stepper Motor

1. Variable reluctance stepper motor.
2. Permanent magnet stepper motor.
3. Hybrid stepper motor.

10.5 BALL SCREWS

Ball screws are employed in feed mechanism of CNC machine tool. The main advantages of employing it are:

1. Load is transmitted through the spherical balls from screw thread to nut which eliminate the direct contact.
2. High transmission efficiency makes larger thrust load to be carried with less torque.
3. Preloading removes backlash and increases axial stiffness of the ball screw.
4. Coefficient of friction will be around 0.004.
5. Accuracy is also high in case of ball screws.

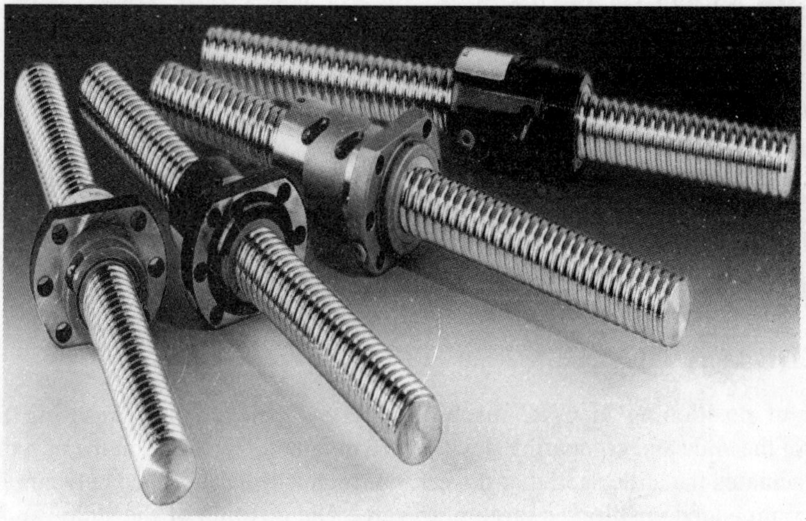

Figure 10.11 The Picture of the Ball Screw.

The need of ball screw arrangement is to avoid rolling friction which occurs in the conventional screws by circulating these balls in the groove. This, in turn, avoids direct contact of screw and nut which reduces the friction. Figure 3.10 helps us understand how the ball screw works, the ball rolls in the groove exit from the trailing end of nut which are again picked by the return tube from outside. The arc formed by these ball screws may be either *gothic or circular arc* as shown in the figure. Gothic arc provides smaller axial clearance while circular arc is characterized by the little axial deformation and greater load carrying capacity. Increase in the number of balls for recirculation decreases friction.

Nuts for ball screw are available in different forms which are as follows:

1. Round flanged nut with embedded tube.
2. Round cylindrical nut with embedded tube.
3. Rectangular shape nut.
4. Small outside diameter flanged nut with tube outside.

These ball screws are mounted in four different methods which are based on the application. They are:

S.No	Application	Mounting Method
1.	Small screw life	One end fixed, other end free
2.	Low speed rotation	One end fixed, other end free
3.	Medium speed rotation	Both ends supported
4.	High accuracy	One end fixed and other end supported
5.	High speed and accuracy	Both ends fixeds

Preloading of ball screw

The removal of back lash is one of the important design criteria for a CNC machine to get high accuracy repeatedly. This is achieved by preloading the ball screw.

When the thrust load is applied, small amount of radial or axial clearance exists in the ball screw which makes the ball and groove to deform slightly. Now the backlash is the sum of the axial clearance and the deformation caused by the axial loading.

Preloading is applied to the two nuts in tension or in compression. A shim plate is inserted between the nuts or between nut and the housing, then they are tightened with a preloading bolt in the direction of compression or pulled with two preloading pins in the direction of tension.

Due to preloading, deformation reduces when axial load is not high which in turn increases the system stiffness. Recommended preload value is one-third of the operating load value. Less preloading reduces the stiffness and heavy preloading decreases the life of the ball screw.

10.6 FEEDBACK DEVICES

The accuracy of positioning in CNC machine tools is achieved by measuring the position/displacement of the slide and comparing it with the commanded position in the part program. The servo system actuates the slide such that the error, which is the difference between programmed position and commanded position, is brought to zero. The position of the slide can be measured directly by fitting a linear transducer to the slide or indirectly by mounting a rotary transducer to the ball screw of the slide or mounting the transducer on the servo-motor shaft. Direct measurement is preferred for more accurate positioning. Basically, positioning feedback systems are of two types. They are:

1. Linear transducers
2. Rotary transducers

Positioning feedback devices

Linear transducers

1. Line grating transducers
2. Ferranti system transducers
3. Inductosyn

Rotary transducers

1. Resolvers
2. Encoders

10.6.1 Linear Transducers

Linear transducer is a linear position measuring device. In the linear transduces the slide moves over a linear wound potentiometer which drops the voltage corresponding to the length of resistance selected by sliding contact.

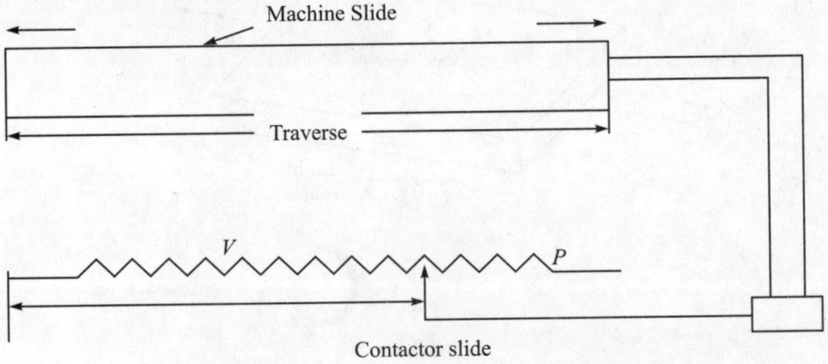

Figure 10.12 Simple Linear Transducer.

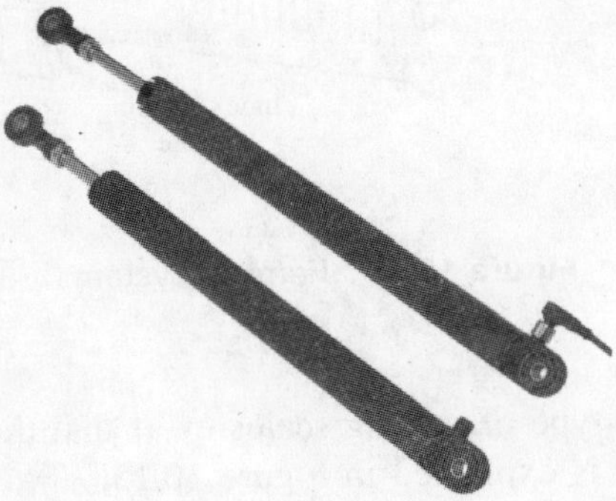

Figure 10.13 The Picture of Linear Transducer.

Glass scale with line grating

Glass scale consists of two scales, one transparent index grating and another a glass scale, both these scales have the same grating. These two glasses are fixed into a single unit. The remaining unit consists of a light source, condenser lens and a photo cell. The slide moves cause the light to pass through the transparent glass which is sensed by photocells. The resistance of photocells changes with the intensity of light falling on it. Since the number of grating lines per unit length is known, the frequency of pulses is converted to get the required length of the movement of slide.

Ferranti system or moiré fringes digitizer

In Ferranti system the index grating is mounted, slightly tilted with respect to line gratings on the scale to produce moiré fringe pattern.

The resistance of photocells changes with the intensity of light falling on it. The periodic signals in the form of pulses are converted proportionately into the displacement of work table by knowing the number of engraved lines per unit length on the linear scale. Usually, a width of 20 microns of the line is used.

This measuring system is shown in Figure 10.14.

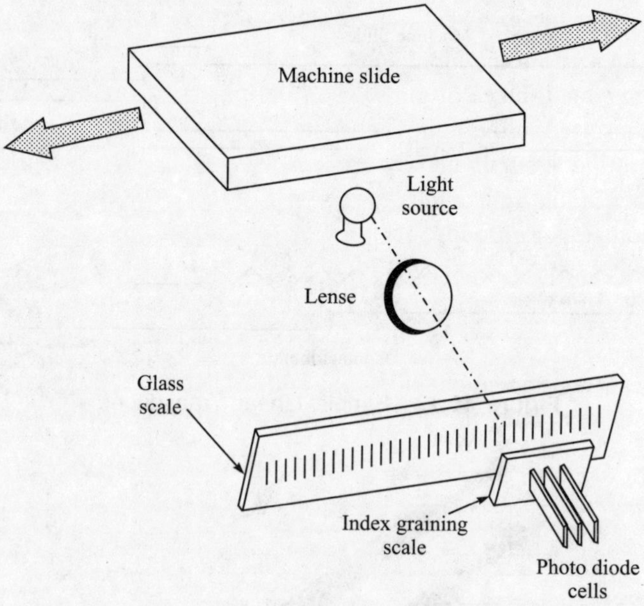

Figure 10.14 Ferranti System.

Ferranti system

Ferranti system uses the same type of grating scales exert that the least count is improved by using moiré effect. This effect is explained in Figure 10.15.

(a) Photo diodes
(b) Glass plate with line grating
(c) Auxiliary grating
(d) Direction of movement of light and dark fringes
(e) Direction of movement of glass plate.

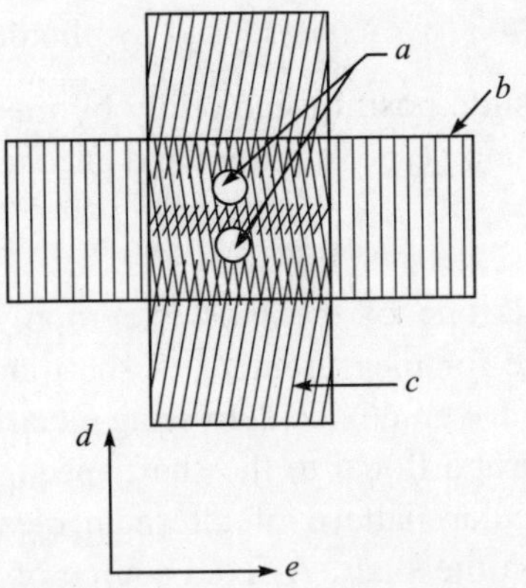

Figure 10.15 The Moiré Effect.

Ferranti system is developed by Ferranti Ltd. It consists of a glass disc marked with a pattern of radial lines and mounted on the input shaft. A lamp and photocell are placed in both sides of the disc. An additional small piece of glass, the fringe grating, with the same pattern of lines as those on the disc, is inserted in the light beam. Moiré fringe patterns are produced by setting the lines on the fringe grating at small angle with disc. Now moiré fringe pattern appears as a series of dark and light zones at the right angles to the marked lines. By rotating the disc, these patterns pass through the photocells to produce the signal. This method is a highly accurate method for measuring small movements.

Inductosyn

It is a precision measuring device developed by Ferrand Corp. It consists of two parts—one is scale and another is slider. They are similar to rotor and stator winding.

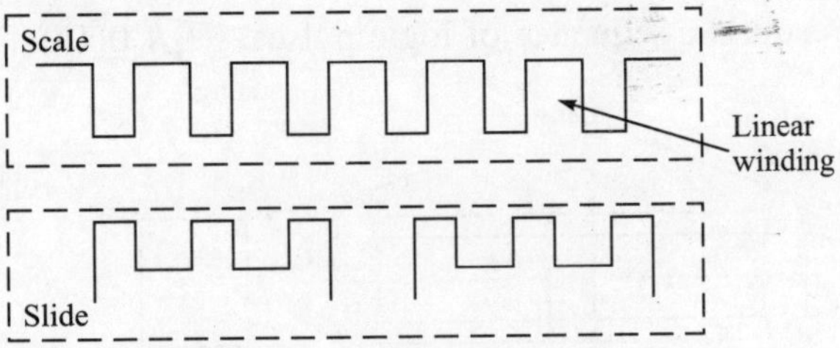

Figure 10.16 Inductosyn.

The coils are in the form of "hairpin turns" etched on a glass, steel or aluminium plate. The coils are wounded around the scale. The space between the coil is 2 mm. When two different voltages are applied with these two sliding coils, an emf is induced, which is corresponding to the position of the slide.

There are two forms of inductosyn—linear and rotary. The accuracy of inductosyn is around 0.5 micron.

10.6.2 Rotary Transducers

Rotary transducers measure the slide position indirectly by measuring the rotational movement of the slide. Resolvers produce analog type of output and encoders display output in digital form.

Encoders

An incremental rotary encoder is one of the most commonly used feedback device in N/C machine tools, since it is suitable for measurement of shaft-angle, or speed.

The rotary encoder is a shaft-driven device delivering electrical pulses at its output terminal. The pulse frequency is directly proportional to the shaft speed. The encoder consists of a glass disc marked with a precise circular pattern of alternate clear and opaque segments on its periphery. The disc is mounted on the shaft. A fixed source of light is provided on one side of the disc and a photocell is placed on the other side of the disc. As the disc rotates, the light is periodically permitted to fall on the photocell, to produce the required output signal which is approximately sinusoidal but is only in the millivolt range. Internal or external amplifiers are supplied to amplify the photocell output to a level suitable for feeding associated logic circuits. The amplified signal is fed to an internal or external schmitttigger, which converts the signal to a proper square wave with suitable rise and fall times.

Optical encoders: The heart of an optical encoder is a pair of light emitters and detectors. These devices transmit and receive infra-red light of about the same wavelength as that used in television remote controls and laptop computer IR links. In an optical encoder, the light sources transmit continuously. The receivers detect light whenever the path between the transmitter and the receiver is not broken by the presence of an object which blocks the light.

A clear plastic mask painted with dark bands is placed between the transmitter and the receiver as shown in Figure 10.17. As the shaft rotates, the dark bands on the mask move across the path from each transmitter to detector. When a detector senses light, it sends out a voltage of zero volts or logic 0; when no light is sensed because the path is blocked, the detector's output is 5 volts or logic. The result is a sequence of logic pulses, 1 0 1 0 1 0. . . , as the shaft rotates.

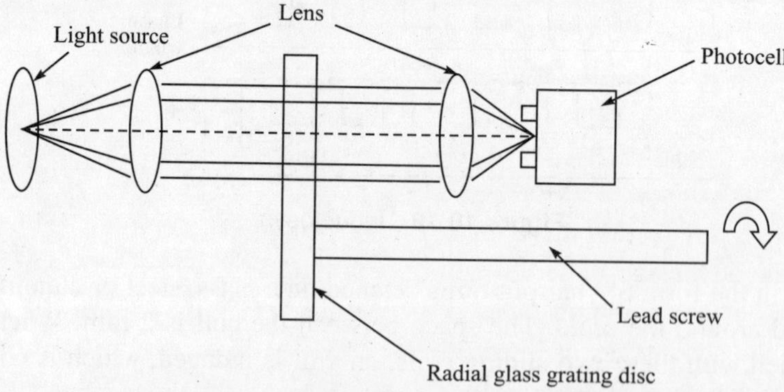

Figure 10.17 Radial Grating Position Transducer.

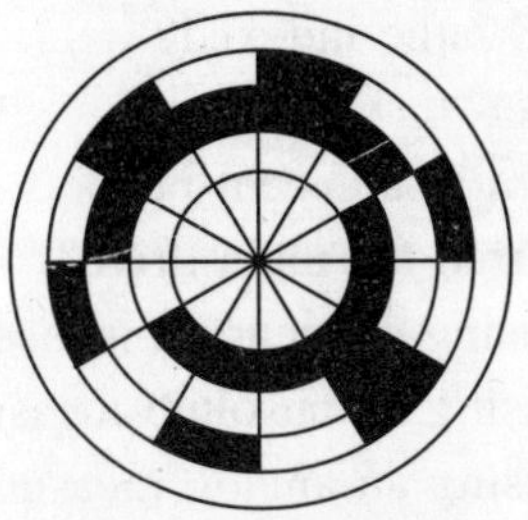

Figure 10.18 Binary Code Encoder.

The rotary encoder is a shaft driven device delivering electrical pulses at its output terminal. Shaft speed is proportional to the pulse frequency. The encoder consists of a glass disc as shown in Figure 10.18 with gratings. By rotating the disc, the light passes through the grating, falls over the photocell and results in output voltage. Two photocells are provided to sense the direction of disc rotation.

Resolvers

It consists of a rotor and a stator, both having two windings at 90° to one another. The output signal is the function of rotating angle obtained by inductive coupling between the stator and the rotor.

Since windings are placed with the phase difference of 90°, the peak value of these components varies sinusoidally, i.e. they vary as sine and cosine value depends on the angular position of the rotor shaft.

REVIEW QUESTIONS

1. What is the most popular type of drive for CMC machines today?
2. Sketch and explain the torque speed characteristics of 3 spindle drive.
3. Why a gearbox is used in the main drive of some CNC machines?
4. Why is a separate cooling motor used with DC drives?
5. List the advantages of AC servo-motors over DC servomotors.
6. Why is a servomotor used for axes feed drive?
7. Why is often a reduction drive using timing belt, employed in feed drives?
8. What is the advantage of servo-motors with built-in encoders?
9. What are the four common types of ball screw mounting arrangements?
10. What are the advantages of linear motion guideways?
11. What type of plastic insert is used in CNC machine slideways?
12. What is the purpose of using LM guides or plastic inserts in slideways?
13. What is the advantage of using slant bed design for turning machines?
14. Why should a brake be used with the X-axis drive in a slant bed lathe?
15. What is the principle of operation of linear motor?

16. How is the turret of a CNC lathe indexed?
17. What is a bidirectional turret?
18. Why axial tools are not kept in adjacent pockets in a tool turret?
19. Why is a poly V belt preferred for main drive?
20. What is the advantage of using a linear transducer for feedback measuring?
21. What is the advantage of using an absolute measuring system?
22. What is the advantage of using an analog measuring system?
23. Describe the principle of working of two analog transducers.
24. Explain the principle of working of two digital incremental transducers.
25. What is a Digital Read Out (DRO)?
26. What is an A/D converter? Where is it used in a CNC machine?
27. Name three types of bearings used in the mounting of spindles of CNC machines.
28. Why is sometimes a spindle oil cooler necessary?
29. Is it necessary to keep a CNC machine in an A/C room?
30. What types of microprocessors are used in modern CNC systems?
31. What is the advantage of multiprocessor CNC system?
32. Name three techniques of interpolation.
33. Give major specifications of a machining centre.
34. Give major specifications of a CNC lathe.
35. What are the differences between a CNC lathe and a turning centre?
36. What is a typical tool changing time in a modern high speed machining centre?
37. How is the slide position referenced?
38. What is meant by accuracy?
39. What is meant by repeatability?
40. How is a CNC machine calibrated?
41. Give the explanation of following G functions for a lathe?
    G04, G09, G27, G54, G7642.
42. Give the explanation of following G functions for a machining centre?
    G17, G30, G81, G99.
43. Give the explanation of following M functions for a machining centre?
    M03,.M19,.M06.

Chapter 11

AUTOMATED GUIDED VEHICLE (AGV)

11.1 INTRODUCTION

Automated Guided Vehicle (AGV) is the programmed vehicle, used to carry load from one location to another in an automated workplace. The most common type of such vehicles normally follows a predetermined path on embedded wires into the ground. These vehicles are called wire *guided* AGVs. Other variety of AGVs is free-ranging AGVs, which offer more flexibility as no pre-specified path is needed.

An AGV is, therefore, a driverless vehicle, which is able to select its own route or path to reach the destination. No human intervention is needed for the guidance, steering or control of the vehicles. An off-board controller is used to send dispatcher commands for the identification of the load, destination of the load, and other instructions related to loading and unloading of the load. Automatic Guided Vehicle Systems (AGVs) are one of the most exciting and dynamic areas in material handling today.

But AGVs is really not so new. Fifty years ago when AGVs was invented, it was then called *driverless systems*. Through the years advances in electronics have led to advances in guided vehicles. The vacuum tub technology gave way to the transistorized technology and now today microprocessor technology. Technological developments may have given AGVs more flexibility and capability, but market acceptance has really given AGVs the application variety to allow it to expand into the standard accepted material handling method it is today.

11.2 HISTORY

Automatic Guided Vehicle Systems (AGVs) have been in existence since 1953. An inventor with a dream devised a method of automating the man-aboard tow truck that had been used in factories for years.

He developed a "track" for the vehicles to follow by imbedding a wire in the factory floor. The first guidance system was created when sensors on the bottom of a tow truck looked for

Figure 11.1

a magnetic field. The field was created by a current running through a wire or series of wires in the floor. Station codes consisted of an array of magnets (north/south combinations) in the floor that a vehicle would read for stopping at a station.

AGV systems existed at this level of technology until the mid-seventies. The advent of solid state controls allowed the systems to expand in capability and flexibility. As a result, applications exploded. Vehicles were not only towing trailers in warehouses, but they were also utilized in unit load delivery, work-in-process production systems, and automotive assembly systems.

Today, the technology continues to evolve. The wire in the floor is still available for the proper application. However, many systems today are being designed without the floor wire. Vehicles are now equipped with sophisticated computers on board to communicate, direct, and manage the system. New vehicle configurations have also proliferated. The basic AGV towing systems still exists. However, automatic call systems, opportunity charging, automatic loading/unloading, and automatic coupling and uncoupling are now available for the right towing application.

Unit load systems have also been redefined from the earlier generations. Multiple load decks, parasitic drive systems, custom fixturing, and opportunity charging are now available.

11.3 FEATURES OF AGV

AGV systems have the following important features:

1. AGV system can easily be interfaced with other modules of flexible manufacturing systems, such as robots, automatic storage and retrieval system (AS/RS), CNC machines, etc. It is an essential element of FMS system, without which effective material handling is perhaps unachievable.
2. AGV system offers considerable flexibility, as it is adaptable to change in product as well as production.
3. AGV system offers great flexibility in the expansion of a manufacturing system. The number of vehicles or area to be covered by the AGV may be easily modified as per the need.

4. AGV material handling system is suitable to computer integrated manufacturing system (CIMS). This is due to on-board controller of the AGV. Through a host computer, the on-board controller can be interfaced with other modules of CIMS. Information linkages of material handling system with other manufacturing activities are easy to achieve.
5. The integrated material handling system facilitates the optimization in the efficiency of the manufacturing systems and improved productivity.
6. AGV system offers elimination or reduction of labour and less transit damages.
7. AGV system has the ability to handle variability in the production rate or changes in product routing.
8. Compared to conventional material handling systems, it has the flexibility to adapt to changes in machines layout and product design without major capital investment.
9. It has the ability to transport low to medium volume unit load material to a reasonable large distance in the factory without many problems.
10. Compared to dedicated conveying systems, AGV system offers fewer restrictions on planning the factory layout. This is due to ease in the movement (backward/forward and left/right) of the AGV on the guide path.
11. It is easy to maintain the AGV system. Individual breakdown of an AGV may be coped with the introduction of another AGV in the same route and removal of the down AGV into the parking area for repair/maintenance.
12. There is low noise and disturbance level with AGV system.
13. Operational safety is more due to non-involvement of labour.
14. AGV system may facilitate the effective implementation of manufacturing information system.
15. AGV system generates relevant information related to inventory control, production, administration, purchase and shipment. Integration of such information is needed in the manufacturing information system.
16. Recently, there has been some emphasis on the research leading to the development of *green production system.* The AGV may act as a supporting link to this system. AGV may carry the jobs to an enclosed area for any operation, which may be environmentally unpleasant.

11.4 BASIC FUNCTIONS OF AGV

Navigation and guidance allows the vehicle to follow a predetermined route which is optimized for the material flow pattern of a given application.

Routing is the vehicle's ability to make decisions along the guidance path in order to select optimum routes to specific destinations.

Traffic management is a system or vehicle's ability to avoid collisions with other vehicles while at the same time maximizing vehicle flow and therefore load movement throughout the system.

Load transfer is the pick-up and delivery method for an AGVs system, which may be simple or integrated with other subsystems.

System management is the method of system control used to dictate system operation.

The proper method selection for each function and its ability to work with the other functions determines in great measure the degree of success of a given system.

11.4.1 Navigation and Guidance

AGVs guide and navigate using one of three principal methods. Although nearly all new systems employ some type of non-wire navigation, some systems still utilize wire guidance. Here the AGV uses a sensor under the vehicle to detect the RF signal emanating from a wire placed in a slot approximately 1 inch below the floor surface.

In the mid-1980s, non-wire guidance and navigation was first introduced using laser target triangulation. Here, reflective targets are mounted above the floor on columns, walls, machines or posts approximately 25 feet apart. Each facility target is surveyed and given a unique 'X,Y' coordinate. These coordinates are loaded into each AGV's memory. On-board each AGV is a rotating laser light beam source and receiver. When the laser light reflects off a facility target, its distance and angle are automatically measured. By processing several laser reflections at a time, and comparing them to the stored target coordinates, the AGV can calculate its position. The AGV then compares its calculated position to a coordinate map of the preplanned path stored in its memory and determines it steering instructions as it proceeds throughout the facility.

A new form of non-wire guidance and navigation was introduced in the mid-1990s. This technology is called *inertial* or *gyro navigation*. Each AGV is equipped with a solid-state gyroscope. This device senses very small deviations in the AGV direction of travel. Like laser navigation, the AGV path is a virtual set of coordinates stored in each AGV's memory. A small marker (magnet) is installed in the floor approximately every 25 feet along the AGV virtual path. The markers are flush with the floor surface and surveyed for their unique X/Y coordinates in the facility. This information is also stored in the AGV's memory. As the AGV negotiates the virtual path, the on-board gyro detects slight change in travel direction and this is compared with the actual stored travel path. The AGV corrects course as needed to stay 'on' the prescribed path in its memory. The infloor markers are used as reference points to correct any slight error accumulated over the distance between markers. Typically, the AGVs track the actual path within 1".

AGVs steering control allows AGVs vehicles to physically maneuver in different ways. There are two basic types of AGVs steering control, "differential speed steer control" and "steered wheel steer control".

- Differential speed control uses two fixed wheel drives and varies the speeds between the two drives on either side of the guide path to permit the vehicles to negotiate a turn; much in the way a tank or a tracked vehicle turns.
- Steered wheel control uses automotive type steering control in which a front steered wheel turns to follow the guide-path.

In either case, the guide paths look the same for most applications. Steered wheel control is used in all type of AGVs vehicles; however, differential steer control is not used in towing applications or on vehicles which have man on-board controls.

In very space limited applications not requiring high throughput volume, differential steer control vehicles are used in pivot steer modes. The vehicle stops on the main line, rotates 90 degrees and proceeds into or out of a station. If sufficient room is present, a normal radius curve is often better from the standpoint of flow and controls simplicity.

Steered wheel type vehicles have excellent guidance tolerance along the guide path. Differential steer vehicles normally have less tracking tolerance and therefore require external guides or positioning devices at load stations when used in automatic load transfer applications.

11.4.2 Routing

AGVs routing is another fundamental control function of an AGVs system. How does the vehicle negotiate the path to take the shortest route from point A to point B? Generally, there are two approaches to the routing function. A simple layout will demonstrate those two concepts. This layout has two locations where the guide path splits into two separate directions. These are called *decision points* or *path branches*. It also has two locations where two guide paths merge together into a single guide path. These are known as *convergences*.

AGVs routing techniques centre on two methods depending on whether the system is wire or non-wire guidance. The wire guided method is known as the *frequency select method* and the non-wire guidance method is known as the *path select method*.

In the frequency select method, the wire guided AGVs vehicle approaches the decision point, reads a marker in the floor, which tells the vehicle what location it is at. This marker is typically a passive code device, which is usually in the form of buried magnets, metal plates, or other code devices. The vehicle uses frequency selection to follow the proper path. When the vehicle is approaching the decision point, it is following a single frequency. At the decision point, two frequencies are present in the same slot. The vehicle depending on which direction it wishes to go selects which frequency to follow and the routing is automatically accomplished.

In the path select method, the non-wire AGVs vehicle approaches a decision point and chooses whether to follow path 1, 2, 3, etc. The choice of path is easily accomplished as it is a virtual path choice made within the vehicle's on-board controller. Since the path map is stored in the AGV's memory, it simply chooses which of the memorized paths to follow based on its ultimate destination.

In our simple system layout, the wire based frequency select method requires that at least two different frequencies be present at a decision point for a two-path decision. If there were three paths to select from at a decision point, then three different frequencies would be required. The multiple frequencies are only required at the decision point or at a convergence point where multiple paths come together. In between divergences and convergences, single frequencies are used. In the frequency select method, the frequencies are used over and over again whenever they are required. Most systems only require two or three frequencies. These frequencies loop through the system in a continuous wire and are always active.

The AGVs vendors using this technique produce frequency layout drawings whereby they design the system so that the necessary frequencies are available at intersections and convergences as required. When the frequency is no longer required, it is routed out of the main guide path slot into an interconnect slot and run to the next needed area in the interconnect slot. This results in some extra slot cutting other than for the main guide path.

In the non-wire path switch select method, the virtual guide path is divided into segments which are stored in the AGVs memory. Since there is no physical wire path installed in the floor, the AGVs simply determine their location on their internally stored maps and determine when they arrive at a path branch which path to take.

11.4.3 Traffic Management

Traffic management is a system or vehicle's ability to avoid collisions with other vehicles while at the same time maximizing vehicle flow and therefore load movement throughout the system.

Different types of traffic management are:

1. Zone control
2. Forward sensing control
3. Combination control.

Zone control

Zone control is easily the most popular and widely used type of traffic management. We can apply zone control to our simplified AGVs layout by segmenting the path into separate zones. The rules of zone control are that only one vehicle is permitted in a given zone at a time.

When a vehicle occupies a zone, the closest a trailing vehicle can get is into the next completely unoccupied zone behind the lead vehicle. The lead vehicle must then proceed into the next zone before the trailing vehicle can move ahead into its next zone. A zone may have multiple stop stations in it. If a vehicle is allowed to occupy a zone, it can proceed to any stop in that zone.

Zone control is accomplished by two general methods.

"Central zone control" accomplishes the zone sequence action through the use of a central controller, which controls each block zone by use of a zone communication point for each zone. The central zone controller regulates the entire network of zone communication points itself. When a vehicle approaches a zone entrance, it communicates from the zone communication point to the central controller. If the central zone controller determines that the vehicle can enter the zone, it communicates to the vehicle at the communication point that it can proceed; otherwise, the vehicle waits for permission to go.

"Vehicle zone control" accomplishes the zone sequence action through the use of the vehicles themselves. Each vehicle communicates to one another without the need for a central zone controller. The vehicles know where they are and communicate that information by radio frequency communication to other vehicles in the area. When these vehicles receive the information, they then decide for themselves, based on their location, whether or not they can enter a given zone. If a vehicle is communicating that it is in a particular zone, any vehicle wishing to enter that zone will automatically stop and wait at the beginning of the occupied zone. When the lead vehicle passes into the next free zone, it changes the information it is transmitted to indicate its new zone position. This new information is transmitted to other vehicles in the system and the vehicles take the appropriate blocking actions.

Briefly summarizing these two principal methods of zone control, they each have distinct advantages and disadvantages.

In a central zone control system, if the central zone controller should go down, there is no ability for the AGVs system to operate. The AGVs cannot operate if the system does not perform traffic control. The central zone control method does allow for flexibility and blocking sophistication, allowing a high degree of vehicle movement for high throughput rates.

The vehicle zone control is based on an intelligent control system onboard the vehicle. Since there is no central, the failure case can be limited only to specific vehicles that fail, and not to a central zone controller failure which would affect all the vehicles.

In any form of zone blocking, the more zones in the system, the greater the degree of vehicle movement permitted. Fifty feet long zones permit much more vehicle movement than do 200 foot long zones, because in a given length of path you can fit four times as many fifty foot zones as you can 200 feet zones. This would permit vehicles to move with a greater degree of freedom from one another and yield higher throughput.

Forward sensing control

Traffic management by forward sensing control uses a sensing system onboard the guided vehicles that detect the presence of a vehicle in front of it. Three types of sensors are used: sonic, which works on a radar principal; optical, which uses an infra-red light sources; and bumper, which uses a physical contact of vehicles to cause traffic control. These methods are not effective in path intersection or convergence areas since the vehicle's forward sensing controls look straight ahead and not to the side for intersecting or converging traffic. Therefore, in those areas these methods have to be augmented by some form of zone traffic control.

In general, forward sensing control is useful where there is a lot of straight paths in the system and where that straight path is not interrupted by intersection curves. In these methods, wherever the sensing system sees a vehicle in front of it, the vehicle goes into a hold, and when the lead vehicle moves out of the range of the sensing device, the trailing vehicle automatically restarts.

One advantage of forward sensing for traffic management is that vehicles can come quite close to one another and there are no fixed holding points as in zone control methods. This permits a greater density of vehicles in a given area. A drawback is that it cannot be used where paths are very dense and complex.

Combination control

Systems can be controlled entirely with one method only, or a system layout can benefit from both types of traffic control known as a *combination control traffic management system.* There may be long runs in which forward sensing control can be used reducing the expense of the traffic management system significantly. In the remaining areas of the path where there are divergence path and convergence path areas, a form of zone control would be used. The vehicles would respond to both types of traffic control depending upon the area in which they are operating.

11.4.4 Load Transfer

AGVs load transfer can be accomplished in many different ways. Please select:

1. Manual load transfer.
2. Automatic couple and uncouple
3. Power roller, belt, chain
4. Power lift/lower
5. Power push/pull

Manual load transfer

Manual methods of load transfer include manually coupling and uncoupling towed trailers, loading and unloading by fork lift, and manually loading and unloading AGVs vehicles.

Manual methods can include uncoupling trailers and moving the trailer off the AGVs system to given work stations, or it can involve simple roller bed transfer from the AGVs to fixed roller stations by manually pushing the load off the vehicles. The most popular conventional AGVs approach uses fork trucks to manually load and unload the trailers towed by an AGVs vehicle. Fork trucks are also used to directly load and unload AGVs unit load carriers.

Automatic couple and uncouple

Automatic uncouple load transfer methods are relatively easy to implement. The vehicle usually stops on a side spur and automatically uncouples trailers, which it is pulling. The AGV then pulls forward and waits or proceeds to another location to receive a new string of trailers to tow. Typically, an operator will manually couple the new trailers to the AGV hugger.

Automatic couple capability is a more challenging capability than automatic uncouple. First, the carts/trailers to which the AGV is coupling must be in a precise position so that the AGV hitch and the trailer hitch engage properly. Second, controls must be provided to insure trailers left on the path are identified so that other AGVs do not collide with them. Third, sophisticated AGV management software and controls need to be provided to support an automatic coupling operation. For example, an operator must 'call' from a terminal when they have some carts ready for an AGV to pickup. This requires a software management program to decide where the closest AGV is that is not currently towing carts, and is available for a new task. If operators move the waiting carts or override the automatic dispatching of AGVs to pickup carts, then the system will not function well. So, increased operator and supervisory discipline needs to be provided.

Power roller, belt, chain

The powered roller belt or chain transfer for guided vehicles has been a standard technique for many years. Unit load carriers can be equipped with these power decks so that they may automatically transfer loads to and from fixed stations. In certain cases, under light load conditions, parasitic drives can be used which allow the AGV unit load carrier vehicle to interface with a non-powered roller stand. Most applications involve the AGV vehicle dropping off to a powered or gravity type conveyor. A roller deck transfer method can be used on AGVs towing systems where the towed trailers are equipped with powered transfer deck mechanisms. In all the cases, the vehicles must precisely align with the given transfer station before the load transfer can be achieved. Normal stopping tolerances are + or –1/2" which is well within the tolerances required at conveyor transfer stations.

Whenever a vehicle is automatically transferring a load, a method or "handshake" logic must be employed. Handshake logic is the method by which the vehicle checks to make sure a load stand is empty before it automatically transfers a load onto the stand. Or when a vehicle comes to pick up a load, it must make sure that there is a load on the stand for pickup before the load transfer occurs. The handshake signals between the vehicle and the load stand allow transfer drives to be activated simultaneously to achieve the load transfer. The signals also turn off the drives when the transfer is complete. The handshake method usually involves sensors on-board the vehicle which interface with sensors on the stand that exchange signals when a load transfer is about to occur. If the proper signals are not exchanged, the vehicle will not attempt the transfer. It can blow its horn or abort the mission and proceed to another destination.

Power lift/lower

There are three general power lift/lower methods of load transfer.

- Pallet fork lift/lower
- Unit load lift/lower
- Fork truck lift/lower

In the pallet truck lift/lower method, the AGVs pallet truck can pick up loads or drop them off directly to the floor. The vehicle must pull straight ahead after depositing a load on the floor so that the pallet forks clear the load before it reenters the main guide path.

The lift/lower unit load carrier can interface with special 'horseshoe' conveyor load stands or with fixed load pickup deposit stations. The vehicle approaches a load transfer area and makes a right angle turn into a horseshoe-shaped load station. It decelerates and pulls into the load station where it accurately stops and lowers or lifts its load deck. The load is then transferred to/from the load stand and the vehicle reverses its direction, and proceeds to its next assignment.

The lift/lower fork truck is a method that allows the guided vehicle to pick loads up off the floor and deposit them on the load stands or in racks. This gives the AGVs system more versatility in larger systems where not all of the conveyors or load stands are at the same height. The vehicle can service an infinite number of different heights within a given lifting range.

Power push/pull

A less frequently used method of automatic load transfer is the push or pull type. Here a vehicle with a non-powered deck or trailers positions itself in front of a fixed station which is equipped with an automatic push or pull mechanism. This mechanism will reach out to the load and push or pull it off the vehicle or trailers. This permits the vehicles to be extremely simple. This method is quite useful if the automatic load transfer is centralized because the cost of multiple load transfer points can be prohibitive. A good example of this approach would be a central storage area supporting manufacturing. Parts baskets can be loaded onto an AGVs train by a powered shuttle automatically. The train proceeds into the manufacturing area. The parts baskets are unloaded manually or by fork truck and are moved to machining centers. Parts baskets are returned to the central storage area on the AGVs train and unloaded automatically by the powered shuttle. Another possible permutation would be to have a powered device load or unload the AGV at certain locations and employ a manual approach to transferring loads at other locations in the system.

11.4.5 System Management

System management is the method of system control used to dictate system operation. It can be accomplished by two functions:

- System monitoring
- Vehicle dispatch

System monitoring

AGVs system monitoring is an important consideration in many systems. Simple systems do no require extensive monitoring or control. Sophisticated systems benefit greatly from monitoring. When a sophisticated system is installed, usually there is a high degree of automation and throughput required. A breakdown or slowdown in the system could cause serious problems if not detected immediately.

There are three approaches to monitoring:

- Locator panel
- CRT colour graphics display
- Central logging and report

A simple monitoring system for AGVs systems can be a locator panel which merely indicates if a vehicle is in a given area of the guide path. It does not identify the vehicle specifically or its condition. A light next to the area a vehicle occupies illuminates to indicate that there is a vehicle in that location. Sometimes a timer is used for each zone to indicate if that zone has been occupied for too long which might indicate that a vehicle problem exists in that location.

A CRT colour graphics display specifically shows where each vehicle is including its status. CRT colour graphics is usually a real-time monitoring type, which can instantly detect a problem, identify specific vehicles, and show the location of the failure on the graphics display. Other useful information includes whether the vehicle is moving or is blocked by the other traffic, whether the vehicle is loaded or empty, if the battery is O.K. and where the vehicle is going. Operators can spot blockages and slowdowns quickly and take corrective action as required.

The CRT colour monitor can show system condition either in a graphic form or in a table form. The table form lists each vehicle ID, its location, destination, condition, mode of control, load status, and alarm condition in a column format listing, which is continually updated.

A central logging and report capability for monitoring an AGVs system is helpful when attempting to develop historical data on the systems performance. Periodically, performance reports can be printed out indicating such things as how long the vehicles were moving, how many loads a given vehicle transported to a given stop station, and where were they taken or even when did a vehicle have a low battery or when did a specific load get picked up and where did it go.

Performance data can be quite useful for keeping system efficiency at the highest possible levels.

Vehicle dispatch

On-board dispatch selector involves a control panel on each vehicle, which is used by an operator to dispatch the vehicle to a single or series of stop stations. He may also select the function he wishes the vehicle to perform at the stop station. This is the most common form of system management and is usually the most flexible type.

An off-board call system can also be used for vehicle management in AGVs systems. These call systems vary in complexity from simple types which involve only a push button at a call location to stop the passing guided vehicle all the way to call terminals which not only can call a specific vehicle, but can also remotely dispatch that vehicle to other destinations without operator interface. Operators can input new destinations into the call box panel at their station, which will communicate those destinations to the guided vehicle after it leaves their stop location. This is useful in systems where load transfer is automatic.

A remote dispatch terminal can be used in systems to give a degree of centralized dispatching to the control of the guided vehicles in the system. The keyboard remote terminal approach allows an operator in a central location to control the individual guided vehicles form a central location. In order to do this, the operator must have some visual status of the vehicles' location and condition so that they might effectively dispatch vehicles to specific areas in the system where they are needed. This is usually a CRT graphic display or locator panel. Many times this approach is used in systems, which cannot justify totally computer-controlled solutions yet, where a high degree of selective movement with automatic transfer is required. For example, if loads are received at a receiving area and taken to selective locations in a storage area with automatic load transfer used by the AGVs vehicles, then a centralized control may be an effective solution. A central operator would use the keyboard to send the vehicles to the receiving area to automatically pickup loads and then dispatch those vehicles to selected storage locations where the loads would be automatically dropped off.

Central computer control for system management is the highest level of control possible. The AGVs vehicles would respond to a central computer's commands for where they should go and what they should do. This would be automating the previous case and eliminating the central dispatcher completely. Most integrated material handling systems incorporate central computer control for AGVs system management. In smaller dedicated systems such as picking up at production machines (injection molding, stretch wrapping, palletizing, machining, etc.), the AGVs are called to the production machine automatically when a load is ready for pickup. The computer control system can select the closest available AGV for the task and then dispatch it automatically. Once a load is picked up, the central control system will communicate a destination to the AGV. This level of control permits automatic tracking of every load in the system. Where the volume is high and automatic load transfer occurs, the computer form of system management is very efficient.

It is also possible to mix these various forms of system management together at the same time in the same system. This depends on the type of control offered by particular vendors, but in many cases systems can be operated in several different modes at the same time. This is particularly useful if you have a sophisticated level of operation in your system and wish to have a backup to it in case the main level of operation control fails. For example, if your system is computer controlled and the central computer fails, then resorting to remote terminal control or onboard vehicle control would allow the AGVs system to continue to operate in the automated mode. Not all central computer control systems permit this type of operation but it is very important in certain situations.

11.5 TYPES OF AGV SYSTEMS

Following AGV systems may be identified on the basis of navigation system, which is the network of guide-paths followed by AGVs: These AGVs follow an inductive guide-path. Floor-embedded wires carry alternating current.

11.5.1 Wire-Guided AGVs

Path layout in guided AGV

The floor embedded wires are arranged to form closed loops. These loops determine the guide-path of the AGV. Each closed loop carries current of different frequencies with low voltage and low amperage. AC current, upto the range of 15 kHz, is used in each loop. In any advanced manufacturing system, simulation study may be needed to justify a path layout.

Figure 11.2 Path layout.

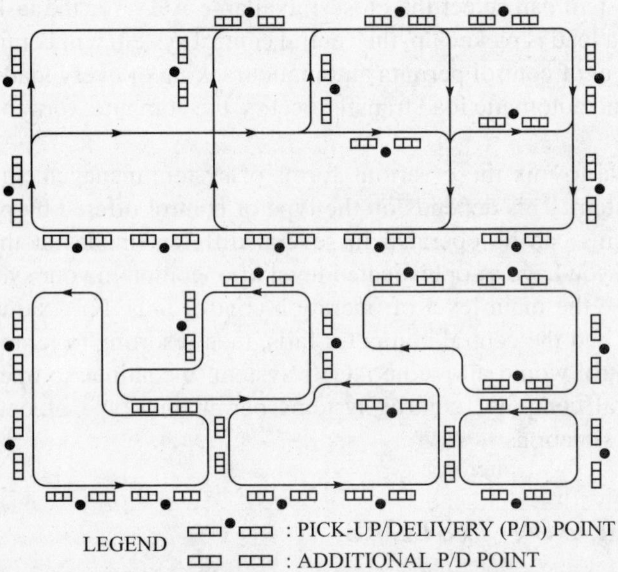

Figure 11.3 Wire guided AGV.

Path layouts for wire-guided AGV

Communication in wire-guided AGV system: Communication of the AGV with the host computer is essential for the shop-floor control and coordination of AGV system.

For this purpose, four main elements interact:

1. On-board controller
2. Floor Embedded wire
3. Reed Sensors
4. Host (or off-board) computer.

An AGV continuously identifies its position and reports its status to the main host computer through the communication cables. At different points in the loop of the guide-path, reed sensors (in the embedded wire) locate AGVs and communicate their status through the wire. Control commands from the host computer reach the onboard controller of AGV via reed sensor on the AGV.

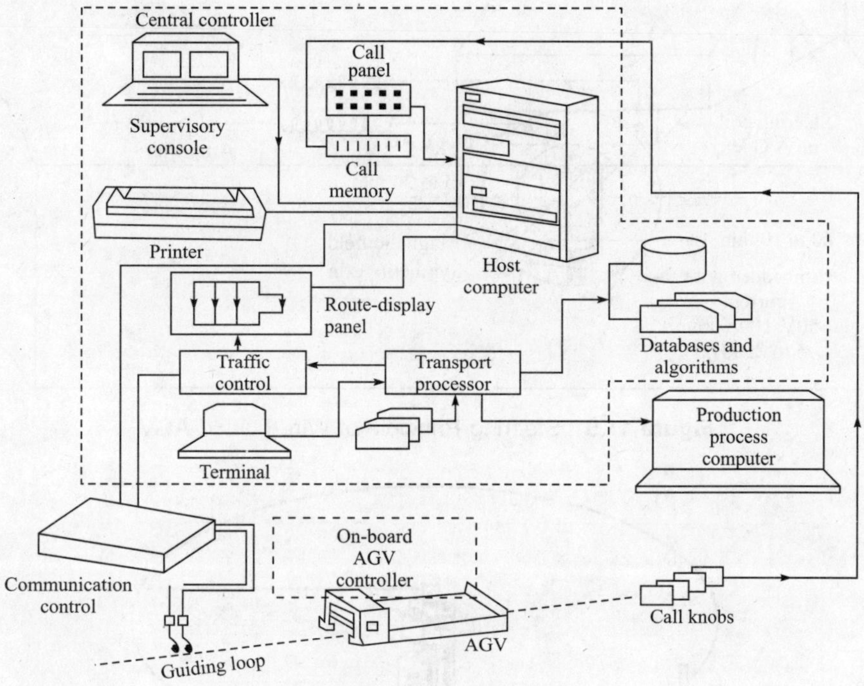

Figure 11.4 Path guided AGV.

AGV system architecture

Steering of vehicle: The principle of steering of AGV on the guide-path is illustrated in Figure 11.4. The frequency generator provides the frequency signal in the embedded wires. A magnetic field is created around the wires. Two coils mounted on the AGV sense this. The embedded wire is mid-way between each coil on the AGV. If in case AGV tilts either way, the intensity of the induced magnetic field becomes different in coils. This differential is sensed and then it is sent to on-board controller of the AGV. Based on this feedback signal, the on-board controller sends signal to put AGV on the right track.

11.5.2 AGVs Having Optical/Chemical/Magnetic Guide-Path

Optically guided AGV

These AGVs use a strip of optical/chemical/magnetic material on their guide-path. It is detectable by a proximity sensor on the vehicle. For example, the chemical guide-path is excited when ultraviolet rays from the AGV fall on the path. They, in turn, re-emit waves in the visible range. Due to this, the contrast ratio of the rays is improved. Location of the guide-path and steering of the vehicle are done through the interpretation of the contrast ratio.

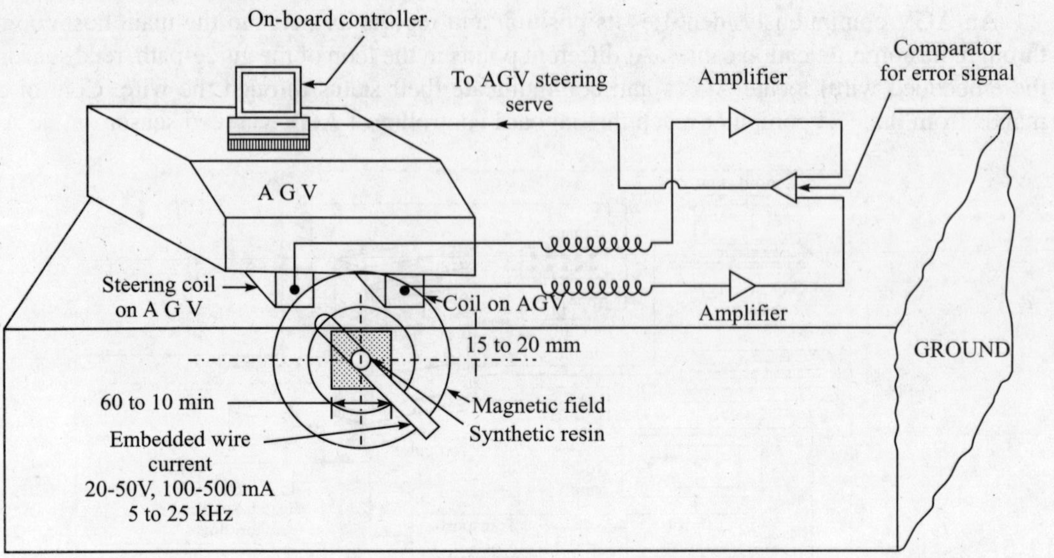

Figure 11.5 Steering Principle of Wire-guided AGV.

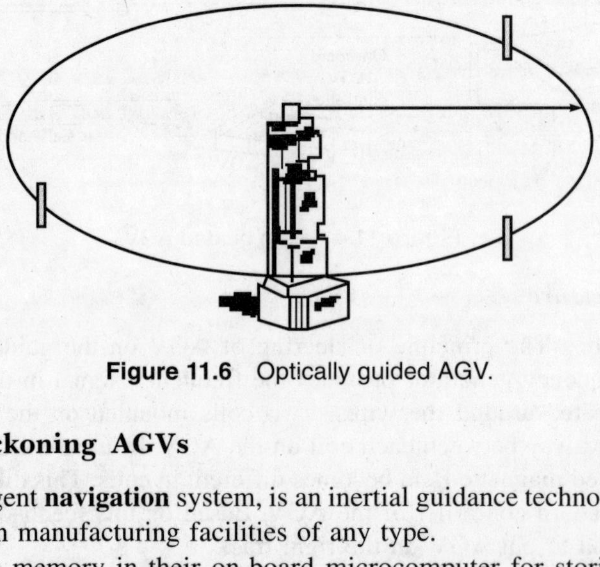

Figure 11.6 Optically guided AGV.

11.5.3 Dead Reckoning AGVs

NAV, AGVP's intelligent **navigation** system, is an inertial guidance technology used by Odyssey AGVs to maneuver in manufacturing facilities of any type.

These AGVs use memory in their on-board microcomputer for storing distance table for stations, acceleration/deceleration information, speed and steering information. It is based on the principle of odometry. The present position of the AGV with respect to the fixed positions in the reference plane in surrounding is initially known. By measuring exact rotation and velocity of each drive wheel, the next position of the vehicle is calculated.

On-board systems are designed to provide instructions to AGV for movement about a known reference point. No guide-path is needed in this system.

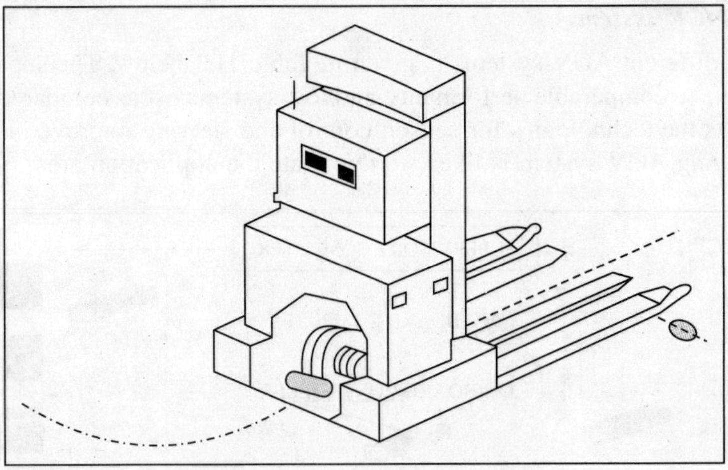

Figure 11.7 Dead Reckoning AGV.

11.5.4 Free Ranging AGV

Unlike wire-guided AGV, the free ranging AGV does not follow any physical guide-path. The guidance of the free ranging AGV may be based on the following techniques:

(i) **Position reference beacons** In this system, some beacons are fixed on the factory locations known to the AGV controller. On-board devices measure exact distance and direction of AGV with respect to beacons. This information is used to measure the position of AGV.

(ii) **Optical imaging of surrounding and stereoscopic vision** Stereoscopic vision system has also been used for the 3-D imaging of the surrounding object. It requires two cameras, separated at known distance (like two human eyes) on the AGV.

The image, captured by each camera at any point, is analyzed to locate an object in surrounding. Much work is, however, needed to integrate the sensor system with AGV system in a cost-effective way. It is an area of tremendous research potential in an autonomous factory.

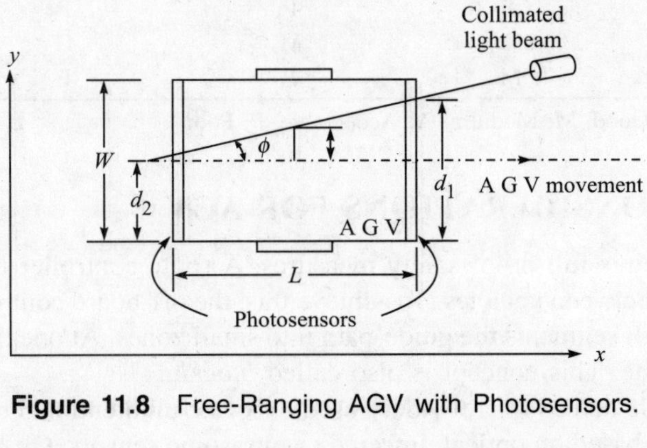

Figure 11.8 Free-Ranging AGV with Photosensors.

Comparison of AGV systems

A comparison of different AGV systems is given in Table 11.1 below. The superiority of wire-guided AGV is due to comparable performance and low system cost as compared to free ranging AGV. However, as the technologies for sensor, control and steering improve in a cost-effective way, the free ranging AGV system is likely to dominate the application area.

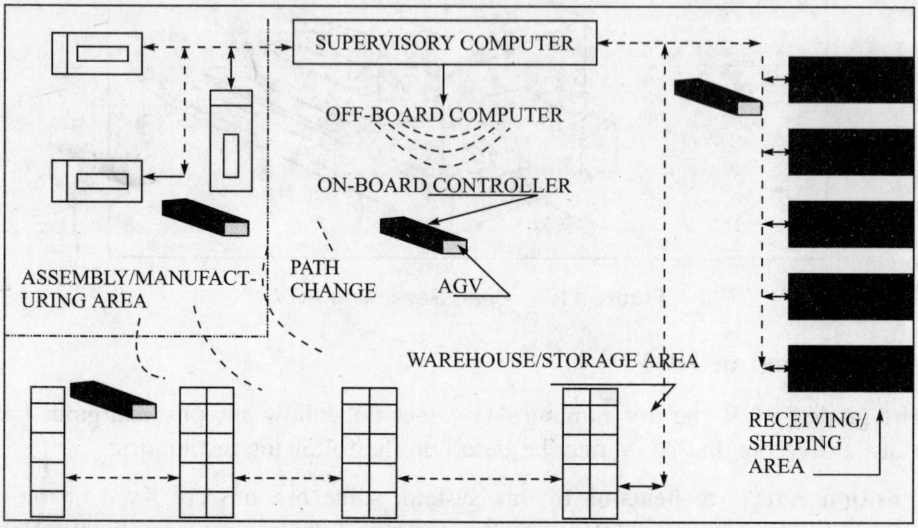

Figure 11.9 Integration of Free Ranging AGVs in the Manufacturing System.

TABLE 11.1 Comparison of different AGV System

Type of AGV	*Wire Guided*	*Optical/ Chemical*	*Dead Reckoning*	*Free Ranging*
Range	G	E	P	C
Accuracy	G	E	A	E
Flexibility	M	G	E	E
Reliability	E	M	G	M
Controllability	E	P	P	A
Vehicle Equipment Cost	G	M	G	P
Station Equipment Cost	M	E	E	G

Legend: E: Excellent, G: Good, M: Medium, A: Acceptable, P: Poor

11.6 SAFETY CONSIDERATIONS FOR AGV

The AGV system requires full-proof safety measures. A route controller normally controls the avoidance of collision between vehicles. To achieve this, the off-board controller is provided with a zone-block software. It segments the guide-path into small zones. At one time, only one vehicle is permitted in one zone. This concept is also called *blocking*.

Sensors for obstacle detection are widely used to locate unintentional obstacles, like human being, pallets, etc. It is based on optical, infrared or ultrasonic sensors. On-board receiver senses the reflection of a beam thrown from the vehicle. The vehicle is programmed to slow down/stop, depending upon the distance of the sensed object. In addition to this, the following safety devices are also used:

Emergency bumper in the front, blinking or rotating lights for warning, warning bell, emergency stop, if the AGV deviates from its path beyond a safe limit, emergency stop-button on the vehicle, limit switch, proximity sensor or photocells.

11.7 DESIGN OF AGV SYSTEM

Various steps in the design of AGV system are shown in Figure 11.10. It is in three distinct phases:

1. Data generating stage
2. Design stage
3. Analysis stage.

The details of each phase are shown in Figure 11.10.

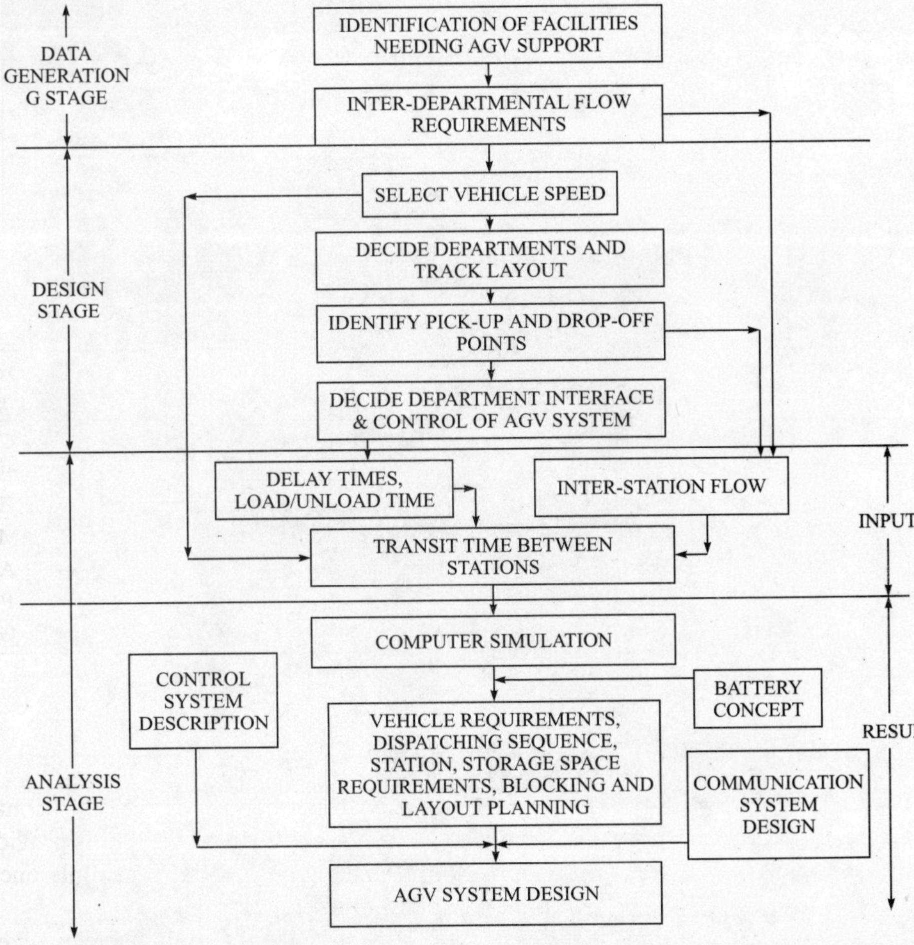

Figure 11.10 AGV Design stages.

REVIEW QUESTIONS

1. Define AGV.
2. What are the salient features of AGV?
3. Describe the basic functions of AGV.
4. Write an advantage of traffic management.
5. Discuss in detail system management.
6. What are the types of AGV system?
7. What do you mean by capital guided AGV?
8. Compare free ranging AGV and dead reckoning AGV.
9. What are the various steps involved in the design of AGV?
10. Explain briefly the historical background of the Automated Guided Vehicle system.

Chapter 12

ROBOTICS

12.1 INTRODUCTION

"An industrial robot is a programmable, multifunctional manipulator designed to move materials, parts, tools, or special devices through variable programmed motions for the performance of a variety of tasks."

—*Robotics Institute of America*

Since the development of the first articulated arm in the 1950s and subsequent developments in computer technology, robots have become more widely used and are now available in a variety of types, styles and sizes. Robots are capable of performing a wide variety of tasks and have become an integral part of automated manufacturing systems. The industrial robot consists of a number of rigid links connected by joints of different types, controlled and monitored by a computer. To a large extent, the physical construction of a robot resembles a human arm. The link assembly mentioned above is connected to the body, which is usually mounted on a base.

This link assembly is generally referred to as a *robot arm*. A wrist is attached to the arm. To facilitate gripping or handling, a hand is attached at the end of the wrist. In robotics terminology, this hand is called an *end-effector*. The complete motion of the end-effector is accomplished through a series of motions and positions of the links, joints and wrist. Applications of robots have increased as manufacturers have moved from hard automation dedicated machines and equipment to soft automation machines that are designed to perform multiple tasks, flexible.

12.1.1 Definitions used in Robotics

Spatial resolution: Ability of the robot to break down its movements into increments #increments = 2n, where n = bits in control memory spatial resolution = range/2n.

Example: A robot controller has 12-bit storage capacity, the full range of the robot = 1.0 m for one joint spatial resolution = 1.0 m/212 = 1.0 m/4096 = 0.244 mm.

Accuracy: How closely the robot arm is able to move to a specific coordinate in the work cell. Accuracy is not constant; it tends to decline when the arm is in the outer range of the work envelope and increase as the arm moves closer to the base; heavier workloads can also cause deflection that degrades accuracy.

Repeatability: Degree to which the robot is able to return to a previously taught position.

Compliance: Displacement of the wrist end in response to a force or torque exerted against it; high compliance—"springy", low compliance—"rigid"; compliance can reduce the robot's precision under load.

Work envelope: Space defined by the geometry of the robot; Cartesian, cylindrical, spherical, revolute.

Degree of freedom: The available axes of movement that a robot has are referred to as the 'degrees of freedom'. This will vary depending upon the robot's configuration and is an important aspect when considering and selecting robots for particular applications.

Every joint or moveable axis on the robot arm is one degree of freedom.

Rotational transverse	movement about a vertical axis
Radial transverse	extension and retraction of the arm
Vertical transverse	up and down motion
Pitch	up and down movement of the wrist
Yaw	side to side movement of the wrist
Roll	rotation of the wrist

Offset: Point of action for the tool mounted to the robot tool plate (or Tool Centre Point—TCP); consider offset in computer programming of the robot.

Duty cycle: Ratio of run time to total operational time that a robot can continuously work at the rated payload without overheating or degrading the robot specifications.

Payload: Mass that the robot is designed to manipulate under the manufacturer's specifications (speed, acceleration/deceleration, duty cycle, etc.) over the entire work envelope. Payload = end-effector weight (or tooling weight) + part weight.

Advantages of machines and robots

- Do not tire or grumble
- Higher quality
- Repeatable performance
- Stronger, faster, more accurate
- More productive
- Work 24 hours each day
- Immune to dangerous environment.

'Quality' here implies a consistent, rather than necessarily higher, standard of achievement.

- Adaptable to problems and environments
- Wide range of sensory inputs, with pattern recognition
- Make decisions, set priorities and define goals
- Investigate new techniques
- Easy to program

Surveys reveal high labour costs are the most usual reason for installing robots, covering about half of all cases.

Social issues

- Unemployment?
- Prosperity = Productivity?
- Elimination of drudgery?
- Workplace safety.

12.2 TYPES OF ROBOT

Robots can be classified in a number of ways, and in this section we look at some of the common characteristics by which we can identify robots. The main characteristics by which a robot can be identified are:

1. Classification based on physical configuration
2. Type of drive
3. Type of end-effectors
4. Reach or work envelope

12.2.1 Classification Based on Physical Configurations

Four basic configurations are identified with most of the commercially available industrial robots. They are:

- Cartesian configuration
- Cylindrical configuration
- Polar configuration
- Jointed-arm configuration

Cartesian configuration

Robots with Cartesian configurations, as shown in Figure 12.1, consist of links connected by linear joints. Because the configuration has three perpendicular slides, they are also called *rectilinear robots*. Another robot that is similar to this configuration is a gantry robot. The structure of a gantry robot resembles a gantry type crane.

- All arm joints are linear.
- Movement along all three axes can occur simultaneously.
- Work envelope is rectangular.

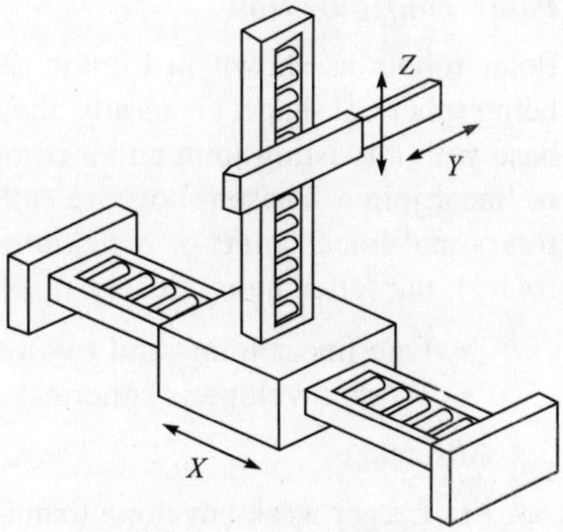

Figure 12.1 Cartesian Robot.

Advantages

- Simple controls
- High degree of mechanical rigidity
- Good accuracy and repeatability
- Consistent payloads throughout work envelope

Disadvantages

- Limited in movement

Applications

- Assembly, surface finishing and inspection.

Cylindrical configuration

In the cylindrical configuration, as shown in Figure 12.2, robots have one rotary joint at the base and linear joints to connect the links. The space in which this robot operates is cylindrical in shape, hence the name cylindrical configuration.

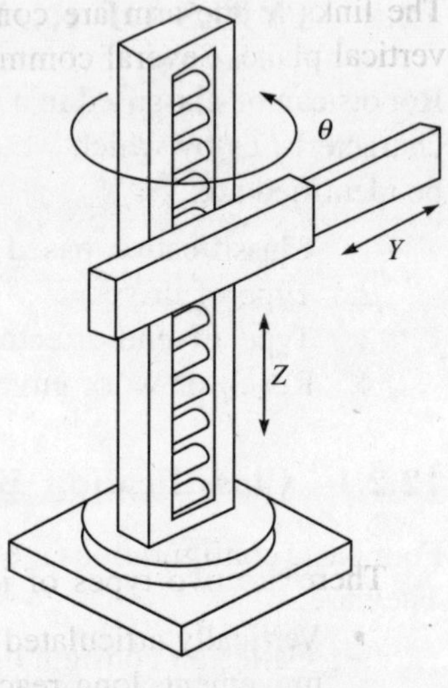

Figure 12.2 Cylindrical Robot.

- Two linear joints and one rotary joint
- Work envelope is cylindrical

Advantages

- Larger work envelope than Cartesian configuration
- Suitable for pick-and-place operations

Disadvantages

- Lower mechanical rigidity
- Repeatability and accuracy lower in direction of rotary movement
- More sophisticated control system required than Cartesian

Applications

- Load/unload, conveyor pallet transfers, material handling

Polar configuration

Polar robots as shown in Figure 12.3 have a work space being spherical shape. Generally, the arm is connected to the base with a twisting joint and a combination of rotary and/or linear joints. These robots are called *spherical* (twisting, rotary and linear joint) or *articulated* (twisting, rotary and rotary), the latter more closely resembles the human arm.

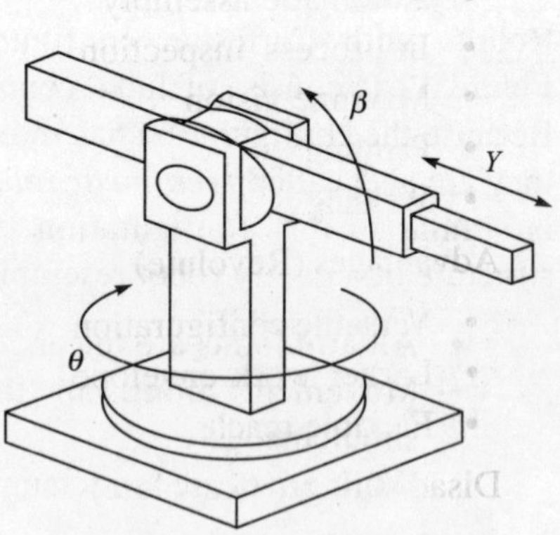

Figure 12.3 Polar Robot.

- One linear joints and two rotary joints
- Work envelope is spherical

Advantages

- Larger work envelope than Cartesian or cylindrical
- Simple design, high payloads

Disadvantages

- Lower mechanical rigidity
- More sophisticated control system than Cartesian or cylindrical, limited vertical movement

Applications

- Injection molding, forging, machine tool loading, material transfer

Jointed-arm configuration

The jointed-arm configuration, as shown in Figure 12.4, is a combination of cylindrical and articulated configurations. The arm of the robot is connected to the base with a twisting joint. The links in the arm are connected by rotary joints. The rotations generally take place in the vertical plane. Several commercially available robots have this configuration.

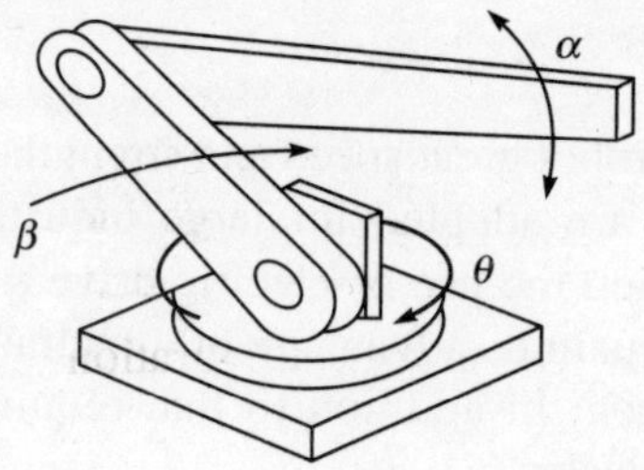

Figure 12.4 Jointed-arm Robot.

There are two types of jointed-arm configuration.

- Vertically articulated (three rotary joints, horizontal movement small relative to vertical movement, long reach makes it suitable for painting applications)
- Horizontally articulated (one linear and two rotary joints, vertical movement small relative to horizontal movement, selective compliance articulated robot arm

Applications

- Automatic assembly
- In-process inspection
- Machine vision
- Painting
- Welding

Advantages (Revolute)

- Versatile configuration
- Larger work envelope
- Flexible reach

Disadvantages (Revolute)

- Sophisticated controller
- Complex programming
- Different locations in work envelope determine accuracy
- Payload and repeatability
- Less stable as arm approaches maximum reach

12.2.2 Types of Robot Drives

An important element of a robot is the drive system. The drive system supplies the power, which enables the robot to move. The dynamic performance of the robot is determined by its drive system, which depends on its application and power requirements. The three types of drive systems that are generally used for industrial robots are as follows.

1. Hydraulic drive
2. Electric drive
3. Pneumatic drive

Hydraulic drive

A hydraulic drive system gives a robot great speed and strength. Because they provide high speed and strength, hydraulic systems are adopted for large industrial robots. Hydraulic robots are preferred in environments in which the use of electric-drive robots may cause fire hazards, for example, in spray painting. The main disadvantage of a hydraulic robot is that it occupies more floor space for ancillary equipment in addition to that required by the robot. Also, there are housekeeping problems such as leaks.

Electric drive

Compared with a hydraulic system, an electric system provides a robot with less speed and strength. Accordingly, electric drive systems are adopted for smaller robots. However, robots supported by electric drive systems are more accurate, exhibit better repeatability, and are cleaner to use. Electrically driven robots are the most commonly available and used industrial robots. Like numerically controlled (NC) machines, electrically driven robots can be classified into two broad categories: stepper motor driven and direct current (DC) servo-motor driven. Most stepper motor-driven robots are of the open loop type, but feedback loops can be incorporated in stepper-driven robots. Servo-driven robots invariably have feedback loops from the driven components back to the driver.

Pneumatic drive

Pneumatic drive systems are generally used for smaller robots. These robots generally have fewer axes of movement and carry out simple pick-and-place material-handling operations, such as picking up an object at one location and placing it at another location. These operations are generally simple and have short cycle times. The pneumatic power can be used for sliding or rotational joints. Pneumatic robots are less expensive than electric or hydraulic robots.

12.2.3 Type of Robot End-effector

The end-effector (commonly known as robot hand) mounted on the wrist enables the robot to perform specified tasks. Various types of end-effectors are designed for the same robot to make it more flexible and versatile. End-effectors are categorized into two major types:

- Grippers
- Tools

Grippers are generally used to grasp and hold an object and place it at a desired location. Grippers can be further classified as mechanical grippers, vacuum or suction cups, magnetic grippers, adhesive grippers, hooks, scoops, and so forth. A double gripper can handle two objects at the same time, and the two gripping devices can be actuated separately. More than two grippers can also be classified as external or internal, depending on whether the part is grasped on its external or internal surface. At times, a robot is required to manipulate a tool to perform an operation on a work part. In such applications, the end-effector is used as a gripper that can grasp and handle a variety of tools and the robot has multi-tool handling function. However, in most robot applications in which only one tool is to be manipulated, the tool is directly mounted on the wrist; here the tool itself acts as the end-effector. Spot welding tools, arc-welding tools, spray painting nozzles and rotating spindles for drilling and grinding are typical examples of tools used as end-effectors.

12.2.4 Robot Reach (Work Envelope)

Robot reach, or more commonly referred to as the 'work envelope', describes the operational range of an industrial robot. All robots have a work envelope or reach capability dependant upon their particular design. For example, Cartesian robots work within a 'cube' as shown in Figure 12.5.

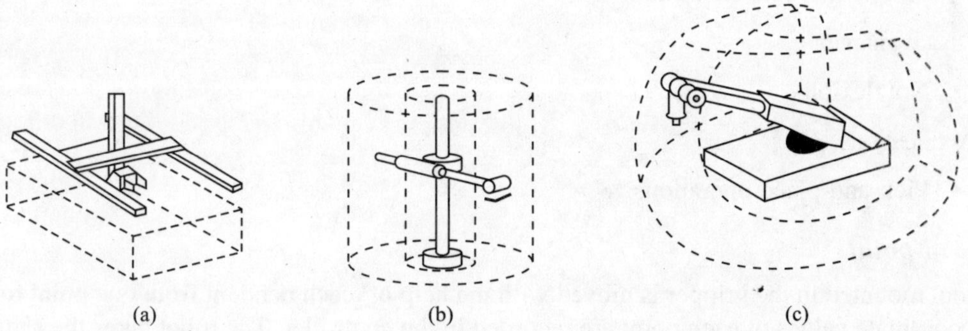

Figure 12.5 (a) Cylindrical robots have a work envelope of a 'cylinder' as shown in (b). Polar and jointed-arm robots extend the flexibility to incorporate parts of 'spheres', as shown in (c).

Robot application

The application of robots to specific tasks is determined by a number of factors, which together identify the total performance capability of the robot. The **capability** of a robot refers to all the factors previously mentioned, i.e. axes of movement, type of drive, end-effectors and the range or envelope of movement. Other factors that can also be included are:

- Speed of operation
- Load capability
- Working environment sensitivity
- Structural rigidity.

Consequently, each application demands a performance capability peculiar to that task. Some typical **applications** for robots can include:

- Welding operations
- Pick-and-place activities
- Clamping for machining
- Transfer and manipulation of parts
- Multiple activities

12.3 BASIC ROBOT MOTIONS

Stop-to-stop

- Open loop
- Position and velocity unknown to controller
- On/off commands stored as valve states
- End travel set by mechanical stops

Advantages

- Simple and reliable system

Disadvantages

- Not flexible

Applications

- Pick-and-place operations

Point-to-point

The tool mounted in the gripper is moved with the help of teach pendant from one point to other. The coordinate values of each point are recorded in the controller. The robot takes the same path when it is programmed. Controller changes each axis at maximum rate.

Advantages

- **Relatively** large and complex problems obtained at moderate cost, proven reliability

Disadvantages

- Lack of straight line control

Application

- Wide variety of operations from loading/unloading to assembly.

Controlled path

- Point-to-point system with added capability of providing control of the end-effector
- Axis actuators driven so that the path between points is a straight line

Advantages

- Path control, complex tasks

Disadvantages

- Increase in controller intelligence

Applications

- Operations requiring path control, welding, etc.

Continuous path

- Number of programmed points is extremely large
- Desired path taught by physically moving the robot arm through its desired motions
- Program button records points until the destination is reached; stores points at a rate of approximately 6 per second, or more.

Advantages

- Very complex tasks can be performed, duplicates movement of the operator

Disadvantages

- Expensive, high level of control

Applications

- Spray painting.

12.4 SENSORS FOR ROBOTS

The addition of sensors to a machine changes it from a dumb machine into an intelligent machine. Humans adapt to changes in their environment in part by using **sensors** to detect what is going on in their environment. The ability to adapt to particular surroundings is one definition of intelligence.

12.4.1 Function of Sensor

Component location—determines identity, position and orientation of parts.

Operation monitoring—confirms progress status of components and machines, checks for jamming, missing parts, successful grasping.

Inspection—examines completed work for blemishes.

Specialized inspection machines are available, including computer control and holography.

12.4.2 Classes of Sensors

Sensors for robots can be divided into three classes: *internal sensors, external sensors,* and sensors such as *interlocks* that fall somewhere in between.

Most sensors are some type of **transducer**.

Transducer: converts one type of physical variable (force, pressure, etc.) into another form (e.g. voltage) that is more usable.

Sensor: transducer used to make a measurement of a physical variable of interest; calibration establishes the relationship between the measured variable and the converted output signal.

There are two basic types of sensors:

- Analog transducers
- Digital transducers

Analog transducers: continuous analog signal (voltage or current); signal is interpreted as the value of the physical variable that is being measured.

Digital transducers: produce a digital output signal (bits or series of pulses); compatible with computers.

12.4.3 Desirable Features of Sensors

1. Accuracy—closeness to "true" value of variable; accuracy = actual value – sensed value; over long average error should be zero
2. Precision—little or no random variability in measured variable
3. Operating range—wide operating range; accurate and precise over entire sensing range
4. Calibration—easy to calibrate; no "drift"—tendency for sensor to lose accuracy over time
5. Reliability—no failures
6. Cost and ease of operation—purchase price, cost of installation and operation

12.5 TYPES OF SENSORS

There are five types of sensors.

1. Position sensors—determine position of robot arm
2. Velocity sensor
3. Tactile sensors—contact sensor, "touch" sensors
4. Proximity and range sensors—non-contact sensors
5. Machine vision systems.

12.5.1 Position Sensors

Position sensors are of three types.

1. Potentiometers (Pots)
2. Resolvers
3. Encoders

Potentiometers

- Analog device whose output voltage is proportional to the position of a wiper; voltage applied across a resistive element
- Voltage between wiper and ground is proportional to (resistance on one side of wiper) total resistance

Resolvers

- Analog device, converts mechanical motion to an electrical sinusoidal format
- Output proportional to angle of rotating element w.r.t. a fixed element
- Simplest form: single winding on rotor and pair of windings on stator.

Encoders

Most common in robotics are optical encoders. They are available as below

- Incremental encoders
- Absolute encoders

Incremental encoder

- It consists of a photo-transmitter, sends light beam through a disk, and signals are captured by photo-receiver
- As disk rotates, light beam is alternately completed and broken
- Output: pulse train whose frequency is proportional to rotational speed of disk
- Count the number of pulses to determine location of shaft or measure time between pulses to determine velocity

Absolute encoder

- It consists of photo-transmitter, sends light beam through disk, and a set of photo-detectors captures the output signal.
- Photo-detectors produce a digital output signal that corresponds to a specific location on the disk; exact position of shaft is known.
- Resolution of encoder is dependent on # tracks; resolution = 2n where n = # tracks.
- It consists of two tracks and two sensors whose output are called channels A and B. They are used to access the direction of rotation by which the channel leads the other. The signal from two channels are ¼(90°) out of phase with each other and are known as **quadrature signal.**

Figure 12.6 Incremental Encoder.

Velocity sensors

The most commonly used velocity sensor is DC Tachometer. Its operation is similar to that of DC generator which provides an output voltage proportional to angular velocity of armature

$$V_0(t) = K_t(t)\,\omega$$

where,

$V_0(t)$ = output voltage of tachometer, V
K_t = tachometer constant, V/rad/s
ω = angular velocity, rad/s

12.5.2 Tactile Sensors

Tactile sensors are of two types:

- *Touch sensors:* Indicate presence or absence of parts either by contact or non-contact type, e.g. limit switches, micro switches.
- *Force sensors:* Measure magnitude of applied force.

Tactile array sensors

Force sensor is composed of a matrix of force-sensing elements. It is used to describe:

(1) Presence of an object
(2) Object's contact area, shape, location and orientation
(3) Pressure and pressure distribution
(4) Force magnitude and location.

It consists of conductive elastomer pads that respond to the force applied by a change in electrical resistance; increase resolution by increasing the number of pads.

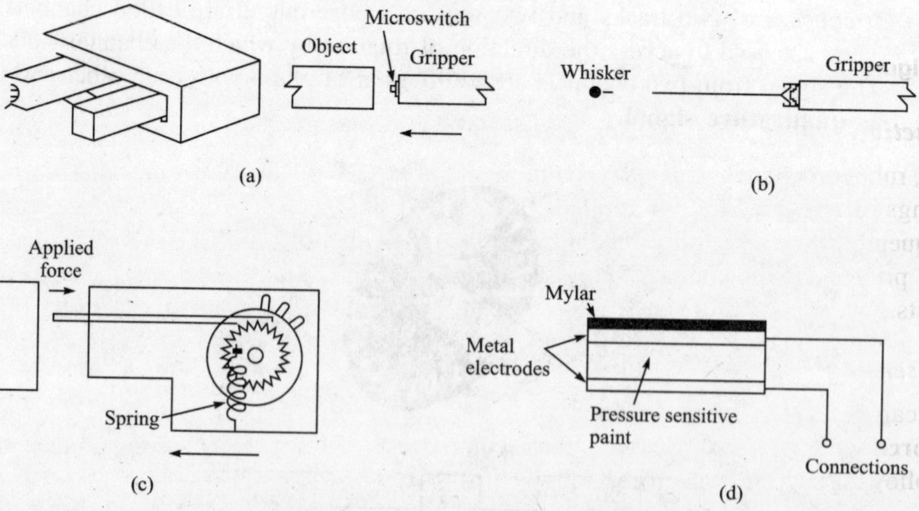

Figure 12.7 (a) A Microswitch being used as a short-range digital touch sensor. (b) A whisker and microswitch being used as a longer-range touch sensor. (c) A spring loaded potentiometer and rod being used as an analog force touch sensor. (d) Pressure-sensitive paint used in an analog touch sensor.

Touch matrix sensor

It is made from conductive foam and laid between row and column wires. It can provide information on the shape of the object touching it, as well as on varying pressure readings at each contact point. The resistance of a row and column junction decreases as the pressure at the point increases.

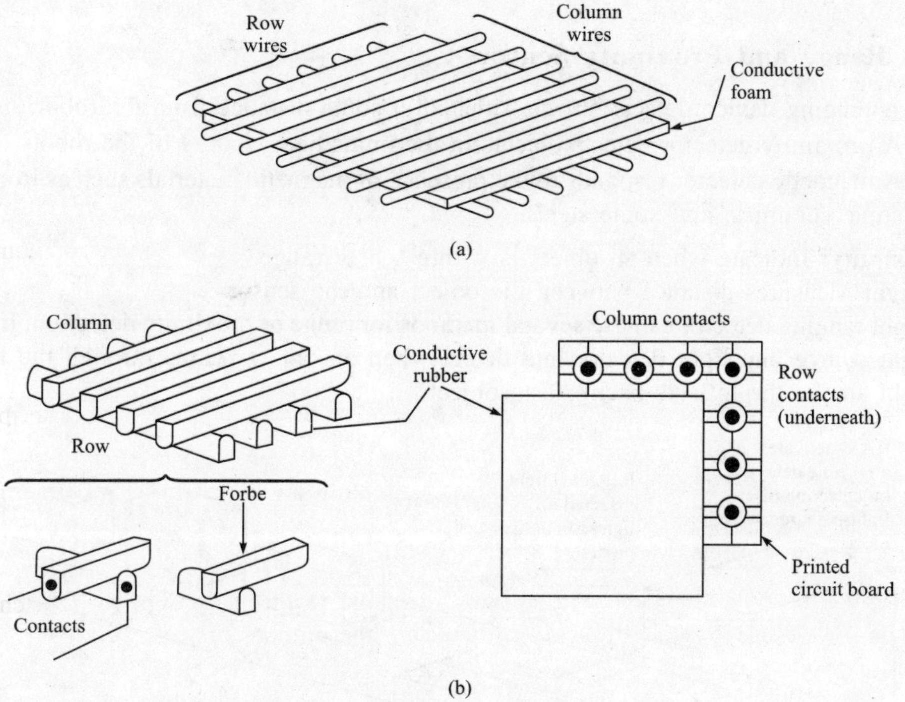

Figure 12.8 (a) A Touch Matrix Sensor. (b) Conductive Rubber Touch Sensor Matrices.

Conductive rubber touch sensor matrices

At left, rubber rods make up one type of touch matrix sensor. The rubber deforms at row-column crossings where pressure is applied. The deforming results in more surface contact and consequently lower resistance readings. At right, a sheet of conductive rubber is placed on the top of printed circuit contacts. Pressure on the rubber decreases the resistance between the contacts.

Force sensing

Force can be sensed in two methods.

Force sensing wrist: Measures three components of force (*Fx*, *Fy*, *Fz*) and their moments. The following are the steps to be followed:

1. Measure force in each axis direction
2. Calculate force offsets
3. Calculate torque applied by each axis to generate force offset at wrist

Joint sensing: (for DC servo-motor applications) Torque exerted by motors is proportional to current flowing through armature; to measure voltage drop across resistor in series with motor.

Disadvantages: Measurements made in joint space—reflect forces and torque required to accelerate links of arm, to overcome friction of joints; if joint friction is high, small forces at tool tip may be masked.

12.5.3 Range and Proximity Sensors

- A ranging device detects objects situated at some distance from the robot.
- A proximity detector detects objects in the immediate vicinity of the robots.
- A magnetic detector responds to the presence of magnetic materials such as iron, cobalt, some ceramics, and some steels.

Proximity: Indicate when an object is within sensor range.

Range: Measures distance between the object and the sensor.

A light ranging detector can use several methods for range or proximity detection. It can use: (a) a light source and light detector and detect when an object passes through the beam, or (b) a light source the reflects back off an object.

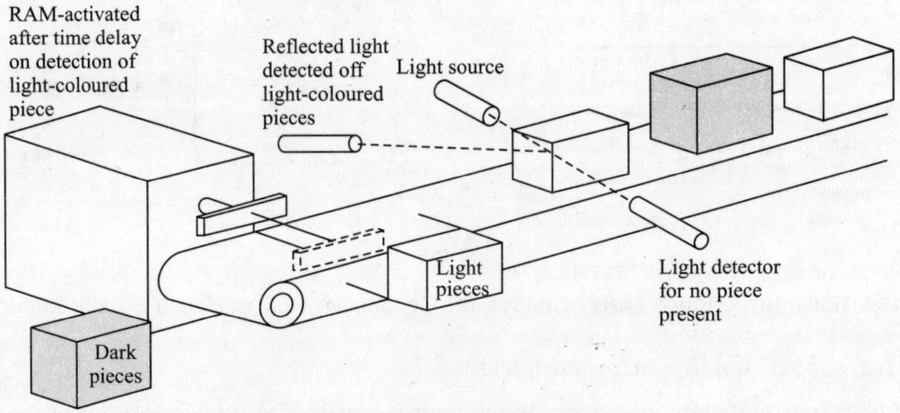

Figure 12.9 Light Ranging Detector.

Photoelectric proximity detectors can be used to sort certain parts. Here, two types of parts are distinguished by colour. Light-coloured parts reflect enough light for the second detector to notice and trigger a moving ram. Dark-coloured parts do not reflect enough light to trigger the detector.

Types of proximity sensors

(1) Inductive : Detects all metals

(2) Capacitive : Detects all materials

(3) Ultrasonic : Detects all materials

Inductive proximity sensor

1. Sensor emits a high frequency radio field projected out from sensor face; field generated by sensing face coil and oscillator circuit.
2. Radio field interrupted and reduced by presence of a metal target.
3. Target absorbs energy from oscillator field.
4. Trigger circuit senses reduction in oscillations and generates an output, it has the repeatability = +/– 1% of sensing distance.

Capacitive proximity sensor

The capacitance of an electrical circuit changes as the dielectric constant of the circuit changes. Some electric lamps equipped with a capacitive proximity detector can be turned on and off by touching them.

1. Capacitor: consists of positive and negative plates separated by an insulating (dielectric) material.
2. Power applied and generates a high frequency field between the plates.
3. Object which passes through the field will change the capacitance.
4. Can "see through" material if contents have higher dielectric constant; water: dielectric constant = 80; air: dielectric constant = 1.

$$C = (D)(A)/d$$

where, C = capacitance
D = dielectric constant of material (target)
A = area of electrode (plates)
d = distance between plates.

Ultrasonic proximity sensor

1. Sensor has acoustic transducer that vibrates at ultrasonic frequencies (> 20,000 Hz).
2. Pulses emitted and aimed at target object.
3. When target appears, pulses reflected back to sensor.
4. Time delay between emitted pulse and echo provides distance information.

12.5.4 Machine Vision

Present-day industrial robots use vision to locate and orient parts. Establishing vision for robots may seem as simple as adding a television camera to robot, but it is much more complicated. All vision systems require at least the following components: an object to be viewed, an illumination source, a vision sensor, an edge detector, a line detector, an object detector, and previously recorded information for interpreting what is being seen.

Vision processing: various steps

Figure 12.10 identifies the major steps in processing vision.

It starts with an object under some type of illumination and ends with understanding.

The illumination source for natural human vision consists of electromagnetic energy in the visible light range.

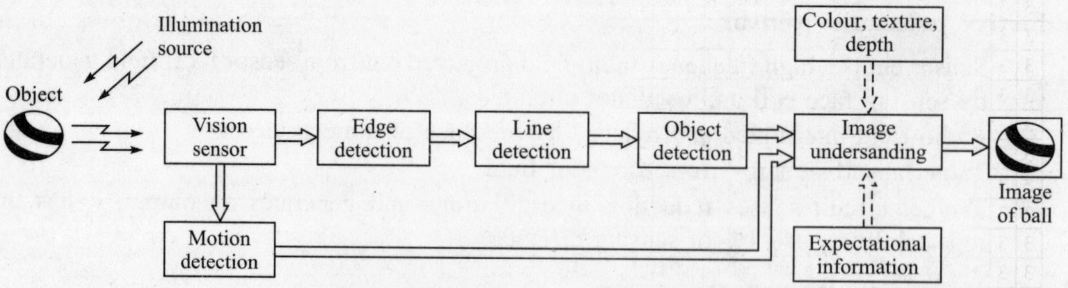

Figure 12.10 Vision Processing System.

High-contrast visual systems usually require placing a high-intensity source of illumination behind the object being viewed. Back lighting only produces a silhouette of the object.

Vision systems that have gray scales—that is, systems that show shades between black and white—generally require placing the primary source of illumination in front of the object.

Figure 12.11 below shows the effect of lighting on the subject being viewed: (a) when a subject is front-lighted, light reflects off the subject and into the vision sensor. This yields the most details about the subject. (b) If the subject is back-lighted, the subject blocks some of the light going to the vision sensor from the illumination source. This produces a high-contrast outline of the subject.

Figure 12.11 Effect of Lighting.

Most machine vision systems use some type of television camera as their vision sensor.

The most common types of solid-state television camera use charge-coupled devices (CCD), change-injection devices (CID), and photo-diode arrays.

Photo-diode cameras may use linear arrays of photo-diodes (called line-scan cameras), or two-dimensional arrays (called diode-matrix cameras).

The output of the vision system must be converted into numerical codes for storage in the system's memory.

Figure 12.12 below shows the visual information in computer memory: (a) The front-lighted ball from Figure 12.11(a) yields a range of values that are represented in computer memory as numbers from 1 to 5. (b) The high back-lighted ball from Figure 12.11(b) can be reduced to values of 1 and 0 in the computer's memory.

3	3	3	3	3	3	3	3	3	3	3	3	3	3	3	3	3
3	3	3	3	3	3	2	1	1	1	2	3	3	3	3	3	3
3	3	3	3	3	1	1	1	1	1	1	1	1	3	3	3	3
3	3	3	3	5	5	1	1	1	1	1	1	1	1	3	3	3
3	3	3	1	4	5	4	1	1	1	1	1	1	1	2	3	3
3	3	3	2	1	5	5	5	1	1	1	1	1	1	1	2	3
3	3	4	5	3	1	5	5	4	1	1	1	1	1	1	2	3
3	3	1	5	5	1	1	4	5	5	4	1	1	1	1	2	3
3	3	1	1	5	5	4	1	4	5	5	5	4	1	1	2	3
3	3	1	1	1	5	5	4	1	1	4	5	5	5	5	3	3
3	3	1	1	1	1	4	5	5	4	1	1	1	3	3	3	3
3	3	3	1	1	1	1	1	5	5	5	5	1	1	3	3	3
3	3	3	3	1	1	1	1	1	1	4	5	5	4	3	3	3
3	3	3	3	3	1	1	1	1	1	1	1	4	3	3	3	3
3	3	3	3	3	3	2	2	1	1	1	2	3	3	3	3	3

(a)

0	0	0	0	0	0	0	0	0	0	0	0	0	0	0	0	0
0	0	0	0	0	0	1	1	1	1	1	0	0	0	0	0	0
0	0	0	0	1	1	1	1	1	1	1	1	0	0	0	0	0
0	0	0	1	1	1	1	1	1	1	1	1	1	0	0	0	0
0	0	0	1	1	1	1	1	1	1	1	1	1	1	0	0	0
0	0	1	1	1	1	1	1	1	1	1	1	1	1	0	0	0
0	0	1	1	1	1	1	1	1	1	1	1	1	1	1	0	0
0	0	1	1	1	1	1	1	1	1	1	1	1	1	0	0	0
0	0	1	1	1	1	1	1	1	1	1	1	1	1	0	0	0
0	0	1	1	1	1	1	1	1	1	1	1	1	1	0	0	0
0	0	1	1	1	1	1	1	1	1	1	1	1	1	0	0	0
0	0	0	1	1	1	1	1	1	1	1	1	1	0	0	0	0
0	0	0	0	1	1	1	1	1	1	1	1	1	0	0	0	0
0	0	0	0	0	1	1	1	1	1	1	0	0	0	0	0	0
0	0	0	0	0	0	0	0	0	0	0	0	0	0	0	0	0

(b)

Figure 12.12 Visual Information.

Edge detection: It is the process of finding the boundaries between areas of different intensities of vision energy.

Figure 12.13 below shows edge detection: (a) This front-lighted figure has visible edge in places other than just around the outside sphere. (b) In this high-contrast figure, only the subject and the edge of the visual field are detectable.

1	1	1	1	1	1	1	1	1	1	1	1	1	1	1	1	1	1
1	0	0	0	0	0	0	1	1	1	1	0	0	0	0	0	0	1
1	0	0	0	0	1	1	0	0	0	0	1	1	0	0	0	0	1
1	0	0	0	0	1	0	0	0	0	0	0	0	1	0	0	0	1
1	0	0	0	1	0	1	0	0	0	0	0	0	0	1	0	0	1
1	0	0	1	1	0	0	1	0	0	0	0	0	0	0	1	0	1
1	0	0	1	0	1	0	0	1	0	0	0	0	0	0	1	0	1
1	0	0	1	0	0	1	0	0	1	1	1	1	1	1	1	0	1
1	0	1	0	1	0	0	1	0	0	0	1	1	0	0	1	0	1
1	0	1	0	0	1	0	0	1	0	0	0	0	1	0	1	0	1
1	0	1	0	0	0	1	0	0	1	0	0	0	0	1	1	0	1
1	0	0	1	0	0	0	1	0	0	1	0	0	0	1	0	0	1
1	0	0	0	1	0	0	0	1	0	0	1	0	1	0	0	0	1
1	0	0	0	0	1	0	0	0	1	1	0	1	1	0	0	0	1
1	0	0	0	0	0	1	0	0	0	0	1	1	0	0	0	0	1
1	1	1	1	1	1	1	1	1	1	1	1	1	1	1	1	1	1

(a)

1	1	1	1	1	1	1	1	1	1	1	1	1	1	1	1	1	1
1	0	0	0	0	0	0	1	1	1	1	1	0	0	0	0	0	1
1	0	0	0	0	1	1	0	0	0	0	1	0	0	0	0	0	1
1	0	0	1	1	0	0	0	0	0	0	0	1	0	0	0	0	1
1	0	0	1	0	0	0	0	0	0	0	0	0	1	0	0	0	1
1	0	0	1	0	0	0	0	0	0	0	0	0	0	1	0	0	1
1	0	1	0	0	0	0	0	0	0	0	0	0	0	1	0	0	1
1	0	1	0	0	0	0	0	0	0	0	0	0	0	1	0	0	1
1	0	1	0	0	0	0	0	0	0	0	0	0	0	1	0	0	1
1	0	1	0	0	0	0	0	0	0	0	0	0	0	1	0	0	1
1	0	1	0	0	0	0	0	0	0	0	0	0	0	1	0	0	1
1	0	0	1	0	0	0	0	0	0	0	0	0	1	0	0	0	1
1	0	0	0	1	0	0	0	0	0	0	0	0	1	0	0	0	1
1	0	0	0	0	1	0	0	0	0	0	1	1	0	0	0	0	1
1	0	0	0	0	0	1	1	1	0	1	0	0	0	0	0	0	1
1	1	1	1	1	1	1	1	1	1	1	1	1	1	1	1	1	1

(b)

Figure 12.13 Edge Detection.

Motion detection: It is the process of detecting things that are moving in the field of vision.

We can detect motion by comparing two edge-detection results. If an edge has moved or a new edge has appeared since the previous edge-detection process was performed, motion has occurred.

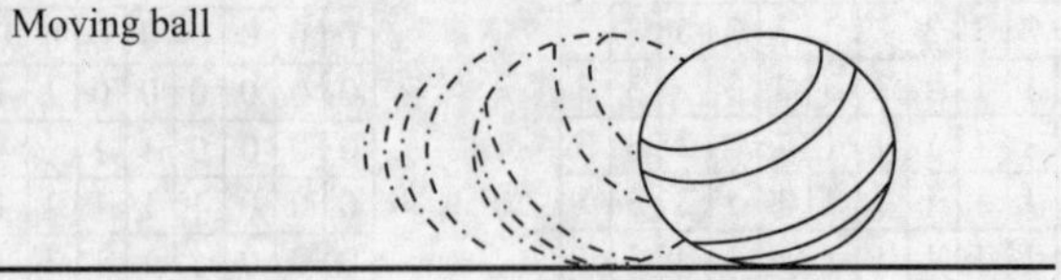

3	3	3	3	2	1	1	1	2	2	3	3	3	3	3	3	3
3	3	3	4	1	1	1	1	1	1	1	3	3	3	3	3	3
3	3	4	5	5	4	1	1	1	1	1	1	3	3	3	3	3
3	3	1	1	5	5	5	5	1	1	1	1	1	3	3	3	3
3	3	3	1	1	1	4	5	5	4	1	1	1	1	3	3	3
3	5	5	5	5	4	1	1	4	5	5	1	1	1	3	3	3
2	1	1	4	5	5	5	4	1	4	5	5	1	1	3	3	3
2	1	1	1	1	4	5	5	4	1	1	5	5	1	3	3	3
2	1	1	1	1	1	1	4	5	5	1	3	5	4	3	3	3
2	1	1	1	1	1	1	1	5	5	5	1	2	3	3	3	3
3	2	1	1	1	1	1	1	1	4	5	4	1	3	3	3	3
3	3	1	1	1	1	1	1	1	1	5	5	3	3	3	3	3
3	3	3	1	1	1	1	1	1	1	1	3	3	3	3	3	3
3	3	3	3	3	2	1	1	1	2	3	3	3	3	3	3	3
3	3	3	3	3	3	3	3	3	3	3	3	3	3	3	3	3

Old picture

3	3	3	3	3	3	3	3	3	2	2	2	2	3	3	3	3
3	3	3	3	3	3	3	3	2	1	1	1	1	5	3	3	3
3	3	3	3	3	3	3	1	1	1	1	1	1	5	1	1	4
3	3	3	3	3	3	1	1	1	1	1	1	4	5	1	1	5
3	3	3	3	3	3	1	1	1	1	1	1	5	5	1	5	5
3	3	3	3	3	2	1	1	1	1	1	4	5	4	1	5	4
3	3	3	3	3	1	1	1	1	1	1	5	5	1	4	5	1
3	3	3	3	3	1	1	1	1	1	4	5	4	1	5	5	1
3	3	3	3	3	1	1	1	1	5	5	4	1	4	5	1	1
3	3	3	3	3	2	1	1	4	5	5	1	4	5	4	1	1
3	3	3	3	3	3	1	5	5	5	1	1	5	5	1	1	1
3	3	3	3	3	3	3	5	4	1	3	5	5	1	1	1	1
3	3	3	3	3	3	3	3	1	2	5	5	1	1	1	1	3
3	3	3	3	3	3	3	3	3	4	1	1	1	1	1	3	3
3	3	3	3	3	3	3	3	3	3	3	3	3	3	3	3	3

New picture

0	0	0	0	1	2	2	2	1	0	–1	–1	–1	0	0	0	0
0	0	0	–1	2	2	2	2	1	0	0	–2	–2	2	0	0	0
0	0	–1	–2	–2	–1	2	0	0	0	0	0	–2	2	–2	–2	1
0	0	2	2	–2	–2	–4	–4	0	0	0	0	3	2	–2	–2	2
0	0	0	2	2	2	–3	–4	–4	–3	0	0	4	4	–2	2	2
0	–2	–2	–2	–2	–2	0	0	–3	–4	–4	3	4	3	–2	2	1
1	2	2	–1	–2	–4	–4	–3	0	–3	–4	0	4	0	1	2	–2
1	2	2	2	2	–3	–4	–4	–3	0	3	0	–1	0	2	2	–2
1	2	2	2	2	0	0	–3	–4	0	4	1	–4	2	2	–2	–2
1	2	2	2	2	1	0	0	–1	0	0	0	2	2	1	–2	–2
0	1	2	2	2	2	0	4	4	1	–4	–3	4	2	–2	–2	–2
0	0	2	2	2	2	2	4	3	0	–2	0	2	–2	–2	–2	–2
0	0	0	2	2	2	2	2	0	1	4	–2	–2	–2	–2	–2	0
0	0	0	0	0	1	2	2	2	1	1	–2	–2	–2	–2	0	0
0	0	0	0	0	0	0	0	0	0	0	0	0	0	0	0	0

New minus old

Figure 12.14 Motion Detection.

Motion detection involves comparing a visual image with a newer one. If the top left number grid represents the older image, and the top right number grid represents the newer image, motion can be detected by subtracting the older image values from the newer image values. The bottom number grid represents the results of this calculation. A minus value represents the cell from which a subject has departed. A plus value represents a cell to which a subject has moved. A zero value represents no change.

Object line detection: It is the process of connecting detected edges into lines.

Line detection involves locating line segments in vertical, horizontal, and both diagonal.

Figure 12.15 shows some of the line detection that could be done on the visual information about the ball.

1	0	0	0	0	0	0	1	1	1	1	1	0	0	0	0	0	1
1	0	0	0	0	0	0	1	1	1	1	1	0	0	0	0	0	1
1	0	0	0	0	0	0	0	0	0	0	1	0	0	0	0	0	1
1	0	0	1	0	0	0	0	0	0	0	0	0	0	0	0	0	1
1	0	0	1	0	0	0	0	0	0	0	0	0	0	0	0	0	1
1	0	0	1	0	0	0	0	0	0	0	0	0	0	1	0	0	1
1	0	1	0	0	0	0	0	0	0	0	0	0	0	1	0	0	1
1	0	1	0	0	0	0	0	0	0	0	0	0	0	1	0	0	1
1	0	1	0	0	0	0	0	0	0	0	0	0	0	1	0	0	1
1	0	1	0	0	0	0	0	0	0	0	0	0	0	1	0	0	1
1	0	1	0	0	0	0	0	0	0	0	0	0	0	1	0	0	1
1	0	0	0	0	0	0	0	0	0	0	0	0	1	0	0	0	1
1	0	0	0	0	0	0	0	0	0	0	0	0	1	0	0	0	1
1	0	0	0	0	0	0	0	0	0	0	0	0	0	0	0	0	1
1	0	0	0	0	0	1	1	1	0	1	0	0	0	0	0	0	1
1	0	0	0	0	0	1	1	1	0	1	0	0	0	0	0	0	1

Vertical

1	1	1	1	1	1	1	1	1	1	1	1	1	1	1	1	1	1
0	0	0	0	0	0	0	1	1	1	1	1	0	0	0	0	0	0
0	0	0	0	0	1	1	0	0	0	0	0	0	0	0	0	0	0
0	0	0	1	1	0	0	0	0	0	0	0	0	0	0	0	0	0
0	0	0	0	0	0	0	0	0	0	0	0	0	0	0	0	0	0
0	0	0	0	0	0	0	0	0	0	0	0	0	0	0	0	0	0
0	0	0	0	0	0	0	0	0	0	0	0	0	0	0	0	0	0
0	0	0	0	0	0	0	0	0	0	0	0	0	0	0	0	0	0
0	0	0	0	0	0	0	0	0	0	0	0	0	0	0	0	0	0
0	0	0	0	0	0	0	0	0	0	0	0	0	0	0	0	0	0
0	0	0	0	0	0	0	0	0	0	0	0	0	0	0	0	0	0
0	0	0	0	0	0	0	0	0	0	0	0	0	0	0	0	0	0
0	0	0	0	0	0	0	0	0	0	0	0	0	0	0	0	0	0
0	0	0	0	0	0	0	0	0	0	0	1	1	0	0	0	0	0
0	0	0	0	0	0	1	1	1	0	0	0	0	0	0	0	0	0
1	1	1	1	1	1	1	1	1	1	1	1	1	1	1	1	1	1

Horizontal

0	1	0	0	0	0	0	0	1	1	1	1	1	0	0	0	0	0
1	0	0	0	0	0	0	1	1	1	1	1	0	0	0	0	0	0
0	0	0	0	0	1	1	0	0	0	0	0	0	0	0	0	0	0
0	0	0	0	1	0	0	0	0	0	0	0	0	0	0	0	0	0
0	0	0	1	0	0	0	0	0	0	0	0	0	0	0	0	0	0
0	0	0	0	0	0	0	0	0	0	0	0	0	0	0	0	0	0
0	0	0	0	0	0	0	0	0	0	0	0	0	0	0	0	0	0
0	0	0	0	0	0	0	0	0	0	0	0	0	0	0	0	0	0
0	0	0	0	0	0	0	0	0	0	0	0	0	0	0	0	0	0
0	0	0	0	0	0	0	0	0	0	0	0	0	0	1	0	0	0
0	0	0	0	0	0	0	0	0	0	0	0	0	1	0	0	0	0
0	0	0	0	0	0	0	0	0	0	0	0	0	0	0	0	0	0
0	0	0	0	0	0	0	0	0	0	0	0	0	1	0	0	0	0
0	0	0	0	0	0	0	0	0	0	0	1	1	0	0	0	0	0
0	0	0	0	0	0	1	1	1	0	1	0	0	0	0	0	0	1
0	0	0	0	0	1	1	1	0	1	0	0	0	0	0	0	1	0

0	0	0	0	0	0	1	1	1	1	1	0	0	0	0	0	1	0
0	0	0	0	0	0	0	1	1	1	1	1	0	0	0	0	0	1
0	0	0	0	0	0	0	0	0	0	0	1	0	0	0	0	0	0
0	0	0	0	0	0	0	0	0	0	0	0	1	0	0	0	0	0
0	0	0	0	0	0	0	0	0	0	0	0	0	1	0	0	0	0
0	0	0	0	0	0	0	0	0	0	0	0	0	0	1	0	0	0
0	0	0	0	0	0	0	0	0	0	0	0	0	0	0	0	0	0
0	0	0	0	0	0	0	0	0	0	0	0	0	0	0	0	0	0
0	0	0	0	0	0	0	0	0	0	0	0	0	0	0	0	0	0
0	0	0	0	0	0	0	0	0	0	0	0	0	0	0	0	0	0
0	0	1	0	0	0	0	0	0	0	0	0	0	0	0	0	0	0
0	0	0	1	0	0	0	0	0	0	0	0	0	0	0	0	0	0
0	0	0	0	1	0	0	0	0	0	0	0	0	0	0	0	0	0
0	0	0	0	0	1	0	0	0	0	0	0	0	0	0	0	0	0
0	0	0	0	0	0	1	1	1	0	1	0	0	0	0	0	0	0
0	0	0	0	0	0	0	1	1	1	0	1	0	0	0	0	0	0

Diagonal

Figure 12.15 Line Detection.

Object shape detection is the process of assembling detected lines to see if they form some known shape.

Object shape detection involves fitting together all the pieces of lines detected earlier. This occasionally requires filling in missing pixels of a line.

Figure 12.16 fits together the pieces from line detection giving object detection.

Image understanding, the final step of vision processing, is where the system assesses whether it is seeing the right things.

0	0	0	0	0	0	0	0	0	0	0	0	0	0	0	0
0	0	0	0	0	0	1	1	1	1	0	0	0	0	0	0
0	0	0	0	1	1	0	0	0	0	1	0	0	0	0	0
0	0	0	1	0	0	0	0	0	0	0	1	0	0	0	0
0	0	1	0	0	0	0	0	0	0	0	0	1	0	0	0
0	0	1	0	0	0	0	0	0	0	0	0	0	1	0	0
0	1	0	0	0	0	0	0	0	0	0	0	0	1	0	0
0	1	0	0	0	0	0	0	0	0	0	0	0	1	0	0
0	1	0	0	0	0	0	0	0	0	0	0	0	1	0	0
0	1	0	0	0	0	0	0	0	0	0	0	0	1	0	0
0	1	0	0	0	0	0	0	0	0	0	0	0	1	0	0
0	0	1	0	0	0	0	0	0	0	0	0	1	0	0	0
0	0	0	1	0	0	0	0	0	0	0	0	1	0	0	0
0	0	0	0	1	0	0	0	0	0	1	1	0	0	0	0
0	0	0	0	0	1	1	1	0	1	0	0	0	0	0	0
0	0	0	0	0	0	0	0	1	0	0	0	0	0	0	0

Figure 12.16 Object Shape Deflection.

Here is a figure (Figure 12.17) doing image understanding for the ball.

0	0	0	0	0	0	1	1	1	1	0	0	0	0	0
0	0	0	0	1	1	0	0	0	0	1	0	0	0	0
0	0	0	1	0	0	0	0	0	0	0	1	0	0	0
0	0	1	0	0	0	0	0	0	0	0	0	1	0	0
0	0	1	0	0	0	0	0	0	0	0	0	0	1	0
0	1	0	0	0	0	0	0	0	0	0	0	0	1	0
0	1	0	0	0	0	0	0	0	0	0	0	0	1	0
0	1	0	0	0	0	0	0	0	0	0	0	0	1	0
0	1	0	0	0	0	0	0	0	0	0	0	0	1	0
0	1	0	0	0	0	0	0	0	0	0	0	0	1	0
0	0	1	0	0	0	0	0	0	0	0	0	1	0	0
0	0	0	1	0	0	0	0	0	0	0	0	1	0	0
0	0	0	0	1	0	0	0	0	0	1	1	0	0	0
0	0	0	0	0	1	1	1	0	1	0	0	0	0	0
0	0	0	0	0	0	0	0	1	0	0	0	0	0	0

0	0	0	0	0	0	1	1	1	1	0	0	0	0	0
0	0	0	0	1	1	0	1	0	0	1	0	0	0	0
0	0	0	1	0	0	0	1	0	0	0	1	0	0	0
0	0	1	0	0	0	1	0	0	0	0	1	1	0	0
0	0	1	0	0	0	1	0	0	0	1	0	0	1	0
0	1	0	0	0	1	0	0	0	0	1	0	0	1	0
0	1	0	0	0	1	0	0	0	1	0	0	0	1	0
0	1	0	0	1	0	0	0	0	1	0	0	0	1	0
0	1	0	0	1	0	0	0	1	0	0	0	0	1	0
0	1	0	1	0	0	0	0	1	0	0	0	0	1	0
0	0	1	1	0	0	0	1	0	0	0	0	1	0	0
0	0	0	1	0	0	0	1	0	0	0	0	1	0	0
0	0	0	0	1	0	1	0	0	0	1	1	0	0	0
0	0	0	0	0	1	1	1	0	1	0	0	0	0	0
0	0	0	0	0	0	0	0	1	0	0	0	0	0	0

0	0	0	0	0	0	1	1	1	1	0	0	0	0	0
0	0	0	0	1	1	1	0	1	0	1	0	0	0	0
0	0	0	1	0	1	0	1	0	0	0	1	0	0	0
0	0	1	0	1	0	0	1	0	0	0	1	1	0	0
0	0	1	0	1	0	1	0	0	0	1	0	0	1	0
0	1	0	1	0	0	1	0	0	0	1	0	0	1	0
0	1	0	1	0	1	0	0	0	1	0	0	1	1	0
0	1	1	0	0	1	0	0	0	1	0	0	1	1	0
0	1	1	0	1	0	0	0	1	0	0	1	0	1	0
0	1	1	0	1	0	0	0	1	0	0	1	0	1	0
0	0	1	1	0	0	0	1	0	0	1	0	1	0	0
0	0	0	1	0	0	0	1	0	0	1	0	1	0	0
0	0	0	0	1	0	0	1	0	0	1	1	0	0	0
0	0	0	0	0	1	1	1	0	1	0	0	0	0	0
0	0	0	0	0	0	0	0	1	0	0	0	0	0	0

Previously recorded images or templates

↖ Compare to

0	0	0	0	0	0	0	0	0	0	0	0	0	0	0	0
0	0	0	0	0	0	1	1	1	1	0	0	0	0	0	0
0	0	0	0	1	1	0	0	0	0	1	0	0	0	0	0
0	0	0	1	0	0	0	0	0	0	0	1	0	0	0	0
0	0	1	0	0	0	0	0	0	0	0	0	1	0	0	0
0	0	1	0	0	0	0	0	0	0	0	0	0	1	0	0
0	1	0	0	0	0	0	0	0	0	0	0	0	1	0	0
0	1	0	0	0	0	0	0	0	0	0	0	0	1	0	0
0	1	0	0	0	0	0	0	0	0	0	0	0	1	0	0
0	1	0	0	0	0	0	0	0	0	0	0	0	1	0	0
0	1	0	0	0	0	0	0	0	0	0	0	0	1	0	0
0	0	1	0	0	0	0	0	0	0	0	0	1	0	0	0
0	0	0	1	0	0	0	0	0	0	0	0	1	0	0	0
0	0	0	0	1	0	0	0	0	0	1	1	0	0	0	0
0	0	0	0	0	1	1	1	0	1	0	0	0	0	0	0
0	0	0	0	0	0	0	0	1	0	0	0	0	0	0	0

Present image

Figure 12.17 Image Understanding.

Image understanding is the final stage of the vision process. It requires previous experience with objects identical or similar to the one being observed. In this figure, the present image will be found to match the leftmost template most closely.

Colour, depth, and texture provide additional information about a viewed scene.

Expectational information is used to process vision towards an expected goal and to ignore all other possibilities until it becomes obvious that the expected goal cannot be reached.

Robot vision is not limited to what humans see, since robots can use radar, sonar, or infrared as an illumination source.

REVIEW QUESTIONS

1. What are the significant advantages of using a robot in a computer integrated manufacturing system?
2. Describe the elements of a robotic system.
3. How do you justify the use of a robot?
4. How do you classify robots?
5. Differentiate between a SCARA and a gantry robot.
6. Define the work envelope of a robot. How does it differ from one type of robot to another?
7. What are the different types of drives used in robots?
8. What are the different types of control systems used in robots?
9. How do you specify a robot?
10. What are the different methods of programming of robots?
11. Discuss the basic types of robot programming languages.
12. How do you integrate CAD data with robot operation?
13. Describe briefly significant application areas of robots.
14. Discuss the importance of proper presentation of work parts to the robots.
15. Determine the effective work envelope of a robot from its specifications.

Chapter 13

ENTERPRISE RESOURCE PLANNING (ERP)

13.1 INTRODUCTION

Enterprise resource planning (ERP) is the industry term used to describe a broad set of activities supported by multi-module application software that helps a manufacturer or other business manage the important parts of its business. These parts can include product planning, parts purchasing, maintaining inventories, interacting with suppliers, providing customer service, and tracking orders. ERP can also include application modules for the finance and human resources aspects of a business. In simple words, we can define ERP as **Enterprise Resource Planning** systems (ERPs) are management information systems (MISs) that integrate and automate many of the business practices associated with the operations or production and distribution aspects of a company engaged in manufacturing products or services.

Enterprise resource planning is a term derived from Manufacturing Resource Planning (MRP II) that followed material requirements planning (MRP). ERP systems typically handle manufacturing, logistics, distribution, inventory, shipping, invoicing, and accounting for a company. Enterprise Resource Planning or ERP software can aid in the control of many business activities, like sales, delivery, billing, production, inventory management, quality management, and human resources management.

ERPs are often called *back office systems* indicating that customers and the general public are not directly involved. ERPs are cross functional and wide in nature, i.e. all the functional departments are integrated into one system including manufacturing, warehouse, information technology, human resource, marketing, etc. The development of new ERP software involves the programmer, the analyst and the user. ERP software made in this method will custom build for that concern, but it is costly one in terms of money and time invalved. Hence, it is better to purchase the commercially available software with required modifications which suite the company.

The consulting is of two types:

- Business consulting
- Technical consulting.

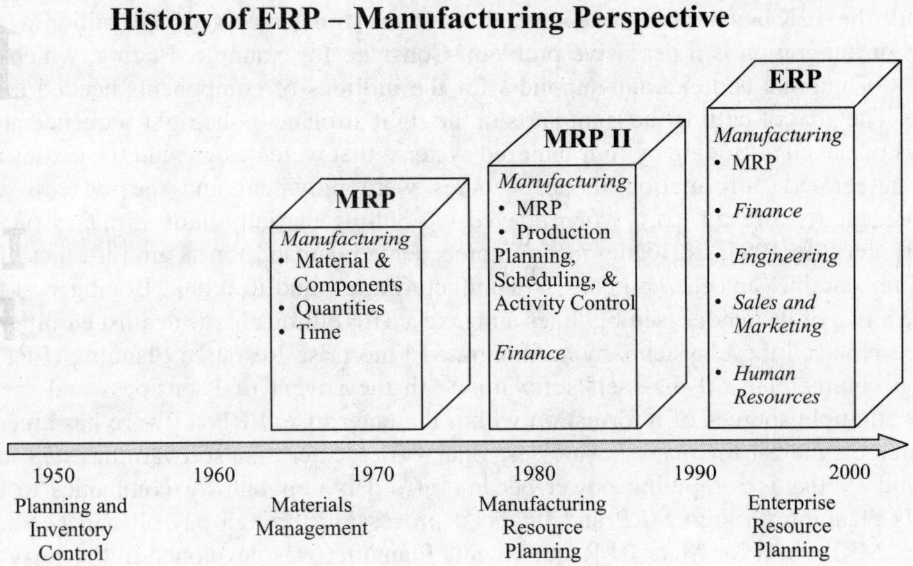

Fig. 13.1 Manufacturing Perspective.

A business consultant goes through the organizations business process completely and implements the same in software thereby he configures it.

Technical consulting often involves programming.

Today, there are also web-based ERP systems. Companies would deploy web-based ERP because they require no client side installation, and is cross-platform and maintained centrally. As long as we have an Internet connection, we can access web-based ERPs through typical web browsers.

In order to implement ERP, we need:

(a) Definition of the project
(b) Identification of the objectives
(c) Project organization
(d) Developing the workplan
(e) Business assessment
(f) Knowledge of the key persons involved
(g) Cost and benefit analysis

13.2 AN OVERVIEW OF ERP

Information in large organizations is often spread across numerous homegrown computer systems, housed in different functions or organizational units. While each of these information islands can able to support a specific business activity, enterprise-wide performance is hampered by the lack of integrated information. Further, the maintenance of these systems can result in substantial costs. For example, many of the older programs cannot properly handle dates beyond the year 2000, and they must be fixed at a steep cost or replaced.

While the Y2K bug has been fixed over time (at an estimated cost of $600 billion worldwide), the lack of integration is a pervasive problem. Consider, for example, Boeing, which relies on hundreds of internal and external suppliers for the millions of components needed to build an airplane. The goal of putting the right parts in the right airplane in the right sequence at the right time was managed at Boeing by four hundred systems that were designed in the sixties and were all but integrated. Information inconsistencies were prevalent and the systems were not synchronized. As a result, parts often arrived late, idling partially-built airplanes on Boeing's assembly lines. In 1997, as Boeing faced unprecedented demand for its aircraft, these problems became unbearable, and the company's manufacturing ground to a halt. Boeing was forced to shut down two of its major assembly lines and take a $1.6 billion charge against earnings. Boeing has since replaced these systems by an integrated Enterprise Resource Planning (ERP) system based on commercial, off-the-shelf software. With the advent of E-business and the need to leverage multiple sources of information within the enterprise, ERP software has emerged as a major area of interest for many businesses. Back-office enterprise software has its roots in the 1960s and 1970s, as computing power became affordable enough for companies to automate materials planning through MRP and financial processing through payroll and general ledger software. MRP, short for Material Requirements Planning, was developed in the early 1960s at IBM and had become the principal production control paradigm in the US.

MRP consists of a set of procedures that convert forecast demand for a manufactured product into a requirements schedule for the components, subassemblies, and raw materials comprising that product. MRP is limited to controlling the flow of components and materials, and does not lend itself to more complete production control and coordination. The next generation of manufacturing software, known as MRP II, was developed to address this shortcoming and to further integrate business activities into a common framework. MRP II divides the production control problem into a hierarchy based on time scale and product aggregation.

MRP II coordinates the manufacturing process, allowing a variety of tasks such as capacity planning, demand management, production scheduling and distribution to be linked together. However, even MRP II is primarily a specialized tool designed to serve the needs of the manufacturing function within a company. Its data and processes are not integrated with those in the rest of the enterprise, such as marketing, finance and human resources.

ERP entered the scene to facilitate information sharing and integration across these different functions, and to operate the enterprise more efficiently and effectively, using a unified data store and consistent processes.

13.2.1 What is ERP?

ERP is a software architecture that facilitates the flow of information among the different functions within an enterprise. Similarly, ERP facilitates information sharing across organizational units and geographical locations. It enables decision-makers to have an enterprise-wide view of the information they need in a timely, reliable and consistent fashion. ERP provides the backbone for an enterprise-wide information system. At the core of this enterprise software is a central database which draws data from, and feeds data into, modular applications that operate on a common computing platform, thus standardizing business processes and data definitions into a unified environment. With an ERP system, data needs to be entered only once. The system provides consistency and visibility or transparency across the entire enterprise. A primary benefit of ERP is easier access to reliable, integrated information. A related benefit is the elimination of redundant data and the rationalization of processes, which result in substantial cost savings.

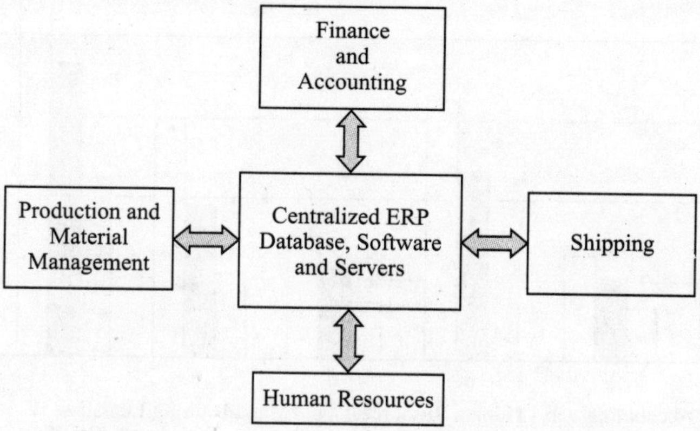

Fig. 13.2 Example of an ERP System.

The integration among business functions facilitates communication and information sharing, leading to dramatic gains in productivity and speed. Cisco Systems, for example, harnessed ERP to help it become the market leader in the global networking industry. Cisco's ERP system was the backbone that enabled its new business model, global Networked Business based on the use of electronic communications, to build interactive, knowledge-based relationships with its customers, business partners, suppliers and employees. In the process, Cisco doubled in size each year and reaped hundreds of millions of dollars in both cost savings and revenue enhancements.

Autodesk, a computer-aided design software company, reported a decrease in its order fulfillment times from two weeks to 24 hours after installing an ERP system. Similar examples abound in today's business environment.

Based on the promise of tightly-integrated corporate functions, globally optimized decisions, and fast and easy access to accurate information, enterprise software has become an essential part of the operations of large businesses in many industries. By 1998, over 20,000 firms around the world spent nearly $17 billion on enterprise software, following annual growth rates that ranged from 30% to 50%. In addition to direct spending on the software itself, companies often spend a multiple of licensing costs on services related to implementation and maintenance. Companies have started to use enterprise software to automate front-office activities such as sales and marketing, call centre operations, product configuration, lead-tracking and customer relationship management.

Following are the ERP software vendors:

- SAP AG
- Oracle
- PeopleSoft
- J.D. Edwards
- Baan

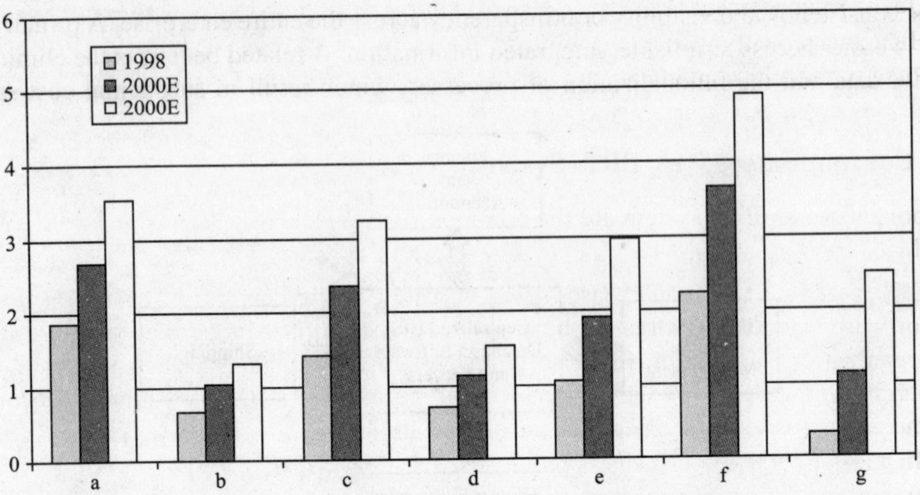

a – Financial Accounting, b – Human Resources, c – Manufacturing/Logistics,
d – Supply Chain Management, e – Customer Management, f – Industry Specific, g – e-commerce

Fig. 13.3 ERP and ERP-related Packaged Application Market 1998–2003.

Classes of Software:

1. Enterprise Resource Planning (ERP)
2. Manufacturing Execution Systems (MES)
3. Line Control Systems (LCS)
4. System Control & Data Access (SCADA)

ERP software includes the following module:

- Project module
- Manufacturing module
- Process module
- Finance module
- Distribution module
- Transporting module
- Tools module
- Service module
- Utilities module
- Materials management module
- Supply chain module

- Quality management
- HR management module
- Plant maintenance module

13.2.2 Components of an ERP System

The components of an ERP system are the common components of a Management Information System (MIS).

- **ERP software** – Module based ERP software is the core of an ERP system. Each software module automates business activities of a functional area within an organization. Common ERP software modules include product planning, parts purchasing, inventory control, product distribution, order tracking, finance, accounting and human resources aspects of an organization.
- **Business processes** – Business processes within an organization fall into three levels—strategic planning management control and operational control. ERP has been promoted as solutions for supporting or streamlining business processes at all levels. Much of ERP success, however, has been limited to the integration of various functional departments.
- **ERP users** – The users of ERP systems are employees of the organization at all levels—from workers, supervisors, mid-level managers to executives.
- **Hardware and operating systems** – Many large ERP systems are UNIX based. Windows NT and Linux are other popular operating systems to run ERP software. Legacy ERP systems may use other operating systems.

13.2.3 How to make ERP Implementation Success?

The implementation of ERP system is a complicated undertaking. According to R. Michael Donovan, only 23 per cent of the users was satisfied with the ERP they currently operated. In this section, we would like to make some suggestions about how to make the ERP in the right way.

1. **Understand business process well before implementation:** In many firms, the business process is not reasonable to follow. For example, many manufacturer fail to realize that they need to redefine the business process to improve their supply chain environment. The purpose of ERP implementation is to improve the business process that supports the company's strategic opportunities.
   Therefore, the firm should cooperate with ERP vendor or implementation consultant to make the process in the right way, such as reorganize the structure of the firm, modify the operating turn, and so on.
2. **Establish project management office:** A successful implementation should be led by Project Management Office (PMO), which is comprised of senior executives of the firm and implementation consultants. "*A PMO can be a central organization with responsibility for management, oversight, and tracking either for a significant project (like ERP implementation) or for all of the projects of a department or an enterprise.*" (Ron. Axam and Darren Jerome, 2003)

The establishment of a PMO can help the firm obtain the success of the project implementation. According to Gartner, it is the best practice to deliver success. "*PMO can design the project, manage the whole project and coordinate every people in the company*."

In the ERP implementation, it plays the following roles.

- Project management solution architecture: PMO will have the right to define the process, technology and others to meet the strategy success of the organization.
- Process companion: It is important for the PMO to deliver the value to the stakeholders based on the feedback of the employees.
- Mentor and coach: It is the PMO's responsibility to promote the understanding of the project to the employees and stakeholders of the organization.
- Facilitator: The PMO should work directly with the project implementation team. Discussing important matters such as resource requirements, project plans, and so on.
- Knowledge broker: PMO must ensure that all-important input and other analytical reports are made available to the stakeholders.

3. **Provide a systematic training:** Systematic training to the customers is very important in the whole implementation process. Every ERP vendor cannot underestimate the use of training. Without detailed training, the "big" system cannot be used properly. Since the ERP is much complicated, ERP vendor must provide training to the people who will operate the system. Different levels of training should be offered to the people who operate at different levels of the system. It is better if every user can input the data and let the data run in the whole system, to see whether the ERP system can generate reports that organizations are willing to get.
4. **Technical suggestion of implementation**
   - Schedule the installation. The firm and ERP implementation team should firstly schedule the installation. Many problems will arise at the start phase. It will always be longer than the schedule time. Both firm and vendor should try to minimize the additional time.
   - Testing before going live. The successful application of ERP system in other firms does not mean that it can be used successfully in your firm. Therefore, testing is a very important part of ERP implementation. ERP vendors should use various tools to test the function of the whole system to ensure that the system works in different environment.
   - Stress specific configuration. Detailed specification includes not only the information about the features and functions but also requirements concerning response times (Theodore Grossman and James Walsh). We can understand that sometimes ERP system will not be able to satisfy some requirements the firm raises. Therefore, the vendor has to program a new code to complement the system. Every firm will also pay more attention to the response time. It is not acceptable that users sit at a computer and wait a long time to get what they need. According to Theodore Grossman and James Walsh, response time frequently is the difference between the success and the failure of an implementation of ERP system.

- Structured methodology of implementation. Successful implementations need a structured plan, covering strategy, people and process. It can provide a detailed guidance to both firm and organization. Even if problems pop up, the implementation team should be able to handle those. Following a sound methodology will increase your percentage of success, although it cannot guarantee the success, it can decrease the effect that the failure brings to the whole project.

13.2.4 Advantages of ERP

In the absence of an ERP system, a large manufacturer may find himself with many software applications that do not talk to each other and do not effectively interface. Tasks that need to interface with one another may involve:

- Design engineering (how best to make the product)
- Order tracking from acceptance through fulfilment
- The revenue cycle from invoice through cash receipt
- Managing interdependencies of complex Bill of Materials
- Tracking the 3-way match between purchase orders (what was ordered), inventory receipts (what arrived), and costing (what the vendor invoiced)
- The accounting for all of these tasks, tracking the revenue, cost and profit on a granular level.

Change how a product is made, in the engineering details, and that is how it will now be made. Effectively, dates can be used to control when the switch over will occur from an old version to the next one, both the date that some ingredients go into effects and the date that some are discontinued. Part of the change can include labelling to identify version numbers.

Computer security is included within an ERP, to protect against both outsider crime, such as industrial espionage and insider crime, such as embezzlement. A data tampering scenario might involve a terrorist altering a Bill of Materials so as to put poison in food products, or other sabotage. ERP security helps prevent abuse as well.

There are concepts of Front office (how the company interacts with customers), which includes CRM or customer relationship management; Back end (internal workings of the company to fulfil customer needs), which includes quality control, to make sure there are no problems not fixed, in the end products; supply chain (interacting with suppliers and transportation infrastructure). All of these can be integrated through an ERP, although some systems have gaps in comprehensiveness and effectiveness. Without an ERP that integrates all these, it can be quite complicated for a manufacturer to handle.

13.2.5 Disadvantages of ERP

Many of the problems that organizations have with ERP systems are due to the inadequate level of investment in ongoing training for all personnel involved, including those implementing and testing changes, as well as a lack of corporate policies protecting the integrity of the data held in the ERP systems and how it is used.

13.2.6 Limitations of ERP

Following are the limitations of ERP.

- Success depends on the skill and experience of the workforce, including training about how to make the system work correctly. Many companies cut costs by cutting training budgets. Privately owned small enterprises are often undercapitalized, meaning their ERP system is often operated by personnel with inadequate education in ERP in general, such as APICS foundations, and in particular ERP vendor package being used.
- Personnel turnover; companies can employ new managers lacking education in the company's ERP system, proposing changes in business practices that are out of synchronization with the best utilization of the company's selected ERP.
- Customization of the ERP software is limited. Some customization may involve changing of the ERP software structure which is usually not allowed.
- Re-engineering of business processes to fit the "industry standard" prescribed by the ERP system may lead to a loss of competitive advantage.
- ERP systems can be very expensive to install.
- ERP vendors can charge sums of money for annual license renewal that is unrelated to the size of the company, using the ERP or its profitability.
- Technical support personnel often give replies to callers that are inappropriate for the caller's corporate structure. Computer security concerns arise, for example, when telling a non-programmer how to change a database on the fly, at a company that requires an audit trail of changes so as to meet some regulatory standards.
- ERPs are often seen as too rigid, and difficult to adapt to the specific workflow and business process of some companies—this is cited as one of the main causes of their failure.
- Systems can be difficult to use.
- The system can suffer from the "weakest link" problem—inefficiency in one department or one of the partners may affect other participants.
- Many of the integrated links need high accuracy in other applications to work effectively. A company can achieve minimum standards, then over time "dirty data" will reduce the reliability of some applications.
- Once a system is established, switching costs are very high for any one of the partners (reducing flexibility and strategic control at the corporate level).
- The blurring of company boundaries can cause problems in accountability, lines of responsibility, and employee morale.
- Resistance in sharing sensitive internal information between departments can reduce the effectiveness of the software.
- There are frequent compatibility problems with the various legacy systems of the partners.
- The system may be over-engineered relative to the actual needs of the customer.

REVIEW QUESTIONS

1. Define ERP.
2. What are the various stages involved in ERP?
3. What are the basic requirements to implement ERP?
4. Discuss the issues related to ERP.
5. Define MRP.
6. List out the various modules involved in ERP.
7. Explain briefly the components of ERP.
8. How do we make ERP implementation a success?
9. Name any three limitations of ERP.
10. What are the merits of ERP?
11. Compare MRP II and MRP.
12. Discuss the application of ERP in auto desk.
13. What is PMO?
14. What is MIS?
15. List out the ERP packages involved in industry.

Chapter 14

PRODUCT DATA MANAGEMENT

14.1 DEFINITION OF PRODUCT LIFE-CYCLE MANAGEMENT (PLM)

Computer Integrated Manufacturing data defines PLM as:

- A strategic business approach that applies a consistent set of business solutions that support the collaborative creation, management, dissemination, and use of product definition information
- Supporting the extended enterprise (customers, design and supply partners, etc.)
- Spanning from concept to end of life of a product or plant
- Integrating people, processes, business systems, and information

It is important to note that PLM is not a definition of a piece, or pieces of technology. It is a definition of a business approach to solving the problem of managing the complete set of product definition information—creating that information, managing it through its life, and disseminating and using it throughout the life cycle of the product. PLM is not just a technology, but is an approach in which processes are as important, or more important than data. It is critical to note that PLM is as concerned with "how a business works" as with "what is being created".

Three core or fundamental concepts of PLM are:

1. Universal, secure, managed access and use of product definition information
2. Maintaining the integrity of that product definition and related information throughout the life of the product or plant
3. Managing and maintaining business processes used to create, manage, disseminate, share and use the information.

While information includes all media (electronic and hard copy), PLM is primarily about managing the digital representation of that information.

Based on user experience over the years, PLM solutions can support a broad range of "products." Examples of "products" include manufactured products, such as automobiles, computers, refrigerators, mobile phones, toys, and airplanes. Many products today also contain software, firmware, and electronic components whose data must be managed. Some organizations have long-lived assets that need to be managed such as utility distribution networks, e.g. power, telecommunications, water, gas, and cable TV, or facilities like plants, drilling rigs, buildings, airports, harbors, railway systems, and logistics warehouses. Other "products" include bridges, highways, and other civil engineering projects. Organizations across many industrial sectors have successfully used PLM solutions to manage product information across the life cycle for all of these "products".

In the 1990s, this life cycle view expanded from managing primarily the mechanical elements of a product's definition to include the electronics and software elements that have become a greater portion of many products. That expansion continued to push the perception of what "design" encompassed. PLM includes management of all product-related information from requirements, through design, manufacturing, and deployment. This information ranges from marketing requirements, product specifications, and test instructions and data, to the as-maintained configuration data from the field. The PLM solution links information from many different authoring tools and other systems to the evolving product configuration. At the same time, the life cycle began to include production-focused attributes and information.

Today, PLM encompasses significant areas of process. It is not just program and project management processes. It is also the processes required to manufacture the product or plant, operate it in the field, and dispose or decommission it at the end of its useful life. PLM solutions help define, execute, measure, and manage key product-related business processes. Manufacturing and operational process plans are also now viewed as an inherent part of PLM. Processes, and the workflow engines that control them, ensure complete digital feedback to both users and other business systems throughout each life cycle stage.

14.2 PDM IMPLEMENTATION—MANAGING A CHANGING ENVIRONMENT

Dynamic young engineers who have just been appointed to lead the project to select and implement a PDM system will know that they have an exciting assignment. To help them succeed, here are a few guidelines to help them through the complex issues of a changing product development and engineering environment. Although some people will look forward to using the new PDM system, what about all those product developers who have trouble using e-mail, spreadsheets, simple computer-based project management tools, and basic word processing systems? It is unlikely they are going to be great PDM users. Similarly, those people who have spent the last thirty years developing (and modifying) product specifications alone in their office using the back of an envelope?

PDM is not a 'tool to improve the product development process'. It is a 'tool to support an improved product development process'. Before it can work well, there must be other changes. Before a company can use PDM effectively, it must have an environment in which it is possible to work well with PDM. Simply buying and installing a system is not going to result in major benefits unless the overall environment (organizational structures, people, organizational methods, infrastructure, etc.) is supportive. To get to this environment requires the application of 'tools for change' such as learning, leadership, communication and the right reward systems. Maybe this sounds too theoretical, but imagine you have to make massive change in your organization—in which most people feel their world is falling apart and they are doing all they can to hang on for dear life. How are you going to get them to stop doing the things they have succeeded with over the last 20 years (such as working with a paper-based document control system), and instead work with a new tool such as PDM?

If people over 40 in the organization are seeing the whole world changing and their company losing ground to its competitors, do you expect them to quickly learn something new? How many people over 50 can see what Internet could mean for the supply chain? How are you going to get someone in engineering to work in a concurrent engineering team led by someone from manufacturing? Do you think it matters to people who decides on their bonus—Engineering, Manufacturing, the team, the team leader? So, what should you do? To introduce change, you have to:

- Understand the need for change
- Accept the need for change and the difficulty of implementing change
- Understand that you personally have to change
- Understand that 'change' is a major activity in its own right
- Find out where you want the changes to lead
- Find out how to carry out the changes
- Carry out change activities

Then implement new tools and techniques such as PDM. The term 'engineering data' includes all data related both to a product and to the processes that are used to specify, develop, produce and support it. Examples of engineering data include specifications, schedules, process plans, technical manuals, project plans, bills of materials, descriptive geometry, formulae, test and quality results, results of analytic experiments and calculations, computer programs, photographs, drawings, sketches, video data and voice data.

Some engineering data (e.g. part geometry) is created within the engineering function, some is created elsewhere (e.g. in field tests). Some of the data (e.g. stress or circuit analysis results) is used within the engineering function, some (e.g. welding instructions) is used elsewhere. Engineering data can be found in many locations and on many media. There is a lot of it (some companies have many hundred thousand documents and many million megabytes of data).

It comes in many forms (e.g. numeric, graphic, alphanumeric). Some of the users of data will be inside a company, others outside (e.g. suppliers and customers).

Engineering data is used in many activities throughout the life cycle of a product.

Engineering data are on a variety of media. In a typical engineering organization, some of the data will be on traditional media such as paper and aperture cards. Some will be on electronic media. Some may be on an Intranet. Some may be on the Web. Some data may be on **mylar or microfilm**. Some may be on **paper tape.** Some data may be on magnetic tape. Some may be on magnetic disk, some on optical disk. Due to the diversity of computers, operating systems, storage devices and storage techniques, there are many differences between the various electronic media. Data on different media have to be managed in different ways.

14.2.1 The Difficulty of Change—PDM

Change is difficult, time-consuming and costly—yet for today's product development organizations it is necessary and offers tremendous potential for improved performance. As they try to respond, many organizations get caught in the trap between fear and greed. They would like the benefits of change, but they are afraid of failure. So they try for low-cost change, exhorting their managers to do their best. The managers nod their heads in agreement, and then go away and do whatever they were doing before—and change does not happen. If you expect a new product to have a significant effect on the organization, you set up a specific project.

It is the same with change. If you want a change to have a significant effect on the organization, you have to set up a change project. Change is an activity in its own right. There is almost a physical barrier to overcome before many people can understand that change will not happen unaided. Although they may understand the need for change, they do not seem to be able to respond. Often, they do not really accept that change for the organization implies that they personally have to change—and they do not understand that 'change' is an activity in its own right.

Does this have anything to do with PDM? Maybe yes, maybe no. In a particular organization it may be that the implementation of a simple PDM system will result in clear benefits without any major changes being necessary—and everyone will be pleasantly surprised how easy the implementation is. Yet in many cases, things are not so simple. An attempt to introduce a system stagnates because it is linked to changes in processes and working procedures, it calls for new skills and interfaces, and it would change old habits. What appears to be the simple introduction of a great new system gradually becomes a major source of problems.

When most companies are faced with a major development activity, they will make sure they have a well-defined objective, clear specifications, a project team, a project manager, a project plan, project phases, test plans, etc.

For mathematicians, the equation for change is easy to write:

$$(New\ state) = (Change\ matrix) \times (Old\ state)$$

14.3 THREE TOOLS OF CHANGE

The three tools of change—communication, learning and reward systems—are often ignored in a PDM implementation, but without them change is unlikely to succeed. Look at them in a general context of change, remembering that change is driven by a very small percentage of the people in the organization. Most people are going to have change thrust upon them—they will be 'the targets of change', 'the victims of change', and 'the people to be changed'. Do not forget that most of these people will be quite happy with the current state of things.

People may complain a lot about the current state, and about their managers and colleagues, but they do accept the current state and live with it. Someone, however, would like to change this stable, generally accepted environment. Even where you are sitting, you can almost hear the screams of the potential victims of change. 'This company has made us do the same thing for the last 20 years and now it asks us to change because apparently that was the wrong thing to do. After 20 years of doing it one way, how can you expect us to change and do it another way?'

'Every few years there is a great new idea from corporate and we are all supposed to change—one year it was quality circles, then it was customer focus. After a few months, these programs always faded away, and everything went back to normal. To overcome the quite legitimate fears and concerns underlying this type of question, it is important to communicate what is changing and why. It is important to give people the opportunity to learn new skills, and it is important to change the reward system so that people can see that changed behaviour will be rewarded and unchanged behaviour will not. Let us look at these three subjects one by one.

It is important to communicate to people why change is needed. Probably the message will be part stick and part carrot—partly describing the dangers of staying in the current 'as-is' state, and partly showing the advantages of moving to the future 'to-be' state. The communication needs to show what the future state will look like, how it will be reached, and what things will look like along the way. It needs to show that the process of change has been clearly thought out, is being led, and is under control.

However, even if the above messages are well communicated and well received, it does not follow that anyone is going to change. People may have done a good job in the past, but that job is going to change in the future—and are they able to do the new job? Do they know how to do their new job? Probably not, so they have to be given the opportunity to learn new skills, new tools, new behaviour and new roles.

However, even if the message is well communicated and well received, and people are given the opportunity to learn so they can behave differently in the new environment, it does not follow that anyone is going to change their behaviour. First, they will see on what basis the carrots and gold stars are going to be distributed. If they see that the rewards are going to be handed out the way they always were, then they are going to behave the way they always did, and they are not going to change.

So, communication is good and necessary, but it is not enough. Communication and learning are good and necessary, but they are not enough. Communication, learning and new reward systems are all needed.

It is important to identify each activity in the product development process, its objective, and its position in the overall flow. If necessary, the activity should be broken down into its constituent tasks. Identify who is involved in the activity, how long it takes, and how often it is carried out. For each activity, the information input created is used and output should be described, as should the sources of information and the definitions of information.

Any information systems used in the activity should be identified, their information requirements described, and their interfaces with other systems described. Management style, procedures and performance measures associated with the activity should be described. The power structure and the culture of the people in the activity should be described. Problems with current operations or non-conformance with documented procedures should be noted, as should suggestions for improvement from the people involved in the activity.

As the individual activities are examined, the information requirements of each activity will become clear. The main parameters about the volume of information will become clear. These will include the number of existing products and parts, the annual number of new products and parts, the number of modifications, and the number of new and modified drawings and other documents.

It will also include the number of drawings released daily, typical design times, the number of engineering changes, the timing of engineering changes, the time taken to process engineering changes, and the number of levels and constituents of bills of materials. The amount of data created each year by computer-based systems needs to be calculated.

The sources and users of engineering/product information will be identified. The way in which users create, access, modify, store and communicate information should be described.

The access needs and rights of users and groups of users need to be understood. Shared and redundant data needs to be identified.

As the current situation is examined, it will be possible to build a picture of the current organization of the company from the point of view of engineering/product information. This will show the number of users and their location, both geographically and functionally, and the way they store and communicate information.

Besides, looking at information use and flow along the product life cycle, it is also necessary to look at the structure of product life cycles, and identify the typical milestones, events, and management activities associated with them. There will be different structures for different product lines, and, no doubt, some individual projects will follow specific rules. The project management techniques in use should be identified. Project status and review needs should be identified.

The management of engineering/product information, in particular at departmental boundaries, needs to be understood, as do data security and data integrity issues. Existing data management systems need to be examined. Transition rules between different states of information must be classified. The rules vary along the product life, from the product concept, when the information's owner can modify it at will, to the time when the product is in the customer's hands, and information can only be modified if strict conditions are met.

Information repositories need to be identified. They will contain engineering drawings, CAD files, text documents and other forms of engineering/product information. The current methods of creating, numbering, classifying, communicating and storing information need to be understood and quantified. The cost structure for preparation and distribution of information should be understood.

14.4 IMPLEMENTING CIM SOFTWARE

Computer integrated manufacturing (CIM) combines manufacturing hardware and software technologies to integrate product, process, and manufacturing management information into a single interactive network, greatly reducing the number of "transactions" necessary to produce a product. Although only in the initial stages of industrial implementation, CIM systems are significantly increasing productivity and lowering manufacturing costs by linking previously independent portions of the production cycle such as computer-aided design terminals and numerically controlled machine tools (usually integrated manufacturing cells) that actually produce the finished part.

CIM incorporates a number of other technologies that are industries unto themselves, such as CAD/CAM, machines tools, controllers, material handling equipment, data management software, and robotics. CIM support software allows the movement of information between parts of the manufacturing process, and is thus the key to making CIM work.

Special characteristics

European developers have made rapid gains during the past few years and are at rough technological parity with the United States in CIM technology. Japan will probably continue to lag in CIM because of the inability to link the various portions of the manufacturing cycle through sophisticated software. Two key factors affecting the future of CIM technologies will be the battle for dominance of operating systems and the adoption of international data standards. Standardization of manufacturing data would greatly facilitate the exchange of information and, in turn, ease the implementation of CIM.

14.5 EDM/PDM

People often ask about the best way to introduce EDM/PDM. 'Should we do EDM/PDM top-down or bottom-up?' 'Should we do a Big Bang implementation, or should we introduce EDM/PDM step-by-step?' 'Should we implement EDM/PDM only in the Engineering Department, or should we implement it company-wide?' 'Do we have to build a prototype?' 'Should we do a feasibility study?'

The answers to some of these questions are company-specific. They depend on a mixture of factors such as the industry sectors the company competes in, the company's organization and culture, the company's computer systems, its personnel, and the company's experience of implementing EDM/PDM. Answers to other questions are less company-specific.

One of the most important factors in choosing the approach is the level of EDM/PDM understanding in the company. If the company understands very little about EDM/PDM, then a feasibility study will be helpful. If it has a basic, but incomplete, understanding, then a prototype will be helpful. If it really understands how EDM/PDM will be used, then it may need neither feasibility study nor prototype.

In a Big Bang approach, a company would switch overnight from only using manual data management to only using computer-based EDM/PDM. Unless a company is extremely well-organized, and has all its processes in order, a Big Bang approach should not be tried. Mega-projects like this rarely work. A step-by-step approach is much more likely to lead to success. Although a Big Bang approach may appear quick and simple, it is generally the opposite. The Big Bang only comes after a very long preparation and planning phase, during which people lose interest and are demotivated by the lack of results. A step-by-step approach, starting with an overall multi-step plan, is more likely to succeed. As each step is implemented, people can see and appreciate the progress, and should any problems occur, these can be resolved before they get out of hand.

The decision as to whether the initial implementation of EDM/PDM should be in the engineering department, or elsewhere, will vary from one sector to another. In some industry sectors there is enough pressure from competitors and customers to overcome any doubts about getting started with EDM/PDM. In these sectors, engineering is often a strategically important function, and able to act fairly independently. As a result, there is a tendency for EDM/PDM to be implemented bottom-up, starting within the engineering department, and spreading out to other areas of the company. The opposite occurs in sectors where engineering is less important, and the engineering department is comparatively weak. In these cases, top management may be the driving force behind EDM/PDM, and may view initial use in the engineering department as an uninspiring approach.

The choice between implementing EDM/PDM in one department or cross-functionally depends a lot on the current organizational structure. Companies that are used to working with cross-functional teams and projects will find it easier to implement EDM/PDM cross-functionally than those where each department is surrounded by high walls. For the former, a cross-functional project will be a good place to start a prototype of EDM/PDM. In a company where the departments are very separate, cross-functional EDM/PDM will be nearly impossible to implement, so EDM/PDM will probably start within a single department, even though this may make it more difficult to go cross-functional later.

Another factor that weighs heavily in the decision is the skill level in the company. There is no point in suggesting a company-wide or Big Bang approach if the company does not have the people available to support or use it, and cannot afford the money to buy in outsiders to do the work. EDM/PDM requires a lot of efforts for implementation, training, data load, user support, and process improvement.

The requirements at each stage of the product life cycle are different, and different skills are needed at each part of the process. One or two supermen can not do it all alone. Perhaps the most important factor in choosing the approach is the EDM/PDM experience so far. If the general level of EDM/PDM awareness is low, then the approach should be cautious and include a lot of training.

If there has been little support from the management, it is unlikely that a top-down approach will be appropriate. If different departments cannot agree on what they want to do with EDM/PDM, then a company-wide approach will be unsuccessful. If the engineering department has not shown any interest, it would not be sensible to propose starting with EDM/PDM in engineering. If no one has taken the work done so far very seriously, it is probably too early to start thinking about implementing EDM/PDM, and the best approach is to try to increase awareness.

REVIEW QUESTIONS

1. What is product data management?
2. What are the three tools used in PDM?
3. Define engineering data.
4. List out any five CIM softwares.
5. What is the benefit of CIM software?

Chapter 15

ARTIFICIAL INTELLIGENCE AND SIMULATION

15.1 INTRODUCTION

Artificial intelligence (AI) is increasingly becoming an important part of everyday life. It is helping people in every field to make better use of information: in the cruise control of cars, the servers that route our email, and medical tools that can automatically assess electrocardiograms. Commonwealth Scientific and Industrial Research Organization is exploring the scientific basis of thought and intelligent behaviour, and using these qualities in machines. Our involvement in Australian artificial intelligence research has resulted in the development of sensors, electronic circuitry, software and databases. These technologies have found application in areas as diverse as medicine, building design and construction.

In a 1977 article, the late AI pioneer Allen Newell predicted a time when the entire man-made world would be permeated by systems that cushioned us from dangers and increased our abilities: smart vehicles, roads, bridges, homes, offices, appliances, even clothes. Systems built around AI components will increasingly monitor financial transactions, predict physical phenomena and economic trends, control regional transportation systems, and plan military and industrial operations. Basic research on common sense reasoning, representing knowledge, perception, learning, and planning is advancing rapidly, and will lead to smarter versions of current applications and to entirely new applications. As computers become ever cheaper, smaller, and more powerful, AI capabilities will spread into nearly all industrial, governmental, and consumer applications.

Moreover, AI has a long history of producing valuable spin-off technologies. AI researchers tend to look very far ahead, crafting powerful tools to help achieve the daunting tasks of building intelligent systems. Laboratories, whose focus was AI, first conceived and demonstrated such well-known technologies as the mouse, time-sharing, high-level symbolic programming languages (Lisp, Prolog, Scheme), computer graphics, the graphical user interface (GUI), computer games, the laser printer, object-oriented programming, the personal computer, email, hypertext, symbolic mathematics systems (Macsyma, Mathematica, Maple, Derive), and, most recently, the software agents which are now popular on the World Wide Web. There is every reason to believe that AI will continue to produce such spin-off technologies.

Intellectually, AI depends on a broad intercourse with computing disciplines and with fields outside computer science, including logic, psychology, linguistics, philosophy, neuroscience, mechanical engineering, statistics, economics, and control theory, among others. This breadth has been necessitated by the grandness of the dual challenges facing AI—creating mechanical intelligence and understanding the information basis of its human counterpart. AI problems are extremely difficult, far more difficult than was imagined when the field was founded. However, as much as AI has borrowed from many fields, it has returned the favour: through its interdisciplinary relationships, AI functions as a channel of ideas between computing and other fields, ideas that have profoundly changed those fields. For example, basic notions of computation such as memory and computational complexity play a critical role in cognitive psychology, and AI theories of knowledge representation and search have reshaped portions of philosophy, linguistics, mechanical engineering, and control theory.

15.2 HISTORICAL PERSPECTIVE

Early work in AI focused on using cognitive and biological models to simulate and explain human information processing skills, on "logical" systems that perform common sense and expert reasoning, and on robots that perceive and interact with their environment. This early work was spurred by visionary funding from the Defense Advanced Research Projects Agency (DARPA) and Office of Naval Research (ONR), which began on a large scale in the early 1960s and continues to this day. Basic AI research support from DARPA and ONR—as well as support from NSF, NIH, AFOSR, NASA, and the US Army beginning in the 1970s—led to theoretical advances and to practical technologies for solving military, scientific, medical, and industrial information processing problems.

By the early 1980s, an "expert systems" industry had emerged, and Japan and Europe dramatically increased their funding of AI research. In some cases, early expert systems success led to inflated claims and unrealistic expectations, while the technology produced many highly effective systems, it proved very difficult to identify and encode the necessary expertise. The field did not grow as rapidly as the investors had been led to expect, and this translated into some temporary disillusionment. AI researchers responded by developing new technologies, including streamlined methods for eliciting expert knowledge, automatic methods for learning and refining knowledge, and common sense knowledge to cover the gaps in expert information. These technologies have given rise to a new generation of expert systems that are easier to develop, maintain, and adapt to changing needs.

Today, developers can build systems that meet the advanced information processing needs of government and industry by choosing from a broad palette of mature technologies. Sophisticated methods for reasoning about uncertainty and for coping with incomplete knowledge have led to more robust diagnostic and planning systems. Hybrid technologies that combine symbolic representations of knowledge with more quantitative representations inspired by biological information processing systems, have resulted in more flexible, human-like behaviour. AI ideas also have been adopted by other computer scientists—for example, "data mining," which combines ideas from databases, AI learning, and statistics to yield systems that find interesting patterns in large databases, given only very broad guidelines.

15.3 CASE STUDIES

The following four case studies highlight the application areas where AI technology is having a strong impact on industry and everyday life.

15.3.1 Authorizing Financial Transactions

Credit card providers, telephone companies, mortgage lenders, banks, and the US Government employ AI systems to detect fraud and expedite financial transactions, with daily transaction volumes in the billions. These systems first use learning algorithms to construct profiles of customer usage patterns, and then use the resulting profiles to detect unusual patterns and take the appropriate action (e.g. disable the credit card). Such automated oversight of financial transactions is an important component in achieving a viable basis for electronic commerce.

15.3.2 Configuring Hardware and Software

AI systems configure custom computer, communications, and manufacturing systems, guaranteeing the purchaser maximum efficiency and minimum set-up time, while providing the seller with superhuman expertise in tracking the rapid technological evolution of system components and specifications. These systems detect order incompleteness and inconsistencies, employing large bodies of knowledge that describe the complex interactions of system components. Systems currently deployed process billions of dollars of orders annually; the estimated value of the market leader in this area is over a billion dollar.

15.3.3 Diagnosing and Treating Problems

Systems that diagnose and treat problems—whether illnesses in people or problems in hardware and software—are now in widespread use. Diagnostic systems based on AI technology are being built into photocopiers, computer operating systems, and office automation tools to reduce service calls. Stand-alone units are being used to monitor and control operations in factories and office buildings. AI-based systems assist physicians in many kinds of medical diagnosis, in prescribing treatments, and in monitoring patient responses. Microsoft's Office Assistant, an integral part of every Office 97 application, provides users with customized help by means of decision-theoretic reasoning.

15.3.4 Scheduling for Manufacturing

The use of automatic scheduling for manufacturing operations is exploding as manufacturers realize that remaining competitive demands an ever more efficient use of resources. This AI technology—supporting rapid rescheduling up and down the "supply chain" to respond to changing orders, changing markets, and unexpected events—has shown itself superior to less adaptable systems based on older technology. This same technology has proven highly effective in other commercial tasks, including job shop scheduling, and assigning airport gates and railway crews. It also has proven highly effective in military settings—DARPA reported that an AI-based logistics planning tool, DART, pressed into service for operations Desert Shield and Desert Storm, completely repaid its three decades of investment in AI research.

15.4 THE FUTURE

Artificial intelligence began as an attempt to answer some of the most fundamental questions about human existence by understanding the nature of intelligence, but it has grown into a scientific and technological field affecting many aspects of commerce and society.

Even as AI technology becomes integrated into the fabric of everyday life, AI researchers remain focused on the grand challenges of automating intelligence. Work is progressing on developing systems that converse in natural language, that perceive and respond to their surroundings, and that encode and provide useful access to all of human knowledge and expertise. The pursuit of the ultimate goals of AI—the design of intelligent artifacts; understanding of human intelligence; abstract understanding of intelligence (possibly superhuman)—continues to have practical consequences in the form of new industries, enhanced functionality for existing systems, increased productivity in general, and improvements in the quality of life. But the ultimate promises of AI are still decades away, and the necessary advances in knowledge and technology will require a sustained fundamental research effort.

This cluster addresses new research to improve decision support systems and reduce uncertainty in the design process, helping to avoid problems during construction. Research could address improvements to Concurrent Engineering (CE) technology. CE applies expert systems and telecommunications to enable multi-disciplinary engineering teams to function as a virtual unit in real-time without the need for physical co-location of team members. A related area of research focus is computer-aided systems engineering (CASE). The CASE toolbox will allow designers to rapidly evaluate the constructability of an entire new building or renovation (mechanical, electrical, security, plumbing, structural, foundations, etc.) while precisely refining the design. Computer Integrated Knowledge Systems (CIKS) can provide logic and control structure for data management, even in the fragmented construction industry. Identifying embedded or hidden costs of operating and maintaining a structure over its service life provide a truer picture of actual costs than do first-cost-driven methods. Advanced design technology can help identify life cycle cost savings that might be realized with advanced materials and systems.

It is not our purpose here to get into a deep analysis of the meanings and accepted definitions of artificial intelligence and expert systems. Indeed it appears that no universally accepted definitions exist, a fact that is bemoaned by some authors as a source of much confusion in the field and an open invitation to "exploit the hype being generated about the field in popular and quasi-technical media" (Harris and Owens, 1986). In their opinion, the *artificial* half of AI is generally taken to mean *man-made*, but there is much disagreement and lack of definition as regards the term *intelligence* as used in this context.

1. If a machine were to act in a way that one recognized as being intelligent, then the machine will be intelligent (why the need for the prefix *artificial*), but
2. he has not as yet seen a machine that is truly intelligent and doubts if he ever will (on the grounds that, for him, true intelligence has an affective/emotional component and he has never seen machine display emotion).

So, to avoid stepping yet deeper into this semantic quagmire, let us accept a simple functional definition, quoted by Rich (1983), which is quite adequate for our purpose here, despite Harris and Owens's criticism that it intentionally avoids the definition of both *artificial* and *intelligence:*

AI is the study of how to make computers do things at which, for the moment people are better...

Despite the implication of a *fluid* field (once we succeed in matching or exceeding the performance of humans, our computer application suddenly ceases to be intelligent), this definition is particularly pertinent to the theme of this conference as it raises the question of whether, and if so when and in what tasks, will computers be better than teachers.

Following the same functional approach, let us select the definition of *expert system* quoted in the user's manual for the *Personal Consultant* expert system shell produced and marketed by Texas Instruments (1985): Expert Systems are software programs that help you solve complex reasoning tasks that normally require a (human) expert ... In other words, an expert system should help a novice, or partly experienced, problem solver, to match acknowledged experts in the particular domain of problem solving that the system is designed to assist. By the way, the term *shell* used above is another way of saying *expert system development tools* but rather an apt way, as the tools provide an empty (content free) structure into which the system developer may fit the specific content and inference strategy that is necessary to solve problems in the given domain.

To be more specific, expert systems are generally conceptualized as depicted in Figure 15.1. The user makes a consultation through the interface system (the communication hardware and also the software which defines the types of queries and formal language to be used) and the system questions the user through this same interface in order to obtain the essential information upon which a judgment is to be made. Behind this interface are two other sub-systems—the knowledge base, made up of all the domain-specific knowledge that human experts use when solving that category of problems, and the inference engine or system that performs the necessary reasoning and uses knowledge from the knowledge base in order to come to a decision with respect to the problem posed.

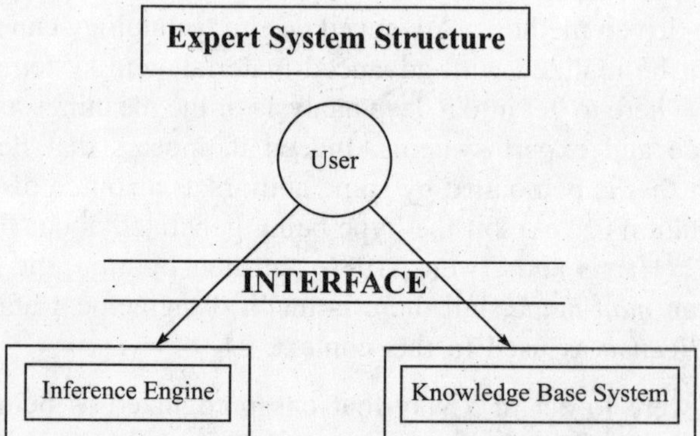

Figure 15.1 Expert System Structure.

In a sense, the knowledge base and inference engine are analogous to the knowledge stored in memory and the reasoning capabilities of the human experts that the system is emulating. However, the analogy is imperfect. The inference engine contains a set of formal logic relationships which may or may not resemble exactly the way that real human experts reach conclusions (but that does not matter as long as the conclusions reached are identical). And the knowledge base is structured in a formal manner, which varies with the *shell*, but most commonly is *rule-based.* The rules have to be defined in a limited number of formal ways. Typically, they may be a set of some hundreds of *if-then* (or *if A and B but not C then D*) types of relationships that describe all the domain-specific knowledge used by the human expert.

And here lies the rub, because human experts do not usually have all their knowledge about a subject neatly structured into a large number of rules of a particular format—life and humanity just is not that well organized. It may be that at the brain-chemistry level, human knowledge is stored in specific standard ways, but we do not know-how for sure and, certainly, we are not personally aware of the exact way that we structure what we know. As a result, the most difficult and time-consuming part of developing an expert system is the extraction of knowledge from the head of an acknowledged expert (or better a group of experts) and then transforming it into a form acceptable to the expert system's knowledge-base structure. This task is so important, complex and time-consuming that it has earned itself the rather grand name of *knowledge engineering* (some prefer *cognitive engineering*, though this is not quite a synonym). Some writers (e.g. Harris and Owens, 1986) query, however, to what extent this activity warrants to be termed engineering, rather than *art or craft.*

Computer Aided Design in general (including Computer-aided Manufacturing (CAM), Engineering (CAE), etc.) has traditionally meant the use of a graphical, interactive system for modelling product or process geometry. The object model is for the most part designed into the system. The possible kinds of relationships that may exist between objects are often built into the internal data model of the system. There are many custom programs that exist for a CAD/CAM/CAE system environment. Such programs are generally procedural in nature, following the way the interactive system works: First do this, and then do that, etc.

The main strengths of CAD/CAM/CAE systems are in the built-in geometry processing and the ease of use of graphical interaction for building and operating on complex geometric models. The growing use of 3-D modelling in design stems from its convenience once the geometry is captured analytically. Designers are able to build and modify a part on a tube much faster and with greater ease than with the use of only a physical model. Moreover, the 3-D models created on computers are reusable and modifiable in far less time and cost than models using "lines and surfaces" representation.

15.5 THE FIELD OF AI IN EDUCATION: ITS STRUCTURE

Artificial intelligence is perhaps the most promising, most misrepresented, and least understood field of computer science. It is certainly the most controversial. Academics from many disciplines—computer science, psychology, philosophy, neurophysiology, mathematics—are involved in the debates over AI. This question remains unanswered: Is it possible for a non-biological entity, a product purely of human invention, a few (million) scratches in a piece of sand (silicon from which computer chips are made) to display intelligent behaviour? Some people take the question a step further and ask: Would we even want such a device? Are the risks too great of **machines** becoming our **superiors**?

In over 20 years of teaching computer science, we have discovered that many students, both majors and non-majors, believe that Al has already arrived and that intelligent, thinking machines exist in one form or another. Yet the realization of true AI, although intriguing, is still a long way off. Much progress has been made in the field, however, especially with the advent of high-speed, highly parallel computer architectures. Here, we explore the foundations of Al and highlight some of the current research areas.

15.6 BASIC ISSUES

So, what is AI? To some, it is the development of machines that simulate intelligent human behaviour. For these performance-oriented researchers, true Al will be attained when computers act and think as we do—or better than we do. These researchers take an engineering approach in which the world of intelligence (animal or machine) is viewed as a collection of discrete modules such as problem-solving ability, learning, language processing, and visual perception. As each of these sub areas is implemented on a computer, it is to be added to the growing arsenal of Al modules whose sum will be an intelligent machine. Some researchers want more than just a smart PC; they want a mobile robot that can interact with the environment. A mobile robot could take an active role in the world, learn new ideas directly, and evolve at a higher rate than one confined to a desktop.

Some people consider the ultimate proof of the existence of performance oriented AI to be the so-called Turing test, named after the mathematician Alan Turing in 1950. He proposed a scenario such as the following.

Give a human (called the interrogator) a computer terminal that is in communication with either a human or a program at the other end. The interrogator is allowed to ask any sequence of questions such as: How is the weather over there? What is the square root of pi? What do you think of the designated hitter rule? In the end, if the interrogator thinks that he/she is communicating with another human when in fact he/she is communicating with a program, then that program shows true AI.

One of the limitations of this test is the presupposition that human intelligence can be evaluated through verbal communication. The ability to communicate verbally (or via computer) is taken as a sign of intelligence by some segments of some societies. However, it is surely not the only sign. Musicians, artists, and inventors all have other, non-verbal skills that are just as suggestive of intelligence. The underlying problem is that there is no scientific definition of intelligence, although researchers in different fields have working theories.

For other researchers, AI is a tool to study the human brain. For these cognitive scientists and neuroscientists, the computer provides a controlled environment with which to model and test (simulate) theories of brain functions. The use of a computer model has the potential to provide feedback and insight into our own mental world. At present, the tool for many brain researchers is the neural network. An artificial neural network (ANN) is a graph in which the nodes represent neurons and the arcs represent axons (the interneuron connections), as shown in Figure 15.2. The figure shows an input layer to which we give a pattern; for instance, two binary numbers to be added or a binary representation of a handwritten character. This layer is connected to one or more hidden layers that manipulate the bits and send results to the output layer.

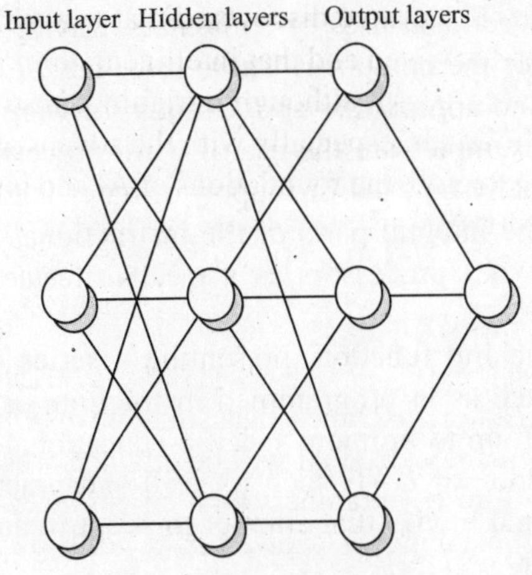

Artificial neural network

0	0	0	0	0	0	0	0	0	0
0	0	0	0	0	0	0	0	0	0
0	0	0	1	1	1	1	0	0	0
0	0	1	0	0	0	0	1	0	0
0	0	1	0	0	0	0	1	0	0
0	0	1	0	0	0	0	1	0	0
0	0	1	0	1	0	0	1	0	0
0	0	1	1	1	1	1	0	0	0
0	0	0	0	0	0	1	0	0	0
0	0	0	0	0	0	0	1	0	0

Digitized letter 'Q'

Figure 15.2 Artificial neural Network.

As we assume to be true for biological neurons, the nodes in the ANN get weighted input from other nodes according to the pattern of connectivity expressed in the graph. The weights represent the importance of the interconnections and in the ANN are initially assigned random values. When the sum of all the weights on a given node exceeds a threshold, the node fires and an output value is propagated through the preset interconnections of the graph.

An ANN can be trained to recognize specific patterns by back propagation. For example, we can train an ANN to recognize digitized handwritten images of the letter '*Q*' (see Figure 15.2) by first deciding what the correct answer is, say the ASCII code for '*Q*.' Then we present the data (the pattern of 1's and 0's) to the input layer and execute the ANN. A pattern of bits emerges at the output layer and is evaluated by computing an error factor, the difference between the output layer and the correct answer. The error factor is sent through the ANN, starting at the output layer and working backwards, adjusting the weights in the process. Similar input data are then applied and executed by the newly adjusted network, a revised error factor is computed, and the error is back propagated. This cycle is repeated hundreds or thousands of times until the ANN adjusts itself to always produce the right answer. This process approximates the behaviour of real neural networks and offers an insight into both human mental activity and the possibilities of AI.

Neural networks are typically explored in software. However, new ANN chips are on the market that have a few hundred neurons built into the circuitry. (No doubt the capacities of such chips will expand greatly as the interest increases.) By comparison, the human brain has about 10^{11} neurons and a total of about 10^{16} interconnections, roughly equivalent to the number of characters in 10 billion 350-page books.

The difficulties of achieving true AI are staggering, especially since intelligence and perception are so poorly understood in biological creatures. Most AI researchers direct their energies to a particular subtopic. These include expert systems, natural-language processing, general problem solving, learning, robotics, and vision.

In order to structure our analysis of the potential areas of application for artificial intelligence-based systems in education, I have borrowed the organizing framework used by Robert Taylor

(1980) when writing in general about the educational uses of computers. Taylor organized the book he edited around three possible roles for the computer. These roles are:

TOOL—the student uses the computer and appropriate applications software to perform a task related to the course of study—typical examples are the use of word processing, statistics packages or spreadsheets—these would normally be used in support of other learning/teaching activities, but in some cases they may become integral parts of the instructional system, as in the case of the ever more common use of word processors as a means of teaching creative writing.

TUTOR—the computer takes on the teaching function, presenting a series of screens of information, test questions and feedback, much as in programmed instruction, or involves the student in a simulation or game situation built up to promote specific educational objectives—this *role* includes most of the common modes of CAI, such as drill-and-practice, tutorial, simulation and also dialogue and conversational modes that attempt to adapt to and learn from the individual learner, much as a human tutor.

TUTEE—the student *teaches* the computer and, in so doing, learns something as well—this mode includes programming the computer and thus learning computer programming and systems analysis, but the best known example of this category is the use of the LOGO language, by which means the student in programming the computer is supposed to gain all manner of insights and *powerful ideas* of high transfer value (Papert, 1980).

However, it is not only the student who uses computers in education. Teachers also may do so, and so may instructional designers, authors and many other groups. They certainly use computers as TOOLS and no doubt often learn a great deal in the process. Less frequently, perhaps, do these groups use computers in the TUTOR or TUTEE roles, specifically to learn something that may help them in their jobs. But it does happen and may become more frequent in the future. So, to fully explore the potential field of applications of AI and expert systems in educating, let us carve up the field into a two by three matrix, as shown in Table 15.1.

TABLE 15.1 AI and Expert System in Education

User	*Tool*	*Tutor*	*Tutee*
Student	1. Commercially produced expert systems used to solve problems (JOB - AID)	2. Intelligent computer-assisted instruction (ICAI)	3. Use of LISP or PROLOG in schools Students construct expert systems for the domain under study
	Student learns by repeated use	Intelligent database search assistant (EMBEDDED TRAINING)	
Teacher/ Instruction Designer/ Author	4. Stand-alone expert systems on instructional design, decision-making, planning, controlling, etc.	5. Intelligent embedded training in school applications software	6. Teacher/Subject matter expert constructs expert systems on specific domains to use in modes 1 and 2
	Intelligent authoring systems for CAI	Intelligent teacher-training systems	

The examples listed in the six cells of this matrix have been chosen to illustrate practical applications of AI systems and in particular *expert systems* which have already been attempted or are currently in the stages of development or design. Some are more of a current reality, whilst others, though technically feasible, may, for a variety of practical or economic reasons, remain in the realm of laboratory research or even pipe dreams for a long time yet. Let us briefly examine each of the six cells of our *field*.

15.7 PROGRESS IN THE APPLICATION OF AI TO EDUCATION

15.7.1 Cell 1: As a TOOL Used by the STUDENT

The most common application of AI so far is in the form of expert systems. Although not yet very common, there are already many examples in commercial use and the market is estimated to grow exponentially during the next few years. It should not be long before commercially developed expert systems will be available in large numbers to assist in all manner of problem solving tasks, including many of interest and relevance in the school room. After all, an expert system is just another form of Job Performance Aid, and as such will surely permeate into school use as slide rules 50 years ago, calculating machines 20 years ago and work processors and spreadsheets yesterday. Many teachers resisted those innovations in their time and no doubt some may resist the entry of expert systems now. I can hear the arguments—"it's all right for the kids to use computers for the repetitive or routine tasks of calculation or writing, but with expert systems on tap, they'll stop thinking for themselves". But will they? And indeed do they think now in the way that the expert software would illustrate? After all, an expert system, to be commercially viable, must represent an exceptionally outstanding level of reasoning, albeit in a tightly defined domain. And most expert system software permits (indeed always should permit) the user to interrogate and analyse the reasoning process. At every stage of consultation, the user may press function keys to ask either why a certain item of information was requested by the computer or how the computer has reached the conclusion that it is recommending. Therefore, the user may follow in detail the reasoning processes built into the expert system and, in so doing, may learn to reason in a similar manner.

This argument is akin to the argument for presenting formulae, worked examples, and all forms of guidelines, algorithms and so on, as learning aids. One may argue that it is not always desirable that students learn by following pre-prepared examples and procedures and that sometimes they should discover the procedures for themselves. But life is too short to discover everything. And anyway, in interacting with an expert system, the user is being shown the reasoning steps that were followed in one particular example problem—it rests with the student to generalize (or not) from the example and discover the general principles of logical reasoning that the system is applying. It may be an important function of the teacher, using expert systems in the classroom, to make sure that the discovery of those general principles really does occur. But to adopt the neo-Luddite attitude of keeping expert systems out is not all that different from attempting to keep students from interacting with real experts, or from watching one of C.P. Snow's excellent television lectures on world history and affairs. The better approach would seem to be to study how the educational potential of interacting with expert systems might best be exploited in the school environment and work towards the sensible integration of these new tools into our instructional systems, before events overtake us and, through out-of-school contact, our students become more expert with expert systems than we are.

15.7.2 Cell 2: As a TUTOR of the STUDENT

In Cell 2, we find the Intelligent Computer Assisted Instruction (ICAI) systems, or Intelligent Tutoring systems that we hear so much about of late. As an area of educational computing research and development, these have been with us for some time. Their history goes as far back as does AI itself and some functional ICAI systems were developed as much as fifteen or more years ago (Sleeman et al., 1975). One early system—SCHOLAR—was a good example of the potential of intelligent tutoring using a structured knowledge-base (in this case a domain in geography, Carbonell, 1970). Many different approaches to the design and development of intelligent CAI have been attempted (see Sleeman and Brown, 1982), but this variety seems to be a function more of the research interests of the developers and the state of knowledge in the AI field, rather than the needs of education and the current state of knowledge about learning and instruction (Part and Seidel, 1986). These two authors draw an interesting set of comparisons between what we may call conventional CAI and ICAI. Some of these contrasts are:

Contrast	**CAI**	**ICAI**
The developers (mainly) ...	Educators ...	Computer scientists
Their goals are to ...	Instruct students ...	Explore AI
They start from ...	Task analysis ...	Knowledge structures
They try to teach ...	Almost any subject ...	Well structured areas
Their methodology is ...	Systems approach ...	Personal, varied
Instructional mode(s) ...	Vary with objective ...	Mainly discovery
Adapts to student's ...	Single responses ...	Response pattern
Validation/evaluation ...	Formative to effectiveness efficiency	Technical opinion and de-bugging

This list, particularly the last observation, would suggest that intelligent tutoring systems have not yet left the laboratory stage. This opinion was shared by Frank Roberts of Control Data Corporation in a Keynote address to the 1986 conference of the Association for the Development of Computer-based Instructional systems (ADCIS), when he remarked that to date there have been about 20 ICAI systems developed, about 15 of these ever reached the stage of practical implementation, only about five have been used on a large scale for a reasonable time and none has been formally evaluated.

Perhaps one reason for the relatively slow progress in the sub-field of intelligent tutoring systems is that the developers are trying to break new ground in several directions at once. The general structure of a typical ICAI system may be represented by the schematic in Figure 15.3. The student interacts with the system by means of some interface system, receiving information and providing responses to questions or sometimes Initiating Dialogue by asking questions—this raises a host of problems associated with natural language communication, two-way graphic communication and so on, which are still in the early stages of research and development. The system then must respond to the student in a manner appropriate to the individual pattern of responses and queries that are received. It must adapt to the individual student's needs, learning style, difficulties, etc. It does this by building and constantly updating a Model of the individual student's characteristics—how to do this is another field for much research, involving the social as well as computer sciences. Information from the student model is then used by the inferencing system to modify the instruction accordingly—there is a need for an adaptive Tutorial Model and the principles on which this adaptation should take place should be drawn from our as yet fledgling instructional science. These adaptations in the tutorial and indeed in its direction, mean that the system must have at its disposal a very expensive and well-structured knowledge-base, or rather two knowledge-bases, one on the subject matter that is to be taught and the other on the teaching methods and tricks associated specifically with that subject. This *expertise system* is a combination of the knowledge of the subject matter expert (our famous SME) and of the expert teacher of that subject.

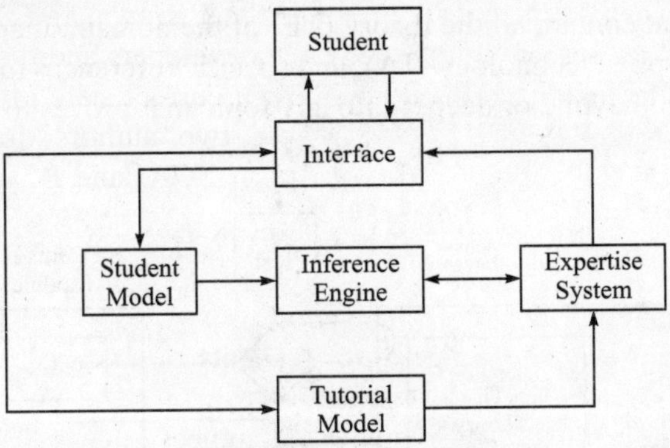

Figure 15.3 Expertise System.

Lest it be thought that we are writing off ICAI as a development which is yet to prove itself and destined to remain for many years the plaything of researchers, let us remember the surprising speed of recent developments in the computer field and not underestimate the cumulative effect of the many developers now working in this field. At this time, some dozens of development projects are underway in the United States alone, many aiming to implement practical military training ICAI systems in the shortest possible time span. We may well have reached a *critical mass* in the body of research that will lead to very fast practical progress from now on.

We must also remember that partial implementation of the model described before may make good sense in many situations. One approach which may cut some of the R & D corners and may have considerable practical application is exemplified by a project which I began to develop a few years ago in Brazil. The overall aims of the project were to develop efficient, individualized updating/retraining systems for electronics technicians from a variety of industries where microchip technology was being introduced (Arce and Romiszowski, 1985). The instructional materials were prepared in printed paper form, but in order to be usable for a variety of differing objectives with a variety of trainees in a subject area in constant development, the materials were written in a highly modularized and very carefully structured manner. Each concept, principle, practical procedure, etc. of digital logic theory, circuit design and troubleshooting was isolated and described in a separate MAP, using the authoring or *mapping* techniques suggested by Robert

Horn of Information Mapping Inc. (Horn 1974). The resultant set of over 1000 maps formed a comprehensive and coherently structured knowledge-base on digital circuits, suitable for both learning the theory and practical application in industry, covering dust about any current need and accessible through the map titles as keywords, in a non-linear and totally user-orientated manner.

In practice, a particular trainee or group follows a sequence of practical laboratory activities, determined by their individual learning needs rather like a conventional, mastery-learning oriented, practical course. Each laboratory activity is separated by a pre/post test which verifies the current state of knowledge of the trainee. This test will direct the trainee to the study of relevant maps in the knowledge base. In the original implementation, this diagnosis of need and prescription of relevant study material was performed by an instructor with the aid of a database management system (set up in Dbase-II), which stored the structure of the knowledge-base (but not the actual materials). The general procedure is shown in Figure 15.4. This shows that the knowledge-base is composed of maps that contain all the theory (TE) of the domain, maps that explain and guide practical application (applied technology–TA), and specific references to library materials, which the trainee may use to go beyond or deeper into any topic that proves to be of particular interest.

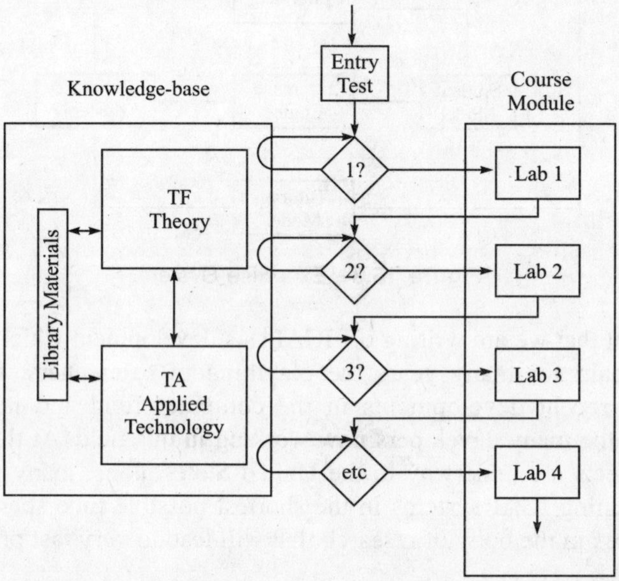

Figure 15.4 KBES [Knowledge Based Expert System].

The evaluation of the trainee and diagnosis of needs, selection of further study material, etc. is performed by the instructor who takes into account the trainee's individual long-term objectives, learning pattern to date, history of errors and difficulties, and so on. Data on all these aspects is accumulated and stored in the same data base (in Dbase-II) that is used to access and select the instructional materials. The instructor uses the data available in order to make more informed, and therefore hopefully more intelligent, decisions concerning the needs of each trainee at each point in the course. By tracking the instructor's decisions over a series of courses, evaluating, and comparing the learning effects of following those decisions, it should be possible, over time, to build up, almost automatically, an expert system to perform the diagnosis and prescription tasks. Thus, by linking up a subject matter knowledge-base (either stored in or merely accessed and managed by means of proprietary database management software) to an expert *instructor* system (developed from practical observations using a proprietary expert system shell), we may construct a useful student evaluation and guidance system which operates with at least a modicum of intelligence (Romiszowski, 1986). This may not be a full ICAI system, but it is not far off and may be realised with existing research know-how and with existing and readily available hardware and software.

Another potential *semi-intelligent* approach, which is an adaptation of the above, is *Automatically Diagnosed Embedded Procedural Training*, or ADEPT (the name and the acronym are my invention). This is an extension of the trend towards the development of computer software packages with *Embedded Training* which in some way forms part of the package itself rather than being a stand-alone training course to be taken before beginning to use the package. One may think of embedded training in the context of computer-related work as a form of in-service training. Horine and Erickson (1986) refer to this as "the explosive growth area in computer based training" and justify it on economic and practical grounds as well as didactically, in that training is taken when required to be applied in practice and only what is needed is learnt. They define two classes of embedded training.

Integrated, which forms part of the software package itself and may be called up by hitting the help key (except that the help you get well designed and appropriate instruction rather than the encounters on help screens).

Concurrent, which is separately packaged and may be written in other programming languages or even stored on, say, videodisc, but which may be accessed rapidly and accurately whenever a difficulty or a new procedure is encountered.

Most embedded training, even when well designed, has to be called up the user/ trainee when experiencing difficulty. Furthermore, the type of difficulty and its origin is not very expertly diagnosed, as it is usually the trainee who is experiencing the difficulty, who has to do the diagnosis. Matters would have been a bit different if an expert user had been peering over the trainee's shoulder and had observed the response patterns that had led up to the difficulty. Chances are that the training needs diagnosis would be much more accurate and profound. Often the expert may even diagnose an operational weakness on the part of the trainee of which the trainee is completely unaware—for example, backspacing by single letters for word processor editing, when there are procedures to return by whole words, lines or paragraphs at a time, or using an inefficient search strategy in using a database. The ADEPT approach is to build that "expert peering over the shoulder" into the applications software, by tracking the user's response pattern, identifying patterns that are suboptimal for the task in hand, diagnosing the probable knowledge deficiency and interrupting the user—"hey, I don't think you're using me right—why not try it this way—do you want me to show you?"

15.8 ARTIFICIAL INTELLIGENCE AND EXPERT SYSTEMS

Artificial intelligence is the simulation of human intelligence processes by machines, especially computer systems. These processes include learning (the acquisition of information and rules for using the information), reasoning (using the rules to reach approximate or definite conclusions), and self-correction. Particular applications of AI include expert systems, speech recognition, and image recognition.

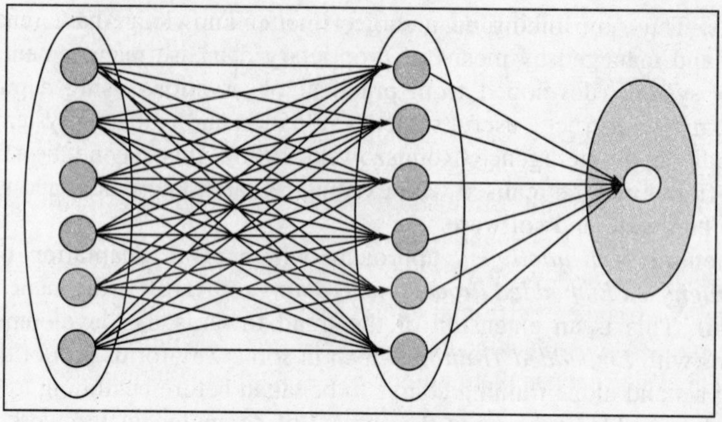

Figure 15.5 AI Application.

The use of soft computing technologies in the Artificial Intelligence (AI) discipline has seen an increase in real-world systems over the past few years. As such, to enhance learning and understanding of scientists and engineers, we have developed an interactive AI tool kit, with a specific focus on soft computing techniques for robotic applications. The core elements of this educational software package are: Comprehensible training modules on three AI soft computing technologies, namely: genetic algorithms, neural networks, and fuzzy logic. Interactive graphical interfaces provides hands-on examples of applying AI techniques to real-world research problems. The three soft computing methodologies currently embedded within the tool kit include Fuzzy logic, genetic algorithm and neural network.

15.9 BRINGING COMMON SENSE, EXPERT KNOWLEDGE, AND SUPERHUMAN REASONING TO COMPUTERS

Artificial Intelligence is the key technology in many of today's novel applications, ranging from banking systems that detect attempted credit card fraud, to telephone systems that understand speech, to software systems that notice when you are having problems and offer appropriate advice. These technologies would not exist today without the sustained federal support of fundamental AI research over the past three decades.

Although there are some fairly pure applications of AI—such as industrial robots, or the *Intellipath* pathology diagnosis system recently approved by the American Medical Association and deployed in hundreds of hospitals worldwide—for the most part, AI does not produce stand-alone systems, but instead adds knowledge and reasoning to existing applications, databases, and environments, to make them friendlier, smarter, and more sensitive to user behaviour and changes in their environments. The AI portion of an application (e.g. a logical inference or learning module) is generally a large system, dependent on a substantial infrastructure. Industrial R&D, with its relatively short time-horizons, could not have justified work of the type and scale that has been required to build the foundation for the civilian and military successes that AI enjoys today. And beyond the myriad of currently deployed applications, ongoing efforts that draw upon these decades of federally-sponsored fundamental research point towards even more impressive future capabilities.

Skills and experience

CSIRO expertise in artificial intelligence includes a number of interrelated fields.

Signal processing uses electronic circuitry to detect and convert signals (such as X-ray, ultrasonic, infra-red) into images or sound to aid in signal interpretation. Subtypes include digital signal processing, analogue, audio, speech and video, and signals from arrayed sensors.

Pattern recognition is a branch of artificial intelligence that can assist in recognising an obscure property from raw input data. A machine capable of pattern recognition can be trained to extract certain features, such as car licence plate numbers, and guide an automated system into making intelligent decisions.

Optical character recognition is the process by which an electronic device examines printed characters on paper and determines their shapes by detecting patterns of dark and light. The shapes are matched against stored sets of characters and used to translate into computer text.

Surveillance systems involve the collection and analysis of data by electronic circuits. The data may enable timely action to be taken with regard to security, border control or the control of a disease.

Decision support systems involve software applications that help analyse data contained within a customised database to support decision-making. This computer technology is used in the management of real estate portfolios, healthcare, cost accounting and budgeting.

Threat and risk analysis applications are electronic-based methods for defining and analysing threats and vulnerabilities associated with a set of assets or capabilities and supply management with information to develop corrective measures.

System diagnostics use CSIRO's unique signal processing method in a number of industries including metal cutting, machine grinding and materials-handling industries.

15.10 SIMULATION

Simulation has long been recognized as a valuable tool for analyzing manufacturing systems. It is effective for assessing the impact of changing system parameters (e.g. reducing processing time) upon system performance measures, and it can also aid decisions concerning system configuration. Currently, it is being used as an analysis tool studying system design and computer communication issues. Future plans are to use simulation as a real-time scheduling and control tool. Due to this ultimate long-term goal of the simulation model use, the model structure is different than traditional manufacturing applications of simulation. Rather than having events associated with workpiece movement and processing drive the simulation, computer communication events drive the model.

- Some complex systems cannot be modeled because of
  - random events,
  - changing operating conditions,
  - too many interactions, and
  - exact solutions do not exist.

- Simulation is used to determine how these systems will behave.
- Simulation typically involves developing a model that includes discrete stations and events that occur with some probable distribution.
- We can then examine the simulation results to evaluate the modeled system. Examples include
  - machine utilization,
  - lead time,
  - down time, etc.
- This is a very effective tool when considering the effect of a change, comparing decision options, or refining a design.
- Some simulation terms include

  System—the real collection of components
  Model—a reasonable, mathematically (simpler) representation of the system
  State—the model undergoes discrete changes. A state is a 'snapshot' of the system
  Entity—a part of the system (e.g. machine tool)
  Attributes—the behaviour of an entity
  Event—something that changes the state of a machine
  Activity—when an entity is going through some activity (e.g. press cycling)
  Delay—a period of time with no activity
- Good approach to simulation
  - Determine what the problem is
  - Set objectives for the simulation
  - Build a model and collect data
  - Enter the model into a simulation package
  - Verify the model then check for validity
  - Design experiments to achieve goals
  - Run simulations and collect results
  - Analyze and make decisions

15.10.1 Model Building

- If we are building a model for a plant floor layout, we will tend to have certain elements, such as
  - material handling paths (transfer)
  - buffers/waiting areas (delays)
  - stock rooms (source)
  - shipping rooms (destination)
  - machine tools (activities)
- Some of these actions can be stated as exact. For example, a transfer time can be approximated and random (manual labour), or exact (synchronous line), or proportional to a distance.
- Some events will occur based on availability. For example, if there are parts in a buffer, a machine tool can be loaded and activity occurs.
- Some activities and events will be subject to probabilities. Consider that the operation time in a press may vary, and there is probability of scrapping a part.

- The random variations can be modeled as,
  — discrete, for individual units
  — continuous, for variations
- Well-known distributions include Normal/Gaussian and Probabilistic Distribution

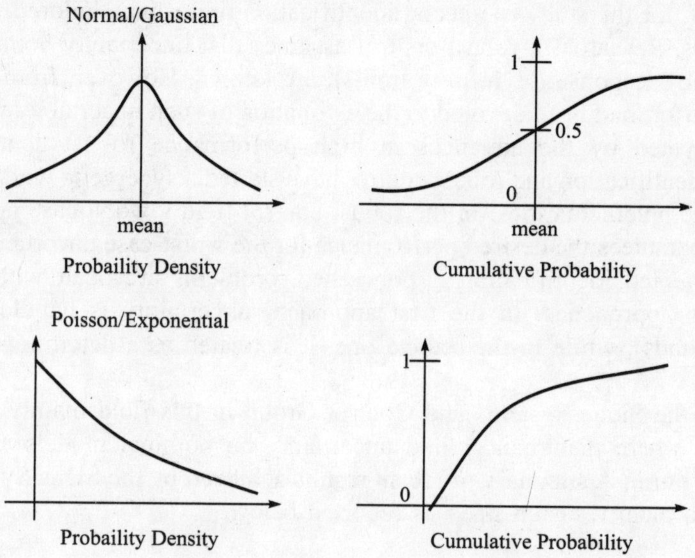

Figure 15.6 Normal/Guassian.

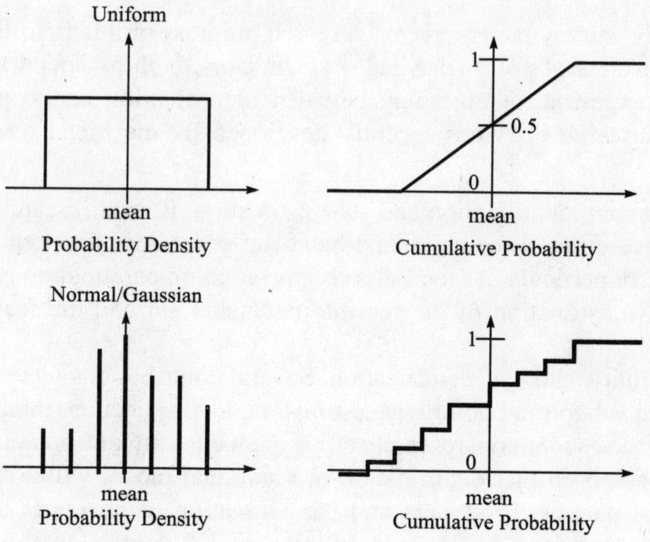

Figure 15.7 Probabilistic Distribution-Simulation.

- This data may be found using data provided by the manufacturer, sampled in-house, etc.

15.11 IDENTIFICATION AND ROBUST CONTROL

Modelling and identification procedures of real processes should be designed having in mind their specific purposes. Although identification techniques have always taken into account typical requirements of control tasks, recent improvements in high performance robust and adaptive control theory ask for the study of special identification procedures tailored to these advanced control techniques. Research in robust control assumes that uncertainty bounds, like tolerance bands on frequency response (H_infinity limits), are known. However, from the identification viewpoint, little effort had been devoted to the estimation of such uncertainty models until a few years ago. Motivated by the advances in high performance robust control, the research communities in identification and robust control have started a synergetic work by stressing their similarities and complementarities. In the robust control field a worst-case approach is usually adopted which guarantees the desired performance for the worst-case uncertainty realization. As far as control oriented identification is concerned, problems are dealt with in the literature according to two approaches. In the first approach, uncertainty is handled as a stochastic quantity (soft bounds), while in the second one, it is treated as a deterministic variable (hard bounds).

Research by the Siena Systems and Control Group in this field mainly refers to the hard bound approach, where disturbances and uncertainty on nominal models are assumed to be bounded in some norm. A summary of recent results achieved by the Siena Systems and Control Group, along two main research lines, is reported below.

15.12 ROBUST CONTROL ORIENTED IDENTIFICATION

As pointed out in the survey papers, recent research progress in this field involves complexity, optimality and robustness of point wise and set estimators, both for identification and filtering; optimal design of experiments; interplay between identification and control. The scientific expertise and the research activities recently developed by the Siena research group can be outlined as follows.

1. Set membership identification and state estimation. Recent research activities deal with the design of optimal recursive and batch set estimators for both parameter and state estimation. In particular, several algorithms based on parallelotopic approximations, for the recursive estimation of the feasible parameter set and the feasible state set, were developed.
2. H_2, H_infinity, and l_1 identification. Several contributions were given in the field of optimal and suboptimal point wise estimators, like projection, interpolatory and central estimators. The common research effort yielded a mixed parametric/non-parametric approach, based on the identification of a nominal model within a given set of finite-dimensional parametric models and the estimation of an upper bound on the model uncertainty, measured by H_2, H_infinity, or l_1 norms. In this context, important results were obtained on the identification of restricted complexity models for real processes, which is considered a key topic in identification since long time. More recently, the problems of optimal orthonormal basis selection and optimal input selection

with respect to the worst-case error provided by central projection estimators have been investigated.

3. Set membership identification of nonlinear models. Motivated by recent developments in hybrid system theory and applications, a new research line focused on identification of nonlinear structures via piecewise affine models has been started. The approaches are based on randomized algorithms or integer programming and yield remarkable performance in terms of both model quality and computational effort. Active research is currently carried out by the Siena research unit on the application of the estimation and identification techniques introduced above, including set membership estimation and filtering for dynamic vision, localization and navigation maps in autonomous mobile robotics.

15.13 ROBUST CONTROL WITH STRUCTURED UNCERTAINTY

The main feature of synthesis methods for unstructured uncertainties is that they allow one to find closed-form solutions to control synthesis problems. The inherent limit of these approaches is that they may be quite conservative in practical applications. Research on techniques and algorithms for problems with structured uncertainties is quite recent. The mu-analysis/synthesis and the parametric robustness analysis are among the most popular approaches to solve a robust control problem with structured uncertainty. Recent results on robust control with structured uncertainties by the Siena research, are described below.

1. Stability and robust performance analysis with parametric and unstructured uncertainty. Contributions were provided on stability and robust performance for linear systems affected by parametric perturbations linearly correlated, polynomially correlated, or mixed with unstructured (H_infinity) uncertainties. An extensive bibliography on this subject is reported in the survey paper.
2. Optimization techniques for control systems with structured uncertainty. The main drawback of analysis and synthesis methods for systems with structured uncertainty is the high computational complexity. In this context, by exploiting convex Linear Matrix Inequality (LMI) optimization techniques, a novel approach has been introduced for solving minimum distance problems from polynomial surfaces. This has allowed for the development of efficient methods for solving complex problems such as the computation of the parametric stability margin of control systems with structured uncertainty or the synthesis of robust controllers with a fixed structure, e.g. PID or lead-lag, for systems with parametric perturbations. An interesting development of the above convex optimization techniques is the construction of polynomial Lyapunov functions, which are aimed at reducing the conservatism introduced by traditional quadratic functions. By combining homogeneous polynomial forms and LMI optimization, the problem of computing polynomial homogeneous Lyapunov functions for systems with structured time-varying uncertainty has been addressed with promising results.
3. Robust model predictive control (MPC). In MPC, at each sampling time an open-loop optimal control problem is solved over a finite horizon. For systems affected by uncertainty, a typical strategy consists of solving a min-max optimization problem,

which, however, turns out to be computationally very demanding for on-line implementation. In order to push MPC to fast-sampling applications (automotive, mechanical, aerospace, etc.), techniques based on multi-parametric programming were recently developed to find the equivalent piecewise affine state-feedback form of the MPC control law, that has tremendous computational advantages over standard on-line optimization solvers.

The book, the workshop 'Robustness in Identification and Control', which was held in Siena in 1998, the Special Issue on 'Robustness in Identification and Control', organized and edited by researchers of Siena, and the Matlab toolbox 'Model Predictive Control', co-authored by some researchers of Siena, testify the effort and the lively interest in this area of research.

REVIEW QUESTIONS

1. What is artificial intelligence?
2. Define expert system in AI.
3. What is KBES?
4. Name any five applications of AI involved in industry.
5. What is simulation?
6. How does simulation work in graphics?
7. What are the terms used in simulation?
8. Draw the characteristic curve of Gaussian simulation.
9. What are the functions in robust control identification?
10. What is MPC?

Chapter 16

PROGRAMMABLE LOGIC CONTROLLER

16.1 INTRODUCTION

A PLC (i.e. Programmable Logic Controller) is a device that was invented to replace the necessary sequential relay circuits for machine control. The PLC works by looking at its inputs and depending upon their state, turning on/off its outputs. The user enters a program, usually via software, that gives the desired results.

Typical PLCs

Figure 16.1 Typical PLC.

PLCs are used in many "real-world" applications. If there is industry present, chances are good that there is a programmable logic control present. If you are involved in machining, packaging, material handling, automated assembly or countless other industries, you are probably already using them. If you are not, you are wasting money and time. Almost any application that needs some type of electrical control has a need for a programmable logic control.

For example, let us assume that when a switch turns on, we want to turn a solenoid on for 5 seconds and then turn it off regardless of how long the switch is on for. We can do this with a simple external timer. But what if the process included 10 switches and solenoids? We would need 10 external timers. What if the process also needed to count how many times the switches individually turned on? We need a lot of external counters.

As you can see, the bigger the process, the more we need a PLC. We can simply program the PLC to count its inputs and turn the solenoids on for the specified time.

16.2 PLC OPERATION

A PLC works by continually ***scanning*** a program. We can think of this scan cycle as consisting of three important steps. There are typically more than three but we can focus on the important parts and not worry about the others. Typically, the others are checking the system and updating the current internal counter and timer values.

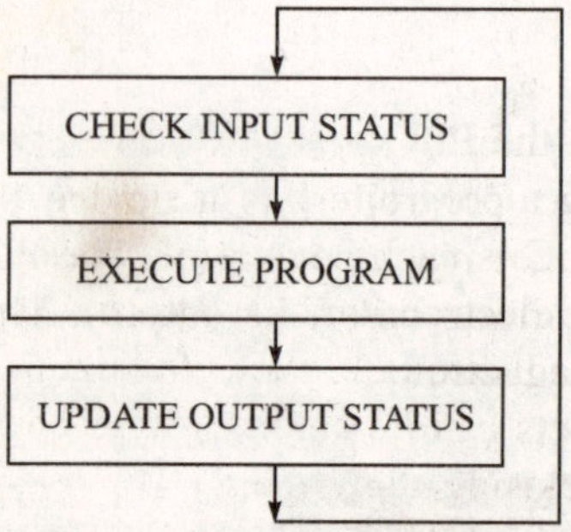

Figure 16.2 Steps in PLC Operation.

Step 1 **Check Input Status** First the PLC takes a look at each input to determine if it is on or off. In other words, is the sensor connected to the first input on? How about the second input? How about the third... It records this data into its memory to be used during the next step.

Step 2 **Execute Program** Next the PLC executes your program one instruction at a time. Maybe your program said that if the first input was on, then it should turn on the first output. Since it already knows which inputs are on/off from the previous step, it will be able to decide whether the first output should be turned on based on the state of the first input. It will store the execution results for use later during the next step.

Step 3 **Update Output Status** Finally, the PLC updates the status of the outputs. It updates the outputs based on which inputs were on during the first step and the results of executing your program during the second step. Based on the example in Step 2, it would now turn on the first output because the first input was on and your program said to turn on the first output when this condition is true. After the third step, the PLC goes back to Step 1 and repeats the steps continuously. One scan time is defined as the time it takes to execute the three steps listed above.

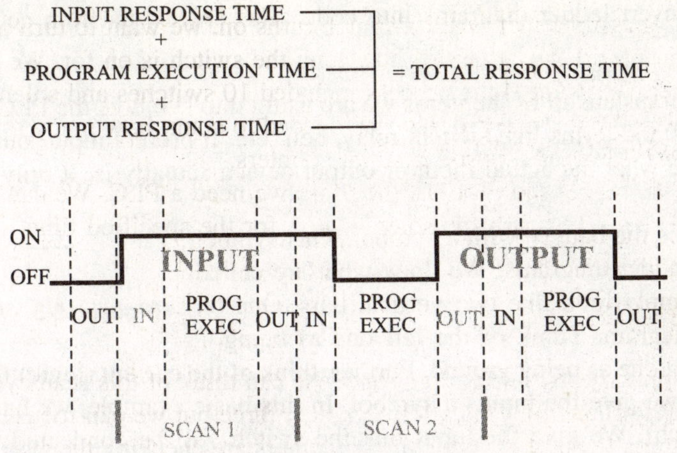

Figure 16.3 Output Status.

16.3 RELAYS

Now that we have understand how the PLC processes inputs, outputs and the actual program, we are almost ready to start writing a program. But first let us see how a relay actually works. After all, the main purpose of a PLC is to replace "real-world" relays.

We can think of a relay as an electromagnetic switch. Apply a voltage to the coil and a magnetic field is generated. This magnetic field sucks the contacts of the relay in, causing them to make a connection. These contacts can be considered to be a switch. They allow current to flow between two points thereby closing the circuit.

Let us consider the following example. Here we simply turn on a bell (Lunch time!) whenever a switch is closed. We have three real-world parts. A switch, a relay and a bell. Whenever the switch closes, we apply a current to a bell causing it to sound.

Notice in Figure 16.4 that we have two separate circuits. The bottom line indicates the DC part. The top line indicates the AC part.

Here we are using a dc relay to control an ac circuit. That is the fun of relays! When the switch is open, no current can flow through the coil of the relay. As soon as the switch is closed, however, current runs through the coil causing a magnetic field to build up. This magnetic field causes the contacts of the relay to close. Now ac current flows through the bell and we hear it.

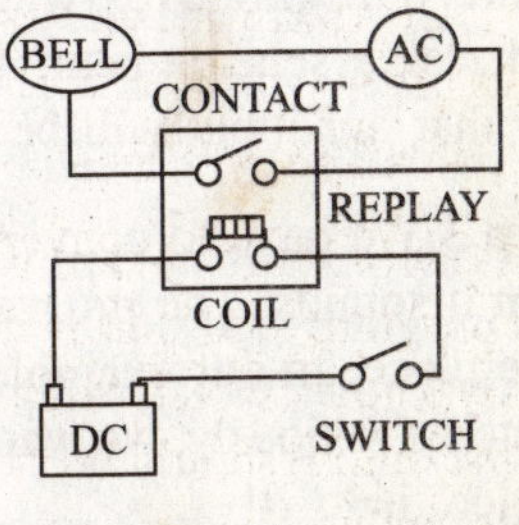

Figure 16.4 Typical Relays.

16.3.1 Replacing Relays

Next, let us use a plc in place of the relay. (Note that this might not be very cost-effective for this application but it does demonstrate the basics we need.) The first thing that is necessary is to create what is called a *ladder diagram*. After seeing a few of these, it will become obvious why it is called a ladder diagram. We have to create one of these because, unfortunately, a plc does not understand a schematic diagram. It only recognizes code. Fortunately, most PLCs have software which convert ladder diagrams into code. This shields us from actually learning the PLC's code.

Step 1 We have to translate all of the items we are using into symbols the PLC understands. The plc does not understand terms like switch, relay, bell, etc. It prefers input, output, coil, contact, etc. It does not care what the actual input or output device actually is. It only cares that it is an input or an output.

First, we replace the battery with a symbol. This symbol is common to all ladder diagrams. We draw what are called *bus bars*. These simply look like two vertical bars. One on each side of the diagram. Think of the left one as being + voltage and the right one as being ground. Further, think of the current (logic) flow as being from left to right. Next, we give the inputs a symbol. In this basic example, we have one real-world input (i.e. the switch). We give the input that the switch will be connected to, to the symbol shown below. This symbol can also be used as the contact of a relay.

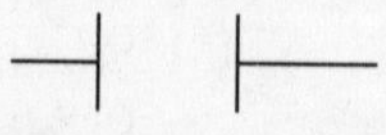

Figure 16.5 A Contact Symbol.

Next, we give the outputs a symbol. In this example, we use one output (i.e. the bell). We give the output that the bell will be physically connected to the symbol shown below. This symbol is used as the coil of a relay.

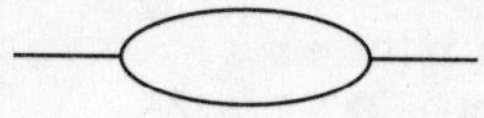

Figure 16.6 A Coil Symbol.

The ac supply is an external supply, so we do not put it in our ladder. The PLC only cares about which output it turns on and not what is physically connected to it.

Step 2 We must tell the PLC where everything is located. In other words, we have to give all the devices an address. Where is the switch going to be physically connected to the PLC? How about the bell? We start with a blank road map in the PLC's town and give each item an address. Could you find your friends if you did not know their address? You know they live in the same town but which house? The PLC town has a lot of houses (inputs and outputs) but we have to figure out who lives where (what device is connected where). We will get further into the addressing scheme later. The PLC manufacturers each do it in a different way! For now, let us say that our input will be called "0000". The output will be called "500".

Step 3 We have to convert the schematic into a logical sequence of events. This is much easier than it sounds. The program we're going to write tells the PLC what to do when certain events take place. In our example, we have to tell the PLC what to do when the operator turns on the switch. Obviously, we want the bell to sound but the PLC does not know that. It is a pretty stupid device, isn't it!

Figure 16.7 is the final converted diagram. Notice that we eliminated the real-world relay from needing a symbol. It is actually "inferred" from the figure.

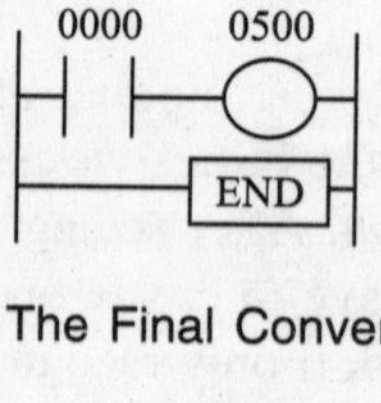

Figure 16.7 The Final Converted Diagram.

16.4 BASIC INSTRUCTIONS

Now let us examine some of the basic instructions in greater detail to see more about what each one does.

Load

The load (LD) instruction is a normally open contact. It is sometimes also called examine if on.(XIO) (as in examine the input to see if it is physically on). The symbol for a load instruction is shown in Figure 16.8.

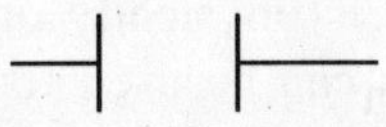

Figure 16.8 A Load (contact) Symbol.

This is used when an input signal is needed to be present for the symbol to turn on. When the physical input is on, we can say that the instruction is True. We examine the input for an on signal. If the input is physically on, then the symbol is on. An on condition is also referred to as logic 1 state.

This symbol normally can be used for internal inputs, external inputs and external output contacts. Remember that internal relays do not physically exist. They are simulated (software) relays.

Load bar

The load bar instruction is a normally closed contact. It is sometimes also called load not or examine if closed. (XIC) (as in examine the input to see if it is physically closed). The symbol for a load bar instruction is shown in Figure 16.9.

Figure 16.9 A Load Not (normally closed contact) sSymbol.

This is used when an input signal does not need to be present for the symbol to turn on. When the physical input is off, we can say that the instruction is True. We examine the input for an off signal. If the input is physically off, then the symbol is on. An off condition is also referred to as logic 0 state.

This symbol normally can be used for internal inputs, external inputs and sometimes, external output contacts. Remember again that internal relays do not physically exist. They are simulated (software) relays. It is the exact opposite of the load instruction.

*Note. With most PLCs, this instruction (load or load bar) 'Must' be the first symbol on the left of the ladder.

Logic State	**Load**	**Load Bar**
0	False	True
1	True	False

Out

The 'out' instruction is sometimes also called an output energize instruction. The output instruction is like a relay coil. Its symbol looks like as shown in Figure 16.10.

Figure 16.10 An OUT (coil) Symbol.

When there is a path of True instructions preceding this on the ladder rung, it will also be True. When the instruction is True, it is physically on. We can think of this instruction as a normally open output. This instruction can be used for internal coils and external outputs.

Outbar

The outbar instruction is sometimes also called an OutNot instruction. Some vendors do not have this instruction. The outbar instruction is like a normally closed relay coil. Its symbol looks like as shown in Figure 16.11.

Figure 16.11 An OUTBAR (normally closed coil) Symbol.

When there is a path of False instructions preceding this on the ladder rung, it will be True. When the instruction is True, it is physically on. We can think of this instruction as a normally closed output. This instruction can be used for internal coils and external outputs. It is the exact opposite of the Out instruction.

Logic State	Out	Outbar
0	False	True
1	True	False

16.5 PLC REGISTERS

We will now take the previous example and change switch 2 (SW2) to a normally closed symbol (load bar instruction). **SW1 will be physically OFF and SW2 will be physically ON initially**. The ladder diagram now looks like as shown in Figure 16.12.

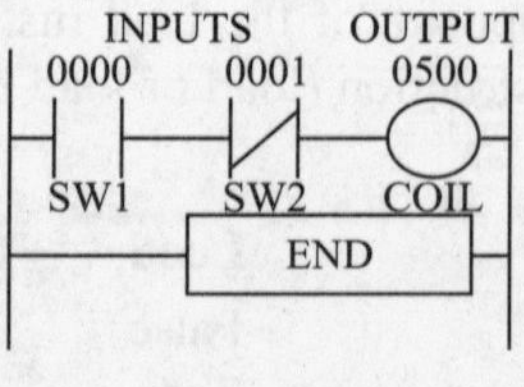

Figure 16.12 Ladder Diagram.

Notice also that we now gave each symbol (or instruction) an address. This address sets aside a certain storage area in the PLC's data files so that the status of the instruction (i.e. true/false) can be stored. Many PLCs use 16 slot or bit storage locations. In the example above, we are using two different storage locations or registers.

REGISTER 00

15	14	16	12	11	10	09	08	07	06	05	04	03	02	01	00
														1	0

REGISTER 05

15	14	16	12	11	10	09	08	07	06	05	04	03	02	01	00
															0

As shown above, we can see that in register 00, bit 00 (i.e. input 0000) was logic 0 and bit 01 (i.e. input 0001) was logic 1. Register 05 shows that bit 00 (i.e. output 0500) was logic 0. Logic 0 or 1 indicates whether an instruction is False or True. **Although most of the items in the register tables above are empty, they should each contain a 0. They were left blank to emphasize the locations we were concerned with.*

LOGICAL CONDITION OF SYMBOL

LOGIC BITS	LD	LDB	OUT
Logic 0	False	True	False
Logic 1	True	False	True

The PLC will only energize an output when all conditions on the rung are TRUE. So, looking at the table above, we see that in the previous example SW1 has to be logic 1 and SW2 must be logic 0. Then and **ONLY** then will the coil be true (i.e. energized). If any of the instructions on the rung before the output (coil) are false, then the output (coil) will be false (not energized).

Let us now look at a truth table of our previous program to further illustrate this important point. Our truth table will show **ALL** possible combinations of the status of the two inputs.

Inputs		Outputs	Register Logic Bits		
SW1(LD)	SW2(LDB)	COIL(OUT)	SW1(LD)	SW2(LDB)	COIL(OUT)
False	True	False	0	0	0
False	False	False	0	1	0
True	True	True	1	0	1
True	False	False	1	1	0

Notice from the chart that as the inputs change their states over time, so will the output. The output is only true (energized) when all preceding instructions on the rung are true.

The program scan

Let us watch what happens in this program scan by scan.

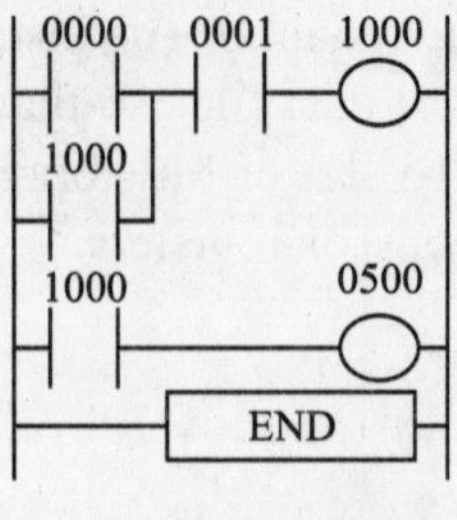

Figure 16.13 The Program Scan.

Initially the tank is empty. Therefore, input 0000 is true and input 0001 is also true.

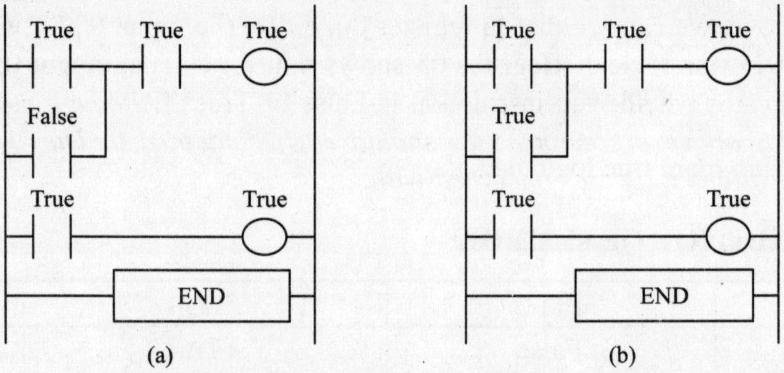

Figure 16.14 (a) Scan 1, (b) Scan 2–100.

Gradually the tank fills because 500(fill motor) is on.
After 100 scans, the oil level rises above the low-level sensor and it becomes open (i.e. false).

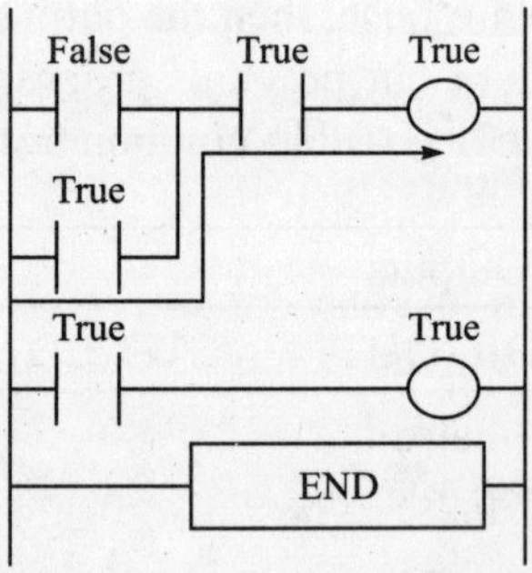

Figure 16.15 Scan 101–1000.

Notice that even when the low-level sensor is false, there is still a path of true logic from left to right. This is why, we used an internal relay. Relay 1000 is latching the output (500) on. It will stay this way until there is no true logic path from left to right (i.e. when 0001 becomes false)

After 1000 scans, the oil level rises above the high-level sensor at it also becomes open (i.e. false).

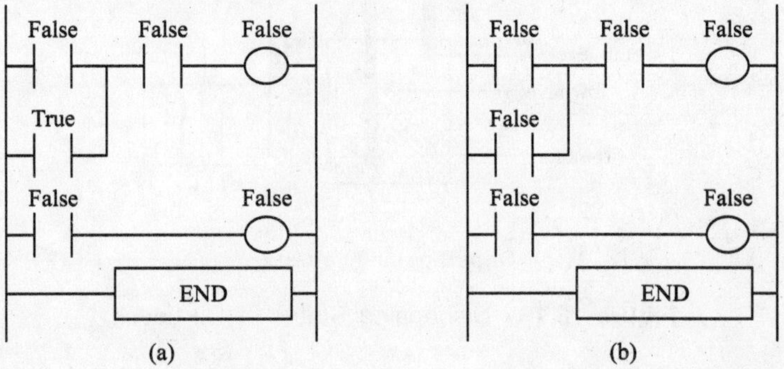

Figure 16.16 (a) Scan 1001, (b) Scan 1002.

Since there is no more true logic path, output 500 is no longer energized (true) and therefore the motor turns off.

After 1050 scans, the oil level falls below the high-level sensor and it will become true again.

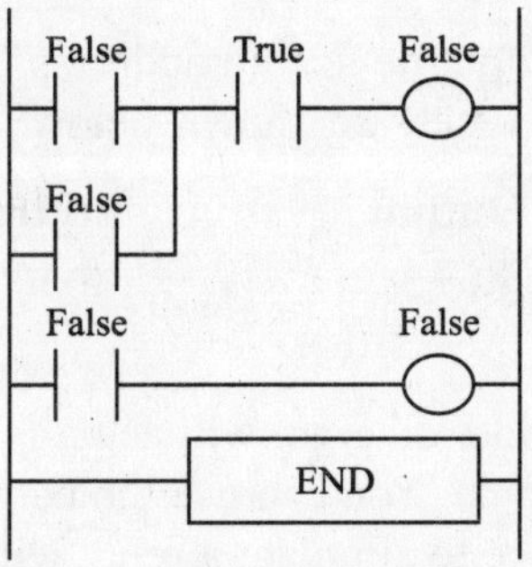

Figure 16.17 Scan 1050.

Notice that even though the high-level sensor became true, there is still NO continuous true logic path and therefore coil 1000 remains false!

After 2000 scans, the oil level falls below the low-level sensor and it will also become true again. At this point, the logic will appear the same as SCAN 1 above and the logic will repeat as illustrated above.

16.6 A LEVEL APPLICATION

Now that we have seen how registers work, let us process a program like PLCs do to enhance our understanding of how the program gets scanned.

Let us consider the following application.

We are controlling lubricating oil being dispensed from a tank. This is possible by using two sensors. We put one near the bottom and one near the top, as shown in Figure 16.18.

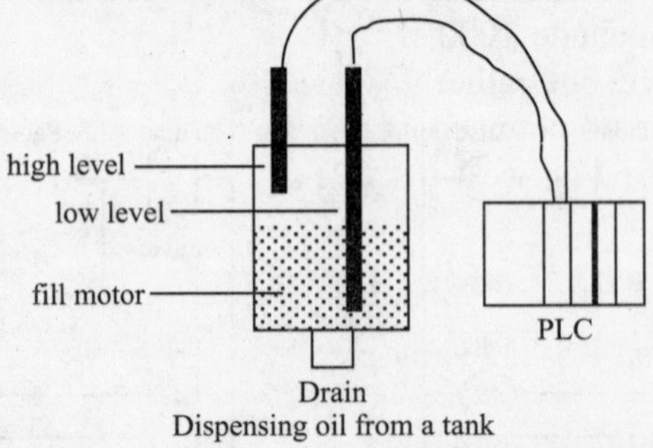

Figure 16.18 Dispensing Status of Oil Level.

Here, we want the fill motor to pump lubricating oil into the tank until the high-level sensor turns on. At that point, we want to turn off the motor until the level falls below the low-level sensor. Then we should turn on the fill motor and repeat the process.

Here we have a need for three I/O (i.e. Inputs/Outputs). Two are inputs (the sensors) and one is an output (the fill motor). Both of our inputs will be **NC (normally closed)** fiber-optic level sensors. When they are NOT immersed in liquid, they will be ON. When they are immersed in liquid, they will be OFF.

We will give each input and output device an address. This lets the PLC know where they are physically connected. The addresses are shown below.

Inputs	Address	Output	Address	Internal Utility Relay
Low	0000	Motor	0500	1000
High	0001			

Figure 16.19 shows how the ladder diagram will actually look like. Notice that we are using an internal utility relay in this example. You can use the contacts of these relays as many times as required. Here they are used twice to simulate a relay with two sets of contacts. Remember, these relays DO NOT physically exist in the PLC but rather they are bits in a register that you can use to simulate a relay.

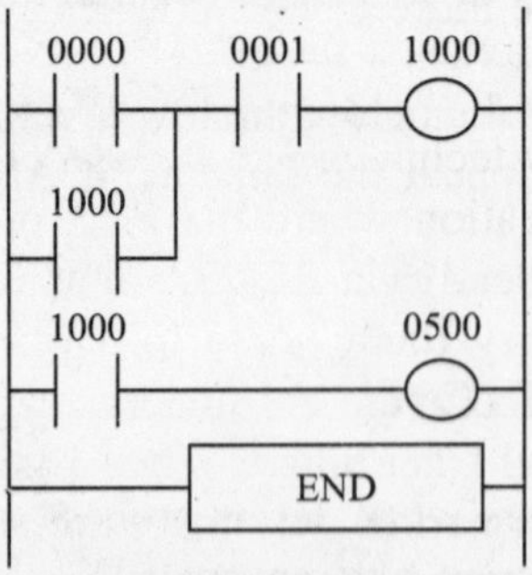

Figure 16.19 Relay Status.

We should always remember that the most common reason for using PLCs in our applications is for replacing real-world relays. The internal utility relays make this action possible.

It is impossible to indicate how many internal relays are included with each brand of PLC. Some include 100's while others include 1000's, while still others include 10's of 1000's! Typically, PLC size (not physical size but rather I/O size) is the deciding factor. If we are using a micro-plc with a few I/O, we do not need many internal relays. If, however, we are using a large PLC with 100's or 1000's of I/O, we will certainly need many more internal relays.

16.6.1 Using RS-232 with Ladder Logic L

Now that we understand what RS-232 is/means, let us see how to use it with our programmable logic control.

We should start out, as always, remembering that a PLC is a PLC is a PLC... In other words, understand the theory first and then figure out how our manufacturer of choice "makes it work". Some manufacturers include RS-232 communication capability in the main processor. Some use the "programming port" for this. Others require you to purchase (i.e. spend extra $'s) a module to "talk RS-232" with an external device. What is an external device, you may be asking? The answer is difficult because there are so many external devices. It may be an operator interface, an external computer, a motor controller, a robot, a vision system, a ... get the point??

To communicate via RS-232, we have to set up a few things. Ask yourself the following questions:

- Where, in data memory, will we store the data to be sent? Essentially we have to store the data we will send... somewhere. Where else but in our data memory!
- Where, in data memory, will we put the data we receive from the external device?
- How will we tell the PLC when it is time to send our data (the data we stored in data memory) out the serial port?
- How will we know when we have received data from our external device?

If you know answers to the above questions, then the rest is easy. If you do not know the above, then make something up and now the rest is easy. Huh??? Simple, pick a memory area to work with and figure out if we can choose the internal relays to use to send and receive data or if the PLC has ones that are dedicated to this purpose. **Sounds too simple??** (No way, it is simple!!)

Before we do it, let us get some more technical terms out of the way so we are on the same playing field.

- **Buffer**-A buffer is a fancy technical word that means a *plastic bag*. In other words, it is a temporary storage location where the PLC or external device stores data it has received (or is waiting to send) via RS-232. When I go to the supermarket to buy my favourite tv dinners, I carry them home in a plastic bag. The plastic bag is not a permanent place for my food (are tv dinners really food??) but rather a temporary storage place for them until I get home. When I get home, I take them out of the bag and cook them. The supermarket was the external device where I got the data (tv dinners) from and my microwave is the PLC. The plastic bag was the buffer (temporary storage place) that was holding my data (tv dinner) until I took them out to use (i.e cook).

- **String**-A string is a cool way of saying "a bunch of characters". The word "hello" is a string. It is a bunch of characters (i.e. h-e-l-l-o) that are connected (strung) together to mean something useful. "43770" is also a string. Although it makes no sense to us, it may mean something valuable to your PLC or external device. It could be a command that tells your robot to send out its current coordinates (or it could simply be the word "hello" upside down).
- **Concatenate**-This word is a mouthful. Simply put, it means to combine two strings together to make one string. An example is combining the two strings "laser" and "jet" together to make one string "laserjet".

With the *mumbo-jumbo* out of the way, let us see it in action. Again, the memory locations and relays vary by manufacturer, but the theory is universal.

1. We assign memory locations DM100 through DM102 to be where we will put our data before we send it out the serial port. Note that many PLCs have dedicated areas of memory for this and only this purpose.
2. We will assign internal relay 1000 to be our send relay. In other words, when we turn on 1000, the PLC will send the data in DM100–DM102 out the serial port to our external device. Note again that many PLCs have dedicated relays (special utility relays) for this and only this purpose. It is great when the manufacturer makes our life easy!

We will send the string "alr" out the PLC serial port to an operator interface when our temp sensor input turns on. This means our oven has become too hot. When the operator interface receives this string, it will display an alarm message for the operator to see. Look back on the ASCII chart and you will see that "alr" is hexadecimal 61, 6C, 72 (a = 61, l = 6C, r = 72). We will write these ASCII characters (in hexadecimal form) into the individual data memory locations. We will use DM100–102. **How?** Remember the LDA or MOV instruction? We will turn on our send relay (1000) when our temperature sensor (0000) turns on. The ladder is shown in Figure 16.20.

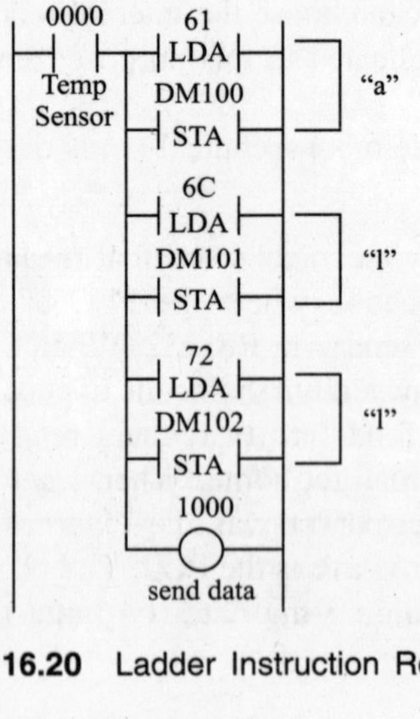

Figure 16.20 Ladder Instruction Register.

Some PLCs may not have dedicated internal relays that send out our data through the RS-232 port. We may have to assign them manually. Further, some PLCs will have a special instruction to tell us where the data is stored and when to send the data. This instruction is commonly called AWT (ASCII Write) or RS. The theory is always the same though. Put the data in a memory location and then turn on a relay to send the data.

Latch instructions

Now that we understand how inputs and outputs are processed by the PLC, let us look at a variation of our regular outputs. Regular output coils are of course an essential part of our programs but we must remember that they are only TRUE when ALL INSTRUCTIONS before them on the rung are also TRUE. What happens if they are not? Then, of course, the output will become false (turn off).

Think back to the lunch bell as an example. What would have happened if we could not find a "push on/push off" switch? Then we would have had to keep pressing the button for as long as we wanted the bell to sound (A momentary switch). The latching instructions let us use momentary switches and program the plc, so that when we push one, the output turns on and when we push another, the output turns off.

Maybe now you are saying to yourself, "What the heck is he talking about?" (It is also what I am thinking!) So let us do a real-world example.

Picture the remote control for your TV. It has a button for ON and another for OFF (Mine does, anyway). When I push the ON button, the TV turns on. When I push the OFF button, the TV turns off. I do not have to keep pushing the ON button to keep the TV on. This would be the function of a latching instruction.

The latch instruction is often called a SET or OTL (output latch). The unlatch instruction is often called a RES (reset), OUT (output unlatch) or RST (reset). Figure 16.21 shows how to use them in a program.

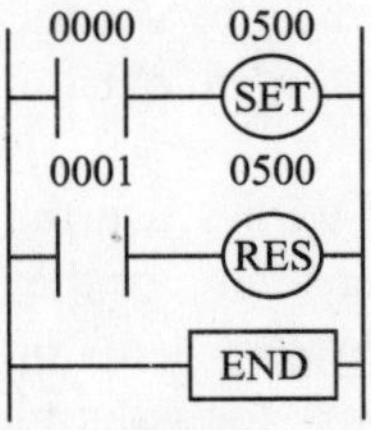

Figure 16.21 Use of SET, RES in a Program.

Here we are using two momentary push button switches. One is physically connected to input 0000 while the other is physically connected to input 0001. When the operator pushes switch 0000, the instruction "set 0500" will become true and output 0500 physically turns on. Even after the operator stops pushing the switch, the output (0500) will remain on. It is latched on. The only way to turn off output 0500 is, turn on input 0001. This will cause the instruction "res 0500" to become true thereby unlatching or resetting output 0500.

Here is something to think about. What would happen if input 0000 and 0001 both turn on at the exact same time.

Will output 0500 be latched or unlatched?

To answer this question, we have to think about the scanning sequence. The ladder is always scanned from top to bottom, left to right. The first thing in the scan is to physically look at the inputs. 0000 and 0001 are both physically on. Next, the PLC executes the program. Starting from the top left, input 0000 is true, therefore it should set 0500. Next, it goes to the next rung and since input 0001 is true, it should reset 0500. The last thing it said was to reset 0500. Therefore, on the last part of the scan when it updates the outputs it will keep 0500 off (i.e. reset 0500).

16.7 COUNTERS

A counter is a simple device intended to do one simple thing—count. Using them, however, can sometimes be a challenge because every manufacturer (for whatever reason) seems to use them in a different way. Rest assured that the following information will let you simply and easily program anybody's counters.

What kinds of counters are there? Well, there are up-counters (they only count up 1,2,3...). These are called CTU (count up), CNT,C, or CTR. There are down counters (they only count down 9,8,7,...). These are typically called CTD (count down) when they are a separate instruction. There are also up-down counters (they count up and/or down 1,2,3,4,3,2,3,4,5,...). These are typically called UDC (up-down counter) when they are separate instructions.

Many manufacturers have only one or two types of counters but they can be used to count up, down or both. *Confused yet*? Can you say "no standardization"? Don't worry, the theory is all the same regardless of what the manufacturers call them. A counter is a counter is a counter...

To further confuse the issue, most manufacturers also include a limited number of high-speed counters. These are commonly called HSC (high-speed counter), CTH (CounTer High-speed?) or whatever.

Typically, a high-speed counter is a "*hardware*" device. The normal counters listed above are typically *software* counters. In other words, they do not physically exist in the PLC, rather they are simulated in software. Hardware counters do exist in the PLC and they are not dependent on scan time.

A good rule of thumb is simply to always use the normal (software) counters unless the pulses you are counting will arrive faster than 2 × the scan time (i.e. if the scan time is 2 ms and pulses will be arriving for counting every 4 ms or longer, then use a software counter. If they arrive faster than every 4 ms (3 ms for example) then use the hardware (high-speed) counters (2 × scan time = 2 × 2 ms = 4 ms).

To use them, we must know three things:

1. Where the pulses that we want to count are coming from. Typically, this is from one of the inputs (a sensor connected to input 0000, for example).
2. How many pulses we want to count before we react. Let us count five widgets before we box them, for example.
3. When/how we will reset the counter so it can count again. After we count five widgets, let us reset the counter, for example.

When the program is running on the PLC, the program typically displays the current or *accumulated* value for us, so we can see the current count value.

Typically, counters can count from 0 to 9999, –32,768 to +32,767 or 0 to 65535. Why the weird numbers? Because most PLCs have 16-bit counters. We will get into what this means in a later chapter, but for now suffice it to say that 0–9999 is 16-bit BCD (binary coded decimal) and that –32,768 to 32767 and 0 to 65535 is 16-bit binary.

Here are some of the instruction symbols we will encounter (depending on which manufacturer we choose) and how to use them. Remember that while they may look different, they are all used basically the same way. If we can set up one, we can set up any of them.

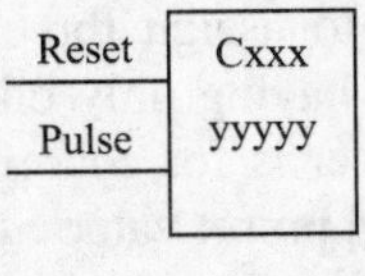

Figure 16.22 A Counter.

In this counter we need two inputs.

One goes before the reset line. When this input turns on, the current (accumulated) count value will return to zero.

The second input is the address where the pulses we are counting are coming from.

For example, if we are counting how many widgets pass in front of the sensor that is physically connected to input 0001, then we would put normally open contacts with the address 0001 in front of the pulse line.

Cxxx is the name of the counter. If we want to call it counter 000, then we would put "C000" here.

yyyyy is the number of pulses we want to count before doing something. If we want to count five widgets before turning on a physical output to box them, we would put five here. If we wanted to count 100 widgets, then we would put 100 here, etc. When the counter is finished (i.e we counted yyyyy widgets), it will turn on a separate set of contacts that we also label Cxxx.

Note that the counter accumulated value ONLY changes at the off to on transition of the pulse input.

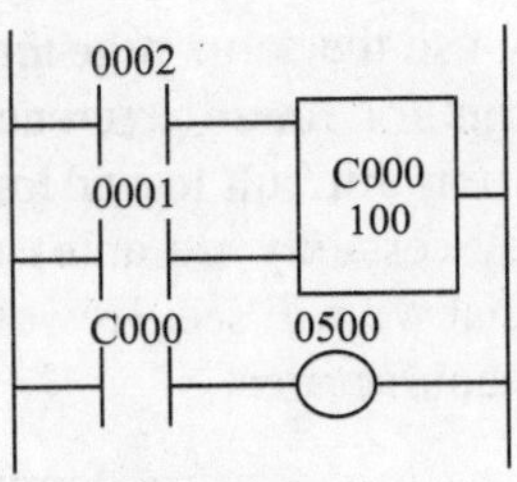

Figure 16.23 Pulse.

Here is the symbol on a ladder showing how we set up a counter (we will name it counter 000) to count 100 widgets from input 0001 before turning on output 500. Sensor 0002 resets the counter.

Figure 16.24 shows the symbol for an up and down counter. We will use the same abbreviation as we did for the example above (i.e. UDCxxx and yyyyy).

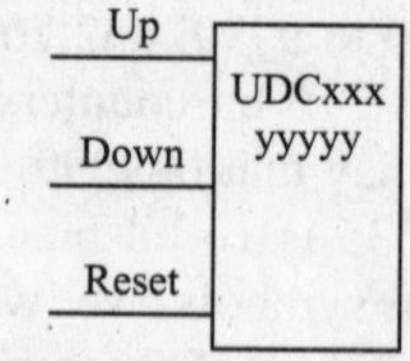

Figure 16.24 Up-down Counter.

In this up-down counter we need to assign three inputs. The reset input has the same function as above. However, instead of having only one input for the pulse counting now has two. One is for counting up and the other is for counting down. In this example, we will call the counter UDC000 and we will give it a preset value of 1000 (We will count 1000 total pulses). For inputs we will use a sensor which will turn on input 0001 when it sees a target, and another sensor at input 0003 will also turn on when it sees a target. When input 0001 turns on, we count up and when input 0003 turns on, we count down. When we reach 1000 pulses, we will turn on output 500. Again note that the counter accumulated value ONLY changes at the off to on transition of the pulse input. The ladder diagram is shown in Figure 16.25.

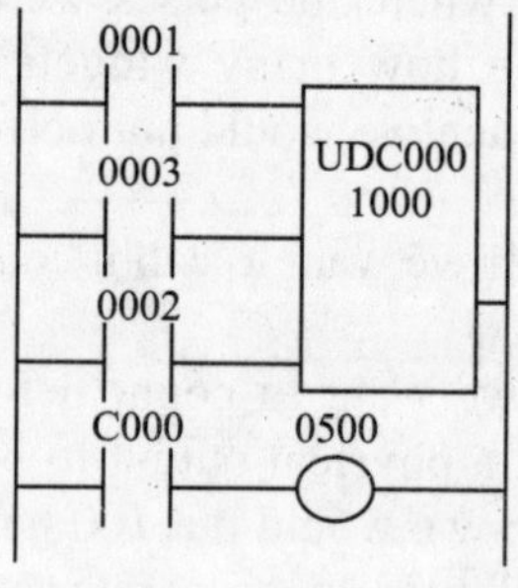

Figure 16.25 Ladder Diagram.

One important thing to note is that counters and timers can not have the same name (in most PLCs). This is because they typically use the same registers. We have not learned about timers yet, but you might make a note of this for future reference because it is pretty important.

Well, the counters above might seem difficult to understand but they are actually quite easy once we get used to using them. They certainly are an essential tool. They are also one of the least *standardized* basic instructions that we will see. However, always remember that the theory is the same from manufacturer to manufacturer!

16.8 TIMERS

Let us now see how a timer works. What is a timer? It is exactly what the word says... it is an instruction that waits a set amount of time before doing something. Sounds simple, does not it?

When we look at the different kinds of timers available, the fun begins. As always, different types of timers are available with different manufacturers. Most of them are discussed here in brief.

- **On-delay timer**-This type of timer simply "delays turning on". In other words, after our sensor (input) turns on, we wait x-seconds before activating a solenoid valve (output). This is the most common timer. It is often called TON (timer on-delay), TIM (timer) or TMR (timer).
- **Off-delay timer**-This type of timer is the opposite of the on-delay timer. This timer simply "delays turning off". After our sensor (input) sees a target, we turn on a solenoid (output). When the sensor no longer sees the target, we hold the solenoid on for x-seconds before turning it off. It is called a TOF (timer off-delay) and is less common than the on-delay type (i.e. few manufacturers include this type of timer).
- **Retentive or accumulating timer**-This type of timer needs two inputs. One input starts the timing event (i.e. the clock starts ticking) and the other resets it. The on/off delay timers above would be reset if the input sensor was not on/off for the complete timer duration. This timer, however, holds or retains the current elapsed time when the sensor turns off in mid-stream. For example, we want to know how long a sensor is on for during one hour. If we use one of the above timers, they will keep resetting when the sensor turns off/on. This timer, however, will give us a total or accumulated time. It is often called an RTO (retentive timer) or TMRA (accumulating timer).

Let us now see how to use them. We typically need to know two things:

- **What will enable the timer?** Typically, this is one of the inputs (a sensor connected to input 0000, for example).
- **How long we want to delay before we react?** Let us wait five seconds before we turn on a solenoid, for example.

When the instructions before the timer symbol are true, the timer starts *ticking*. When the time elapses, the timer will automatically close its contacts. When the program is running on the plc, it typically displays the elapsed or *accumulated* time for us so we can see the current value. Typically, timers can tick from 0 to 9999 or 0 to 65535 times.

Why the weird numbers? Again it is because most PLCs have 16-bit timers. We will get into what this means in a later chapter, but for now suffice it to say that 0–9999 is 16-bit BCD (binary coded decimal) and that 0 to 65535 is 16-bit binary. Each tick of the clock is equal to x-seconds.

Typically, each manufacturer offers several different ticks. Most manufacturers offer 10 and 100 ms increments (ticks of the clock). An "**ms**" is a milli-second or 1/1000th of a second. Several manufacturers also offer 1 ms as well as 1 second increments. These different increment timers work the same manner, but sometimes they have different names to show their time base. Some are TMH (high-speed timer), TMS (super high-speed timer) or TMRAF (accumulating fast timer).

Shown below is a typical timer instruction symbol (Figure 16.26) we will encounter (depending on which manufacturer we choose) and how to use it. Remember that while they may look different, they are all used basically the same way. If we can set up one, we can set up any of them.

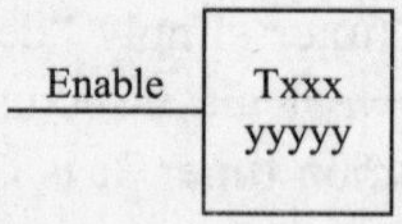

Figure 16.26 Timer Instruction Symbol.

This timer is the on-delay type and is named Txxx. When the enable input is on, the timer starts to tick. When it ticks yyyyy (the preset value) times, it will turn on its contacts that we will use later in the program. Remember that the duration of a tick (increment) varies with the vendor and the time base used. (i.e. a tick might be 1 ms or 1 second or...).

Below is the symbol shown on a ladder diagram (Figure 16.27).

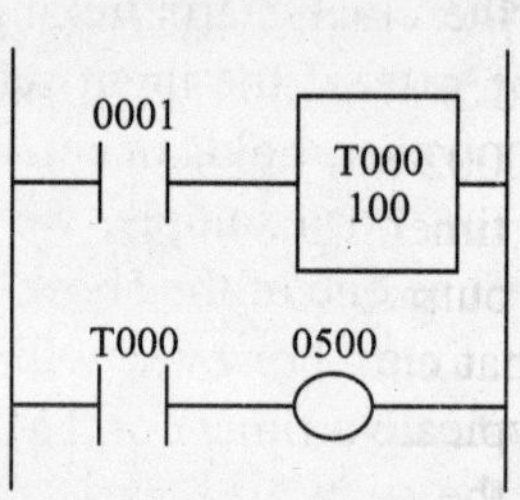

Figure 16.27 Ladder Diagram.

In Figure 16.27 we wait for input 0001 to turn on. When it does, timer T000 (a 100 ms increment timer) starts ticking. It will tick 100 times. Each tick (increment) is 100 ms so the timer will be a 10000 ms (i.e. 10 second) timer. 100 ticks × 100 ms = 10,000 ms. When 10 seconds have elapsed, the T000 contacts close and 500 turns on. When input 0001 turns off(false), the timer T000 will reset back to 0 causing its contacts to turn off(become false), thereby making output 500 turn back off.

An accumulating timer would look similar to Figure 16.28:

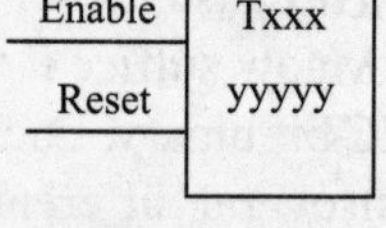

Figure 16.28 Accumulating Timer.

This timer is named Txxx. When the enable input is on, the timer starts to tick. When it ticks yyyyy (the preset value) times, it will turn on its contacts that we will use later in the program. Remember that the duration of a tick (increment) varies with the vendor and the time base used (i.e. a tick might be 1 ms or 1 second or...). If, however, the enable input turns off before the timer has completed, the current value will be retained. When the input turns back on, the timer will continue from where it left off. The only way to force the timer back to its preset value to start again is to turn on the reset input. The symbol is shown in the ladder diagram (Figure 16.29).

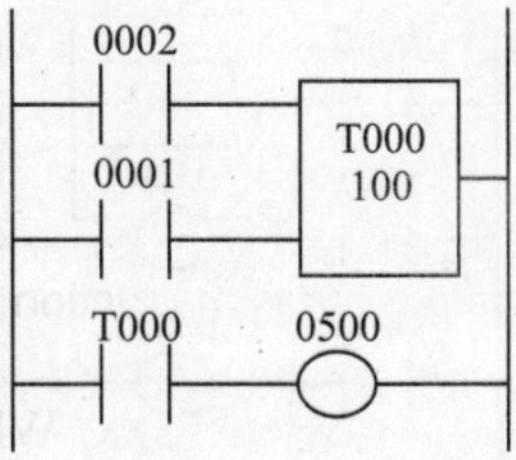

Figure 16.29 Ladder Diagram.

In this diagram, we wait for input 0002 to turn on. When it does, timer T000 (a 10 ms increment timer) starts ticking. It will tick 100 times. Each tick (increment) is 10 ms so the timer will be a 1000 ms (i.e. 1 second) timer. 100 ticks × 10 ms = 1,000 ms. When 1 second has elapsed, the T000 contacts close and 500 turns on. If input 0002 turns back off, the current elapsed time will be retained. When 0002 turns back on, the timer will continue where it left off. When input 0001 turns on (true), the timer T000 will reset back to 0 causing its contacts to turn off (become false), thereby making output 500 turn back off.

One important thing to note is that **counters and timers cannot have the same name** (in most PLCs). This is because they typically use the same registers.

Always remember that although the symbols may look different, they all operate the same way. Typically, the major differences are in the duration of the ticks (increments).

16.9 MASTER CONTROLS

Let us now look at what are called master controls. Master controls can be thought of as *emergency stop switches*. An emergency stop switch typically is a big red button on a machine that will shut it off in cases of emergency. Next time you are at the local gas station, look near the door on the outside to see an example of an e-stop. We are not implying that this instruction is a substitute for a "hard wired" e-stop switch. There is no substitute for such a switch! Rather, it is just an easy way to get to understand them.

The master control instruction is typically used in pairs with a master control reset. However, this varies with manufacturer. Some use MCR in pairs instead of teaming it with another symbol. It is commonly abbreviated as MC/MCR (master control/master control reset), MCS/MCR (master control set/master control reset) or just simply MCR (master control reset).

Here is one example of how a master control symbol looks (Figure 16.30).

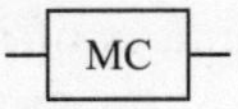

Figure 16.30 Appearance of Master Control Symbol.

Figure 16.31 An Example of a Master Control Reset.

To make things interesting, many manufacturers make them act differently. Let us now take a look at how it is used in a ladder diagram. Consider the following example.

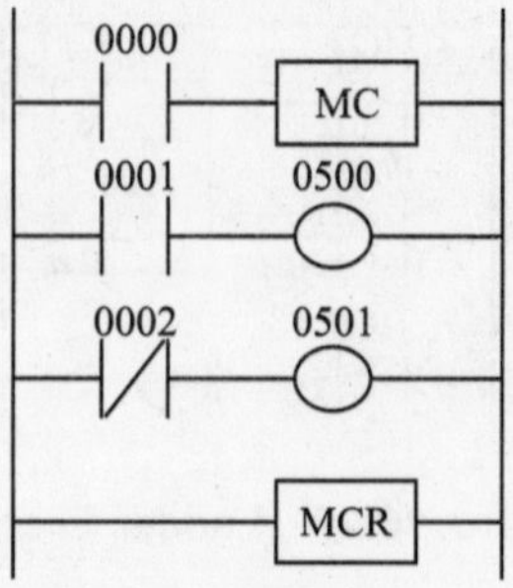

Figure 16.32 Use of MC and MCR in a Ladder Diagram.

Here is how different PLCs will run this program.

Manufacturer X-In this example, rungs 2 and 3 are only executed when input 0000 is on (true). If input 0000 is not true, the PLC pretends that the logic between the mc and mcr instructions does not exist. It would, therefore, bypass this block of instructions and immediately go to the rung after the MCR instruction.

Conversely, if input 0000 is true, the PLC would execute rungs 2 and 3 and update the status of outputs 0500 and 0501 accordingly. So, if input 0000 is true, program execution goes to rung 2. If input 0001 is true, 0500 will be true and hence it will turn on when the PLC updates the outputs. If input 0002 is true (i.e. physically off), 0501 will be true and, therefore, it will turn on when the PLC updates the outputs.

MCR just tells the PLC "that's the end of the MC/MCR block".

In this PLC, scan time is not extended when the MC/MCR block is not executed, because the PLC pretends the logic in the block does not exist. In other words, the instructions inside the block are not seen by the PLC and, therefore, it does not execute them.

Manufacturer Y-In this example, rungs 2 and 3 are always executed regardless of the status of input 0000. If input 0000 is not true, the PLC executes the MC instruction (i.e. MC becomes true). It then forces all the input instructions inside the block to be off. If input 0000 is true, the MC instruction is made to be false.

Then, if input 0000 is true, program execution goes to rung 2. If input 0001 is true, 0500 will be true and hence it will turn on when the PLC updates the outputs. If input 0002 is true (i.e. physically off), 0501 will be true and, therefore, it will turn on when the PLC updates the outputs. MCR just tells the PLC "that's the end of the MC/MCR block". When input 0000 is false, inputs 0001 and 0002 are forced off regardless of if they are physically on or off. Therefore, outputs 0500 and 0501 will be false.

The difference between manufacturers X and Y above is that in the Y scheme, the scan time will be the same (well close to the same) regardless of if the block is on or off. This is because the plc sees each instruction whether the block is on or off.

Most of the manufacturers will make a previously latched instruction (one that is inside the MC/MCR block) retain its previous condition.

If it was true before, it will remain true.

If it was false before, it will remain false.

Timers should not be used inside the MC/MCR block because some manufacturers will reset them to zero when the block is false, whereas other manufacturers will have them retain the current time state.

Counters typically retain their current counted value.

Here is the part to note most of all. When the MC/MCR block is off (i.e. input 0000 would be false in the ladder example shown previously), an OUTB (OutBar or OutNot) instruction would not be physically on. It is forced physically off.

Figure 16.33 OutBar Instruction.

16.10 SHIFT REGISTERS

In many applications, it is necessary to store the status of an event that has previously happened. As we have seen in previous chapters, this is a simple process. But what do we do if we must store many previous events and act upon them later. The answer is, we call upon the shift register instruction.

We use a register or group of registers to form a train of bits (cars) to store the previous on/off status. Each new change in status gets stored in the first bit and the remaining bits get shifted down the train. Huh? Read on.

The shift register goes by many names. SFT (ShiFT), BSL (Bit Shift Left), SFR (Shift Forward Register) are some of the common names. These registers shift the bits to the left. BSR (Bit Shift Right) and SFRN (Shift Forward Register Not) are some examples of instructions that shift bits to the right. We should note that not all manufacturers have shift registers that shift data to the right, but most all do have left shifting registers.

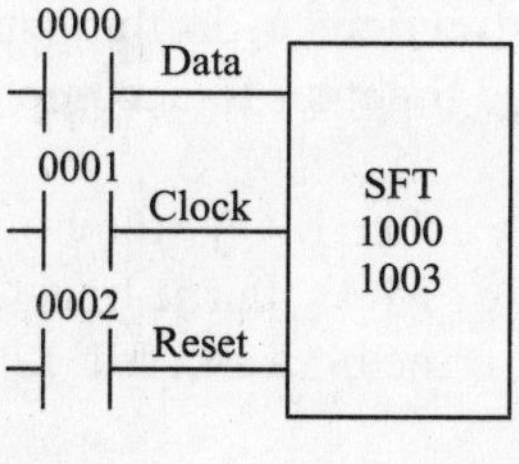

Figure 16.34 The Shift Register.

A typical shift register instruction has a symbol like that shown above. Notice that the symbol needs three inputs and has some data inside the symbol.

The reasons for each input are as follows:

- **Data**-The data input gathers the true/false status that will be shifted down the train. When the data input is true, the first bit (car) in the register (train) will be 1. This data is only entered into the register (train) on the rising edge of the clock input.

- **Clock**-The clock input tells the shift register to *do its thing*. On the rising edge of this input, the shift register shifts the data one location over inside the register and enters the status of the data input into the first bit. On each rising edge of this input, the process will repeat.
- **Reset**-The reset input does just what it says. It clears all the bits inside the register were using to 0.

The 1000 inside the shift register symbol is the location of the first bit of our shift register. If we think of the shift register as a train (a choo-choo train that is), then this bit is the locomotive. The 1003 inside the symbol above is the last bit of our shift register. It is the caboose. Therefore, we can say that 1001 and 1002 are cars in between the locomotive and the caboose. They are intermediate bits. So, this shift register has 4 bits (i.e. 1000,1001,1002,1003).

Getting and moving data

Let us now start working with some data. This is what can be considered to be getting into the *advanced* functions of a PLC. This is also the point where we will see some marked differences between many of the manufacturer's functionality and implementation. On the lines that follow we will explore two of the most popular ways to get and manipulate data.

Why do we want to get or acquire data? The answer is simple. Let us say that we are using one of the manufacturer's optional modules. Perhaps it is an A/D module. This module acquires analog signals from the outside world (a varying voltage or current) and converts the signal to something the PLC can understand (a digital signal, i.e. 1's and 0's). Manufacturers automatically store this data into memory locations for us. However, we have to get the data out of there and move it somewhere else; otherwise, the next analog sample will replace the one we just took. In other words, move it or lose it! Something else we might want to do is store a constant (i.e. fancy word for a number), get some binary data off the input terminals (maybe a thumbwheel switch is connected there, for example), do some math and store the result in a different location, etc.

As was stated before, there are typically two common instruction *sets* to accomplish this. Some manufacturers use a single instruction to do the entire operation, while others use two separate instructions. The two are used together to accomplish the final result. Let us now look briefly at each instruction.

The single instruction is commonly called MOV (move). Some vendors also include a MOVN (move not). It has the same function of MOV but it transfers the data in inverted form (i.e. if the bit was a 1, a 0 is stored/moved, or if the bit was a 0, a 1 is stored/moved). The MOV typically looks like as shown in Figure 16.35.

MOV
XXXX
YYYY

Figure 16.35 MOV Instruction Symbol.

The paired instruction typically is called LDA (LoaD Accumulator) and STA (STore Accumulator). The accumulator is simply a register inside the CPU where the PLC stores data temporarily while it is working. The LDA instruction typically looks like that shown below, while the STA instruction looks like that shown below to the right.

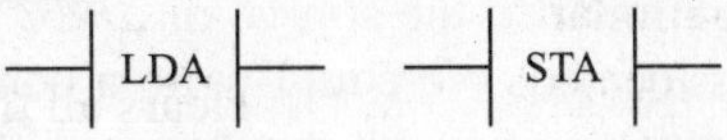

Figure 16.36 LoaD Accumulator (LDA) and STore Accumulator (STA).

Regardless of whether we use the one symbol or two symbol instruction set (we have no choice as it depends on whose PLC we use), they work the same way.

Let us see the single instruction first. The MOV instruction needs to know two things from us.

- **Source (xxxx)**-This is where the data we want to move is located. We could write a constant here (2222, for example). This would mean our source data is the number 2222. We could also write a location or address where the data we want to move is located. If we wrote DM100, this would move the data that is located in data memory 100.
- **Destination (yyyy)**-This is the location where the data will be moved to. We write an address here. For example, if we write DM201 here, the data would be moved into data memory 201. We could also write 0500 here. This would mean that the data would be moved to the physical outputs. 0500 would have the least significant bit, 0501 would have the next bit... 0515 would have the most significant bit. This would be useful if we had a binary display connected to the outputs and we wanted to display the value inside a counter for the machine operator at all times (for example).

The ladder diagram to do this would look similar to that shown in Figure 16.37.

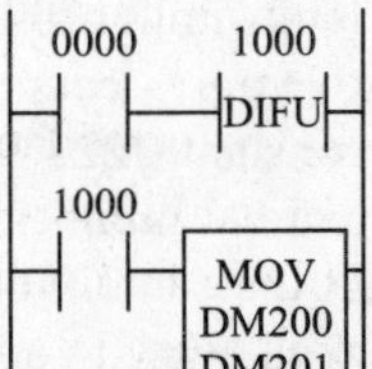

Figure 16.37 Ladder Diagram.

Notice that we are also using a "difu" instruction here. The reason is simply because if we did not, the data would be moved during each and every scan. Sometimes this is a good thing (for example, if we are acquiring data from an A/D module), but other times it is not (for example, an external display would be unreadable because the data changes too much).

The ladder shows that each time real-world input 0000 becomes true, difu will become true for only one scan. At this time, LoaD 1000 will be true and the PLC will move the data from data memory 200 and put it into data memory 201. Simple but effective. If, instead of DM200, we had written 2222 in the symbol, we would have moved (written) the number (constant) 2222 into DM201.

The two symbol instruction works in the same method but looks different. To use them, we must also supply two things, one for each instruction.

- **LDA**-This instruction is similar to the source of a MOV instruction. This is where the data we want to move is located. We could write a constant here (2222, for example). This would mean our source data is the number 2222. We could also write a location or address of where the data we want to move is located. If we wrote DM100, this would move the data that is located in data memory 100.
- **STA**-This instruction is similar to the destination of a MOV instruction. We write an address here. For example, if we write DM201 here, the data would be moved into data memory 201. We could also write 0500 here. This would mean that the data would be moved to the physical outputs. 0500 would have the least significant bit, 0501 would have the next bit... 0515 would have the most significant bit. This would be useful if we had a binary display connected to the outputs and we wanted to display the value inside a counter for the machine operator at all times (for example).

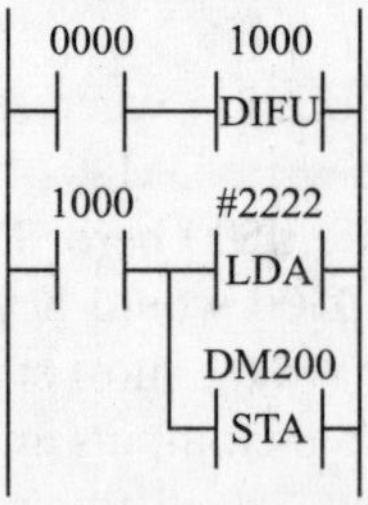

Figure 16.38 Ladder Diagram.

The ladder diagram to do this would look similar to that shown above. Here again, we notice that we are using a one-shot so that the move only occurs once for each time input 0000 becomes true. In this ladder, we are moving the constant 2222 into data memory 200. The "#" is used by some manufacturers to symbolize a decimal number. If we just used 2222, this PLC would think it meant address 2222. PLCs are all the same... but they are all different.

We can think of this instruction as the gateway to advanced instructions. I am sure you will find it useful and invaluable. Many advanced functions are impossible without this instruction.

Boolean math

Let us now take a look at some simple "Boolean math". Boolean math lets us do some very basic functions with the bits in our registers. These basic functions typically include AND, OR and EXOR functions. Each is described below.

- **AND**-This function enables us to use the truth table in next page. Here, we can see that the AND function is very much related to multiplication. We see this because the only time the result is true (i.e. 1) is when both operators A AND B are true (i.e. 1).
  The AND instruction is useful when your PLC does not have a masking function. Oh yeah, a masking function enables a bit in a register to be "left alone" when working on a bit level. This is simply because any bit that is ANDed with itself will remain the value

it currently is. For example, if you wanted to clear (make them 0) only 12 bits in a 16-bit register, you might AND the register with 0's everywhere except in the 4 bits you wanted to maintain the status of clear.
See the truth table below to figure out what we mean. (1 AND 1 = 1, 0 AND 0 = 0)

Result = A *AND* B

A	B	Result
0	0	0
1	0	0
0	1	0
1	1	1

- **OR**-These functions are based upon the truth table below. Here, we can see that the OR function is very much related to addition. We see this because the only time the result is true (i.e. 1) is when operator A OR B is true (i.e. 1). Obviously, when they are both true, the result is true. (If A OR B is true...)

Result = A *OR* B

A	B	Result
0	0	0
1	0	1
0	1	1
1	1	1

- **EXOR**-This function enables us to use the truth table below. Here, we can see that the EXOR (XOR) function is not related to anything. An easy way to remember the results of this function is to think that A and B must be one or the other case, exclusively. In other words, they must be opposite of each other. When they are both the same (i.e. A=B), the result is false (i.e. 0).
This is sometimes useful when you want to compare bits in two registers and highlight which bits are different. It is also needed when we calculate some checksums. A checksum is commonly used as error checking in some communications protocols.

Result = A *XOR* B

A	B	Result
0	0	0
1	0	1
0	1	1
1	1	0

The ladder logic instructions are commonly called AND, ANDA, ANDW, OR, ORA, ORW, XOR, EORA XORW.

As we saw with the MOV instruction, there are generally two common methods used by the majority of PLC makers. The first method includes a single instruction that asks us for a few key pieces of information. This method typically requires:

- **Source A**-This is the address of the first piece of data we will use. In other words, it is the location in memory of where A is.

- **Source B**-This is the address of the second piece of data we will use. In other words, it is the location in memory of where B is.
- **Destination**-This is the address where the result will be put. For example, if A AND B = 0, the result (0) would automatically be put into this *destination* memory location.

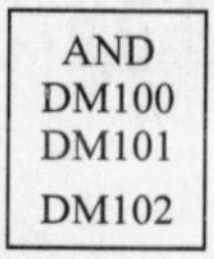

Figure 16.39 AND Symbol.

The instructions above typically have a symbol that looks like as shown in Figure 16.39. Of course, the word AND would be replaced by OR or XOR. In this symbol, the source A is DM100, the source B is DM101, and the destination is DM102. Therefore, we have simply created the equation DM100 AND DM101 = DM102. The result is automatically stored into DM102. The Boolean functions on a ladder, diagram are shown in Figure 16.40.

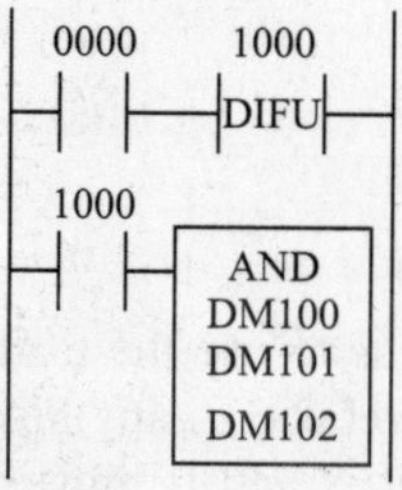

Figure 16.40 The Boolean Functions on a Ladder Diagram.

Please note that once again we are using a one-shot instruction. As we have seen before, this is because if we did not use it, we would execute the instruction on every scan. Odds are good that would only want to execute the function one time when input 0000 becomes true.

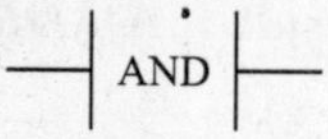

Figure 16.41 AND Symbol (Dual Instruction Method).

The dual instruction method would use a symbol similar to that shown above (Figure 16.41). In this method, we give this symbol only the source B location. The source A location is given by the LDA instruction. The destination would be included in the STA instruction.

The ladder diagram (Figure 16.42) explain in the above said.

The results are the same as the single instruction method shown above. It should be noted that although the symbol and ladder diagrams above show the AND instruction, OR or EXOR can be used as well. Simply substitute the word "AND"

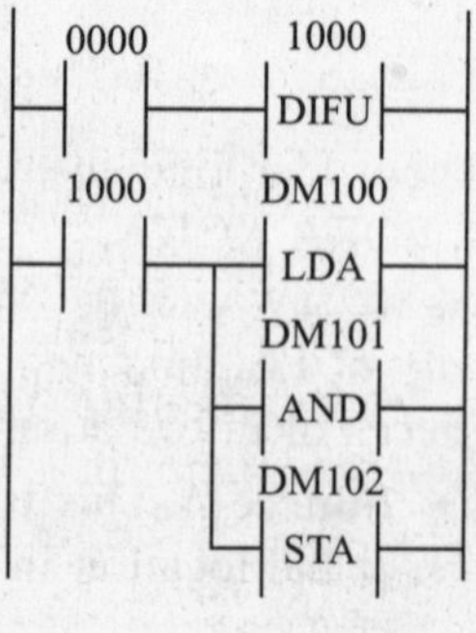

Figure 16.42 Ladder Diagram.

within the instruction to be either “OR” or “EXOR”. The results will be the same as shown in their respective truth tables.

We should always remember that the theory is most important. If we can understand the theory of why things happen as they do, we can use anybody’s PLC. If we refer to the manufacturers documentation, we can find out the details for the particular PLC we are using. Try to find the theory in that documentation and you might come up short. The details are insignificant while the theory is very significant.

16.11 MATH INSTRUCTIONS

Let us now look at using some basic math functions on our data. Many times in our applications we must execute some type of mathematical formula on our data. It is a rare occurrence when our data is actually *exactly* what we needed.

As an example, let us say we are manufacturing widgets. We do not want to display the total number we have made today, but rather we want to display how many more we need to make today to meet our quota. Let us say our quota for today is 1000 pieces. We will say X is our current production. Therefore, we can figure that 1000–X = widgets left to make. To implement this formula, we obviously need some math capability.

In general, PLCs almost always include these math functions:

- **Addition**-The capability to add one piece of data to another. It is commonly called ADD.
- **Subtraction**-The capability to subtract one piece of data from another. It is commonly called SUB.
- **Multiplication**-The capability to multiply one piece of data by another. It is commonly called MUL.
- **Division**-The capability to divide one piece of data from another. It is commonly called DIV.

As we saw with the MOV instruction, there are generally two common methods used by the majority of PLC makers. The first method includes a single instruction that asks us for a few key pieces of information. This method typically requires:

- **Source A**-This is the address of the first piece of data we will use in our formula. In other words, it is the location in memory of where the first “number” is that we use in the formula.
- **Source B**-This is the address of the second piece of data we will use in our formula. In other words, it is the location in memory of where the second “number” is that we use in the formula. Note that typically we can only work with two pieces of data at a time. In other words, we cannot work directly with a formula like 1 + 2 + 3. We would have to break it up into pieces. Like 1 + 2 = X then X + 3 = our result.
- **Destination**-This is the address where the result of our formula will be put. For example, if 1 + 2 = 3, (I hope it still does!), 3 would automatically be put into this destination memory location.

The instructions above typically have a symbol that looks like Figure 16.43. Of course, the word ADD would be replaced by SUB, MUL, DIV, etc. In this symbol, the source A is DM100, the source B is DM101, and the destination is DM102. Therefore, the formula is simply whatever value is in DM100 + whatever value is in DM101. The result is automatically stored into DM102.

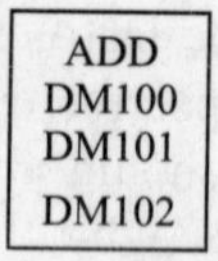

Figure 16.43 ADD Symbol.

Figure 16.44 shows how to use math functions on a ladder diagram. Please note that once again we are using a one-shot instruction. As we have seen before, this is because if we did not use it, we would execute the formula on every scan. Odds are good that would only want to execute the function one time when input 0000 becomes true. If we had previously put the number 100 into DM100 and 200 into DM101, the number 300 would be stored in DM102 (i.e. 100 + 200 = 300, right??).

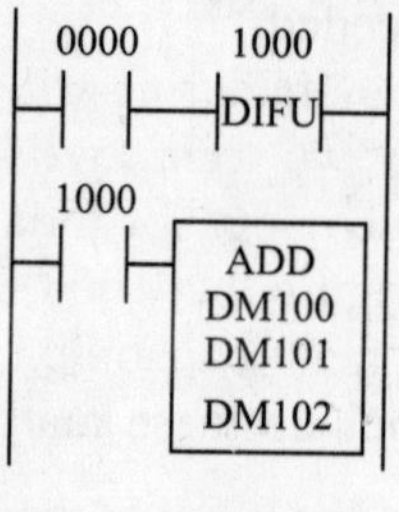

Figure 16.44 Math Functions on a Ladder Diagram.

The dual instruction method would use a symbol similar to that shown in Figure 16.45. In this method, we give this symbol only the source B location. The source A location is given by the LDA instruction. The destination would be included in the STA instruction.

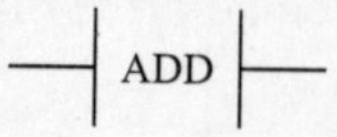

Figure 16.45 ADD Symbol (Dual Method).

Shown below (Figure 16.46) is a ladder diagram showing what we mean.

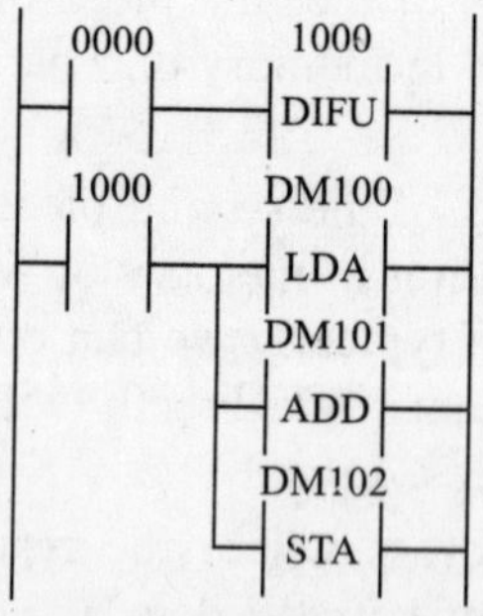

Figure 16.46 Ladder Diagram.

The results are the same as the single instruction method shown before.
Many PLCs also include other math capabilities. Some of these functions could include:

- Square roots
- Scaling
- Absolute value
- Sine
- Cosine
- Tangent
- Natural logarithm
- Base 10 logarithm
- X^Y (X to the power of Y)
- Arcsine (tan, cos)
- and more....check with the manufacturer to be sure.

Some PLCs can use floating point math as well. Floating point math is simply using decimal points. In other words, we could say that 10 divided by 3 is 3.333333 (floating point). Or we could say that 10 divided by 3 is 3 with a remainder of 1(long division). Many micro/mini PLCs do not include floating point math. Larger systems typically do.

DC inputs

Let us now take a look at how the input circuits of a PLC work. This will give us a better understanding of how we should wire them up. Typically, dc input modules are available that will work with 5, 12, 24 and 48 volts. Be sure to purchase the one that fits your needs based upon the input devices you will use.

We will first look at how the dc inputs work. DC input modules allow us to connect either PNP (sourcing) or NPN (sinking) transistor type devices to them. If we are using a regular switch (i.e. toggle or pushbutton, etc.), we typically do not have to worry about whether we wire it as NPN or PNP. We should note that most PLCs will not let us mix NPN and PNP devices on the same module. When we are using a sensor (photo-eye, prox, etc.), we do, however, have to worry about its output configuration. **Always verify whether it is PNP or NPN** (Check with the manufacturer when unsure).

The difference between the two types is whether the load (in our case, the PLC is the load) is switched to ground or positive voltage. An NPN type sensor has the load switched to ground whereas a PNP device has the load switched to positive voltage. Figure 16.47 shows is what the outputs look like for NPN and PNP sensors.

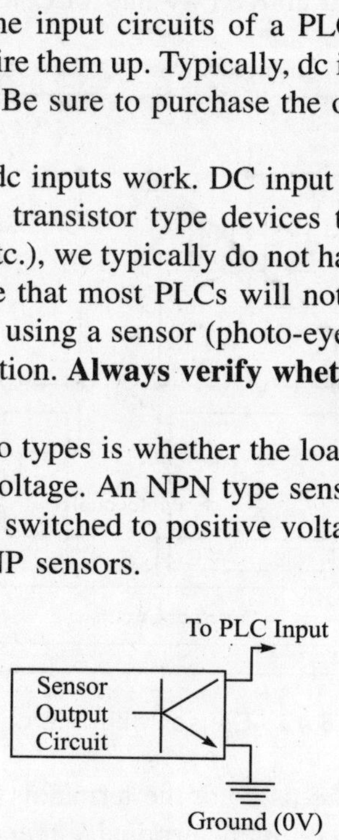

Figure 16.47 NPN (Sinking) Sensor.

On the NPN sensor we connect one output to the PLCs input, and the other output to the power supply ground. If the sensor is not powered from the same supply as the PLC, we should connect both grounds together. NPN sensors are most commonly used in North America.

Many will say that PNP is better (i.e. safer) because the load is switched to ground, but whatever works for you is best. Just remember to plan for the worst.

On the PNP sensor we connect one output to positive voltage and the other output to the PLC's input. If the sensor is not powered from the same supply as the PLC, we should connect both V+'s together. PNP sensors are most commonly used in Europe.

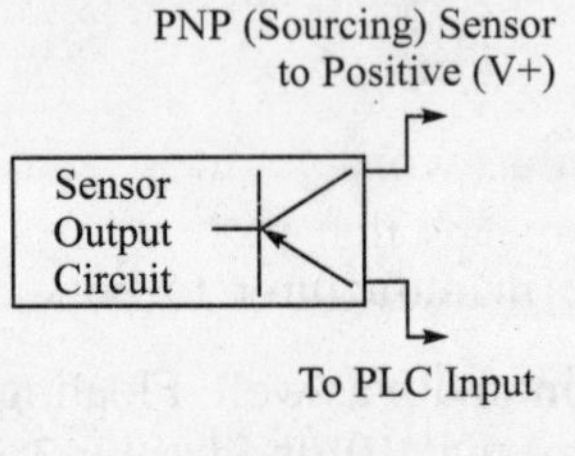

Figure 16.48 PNP (Sourcing) Sensor.

Inside the sensor, the transistor is just acting as a switch. The sensor's internal circuit tells the output transistor to turn on when a target is present. The transistor then closes the circuit between the two connections shown above (V+ and PLC input).

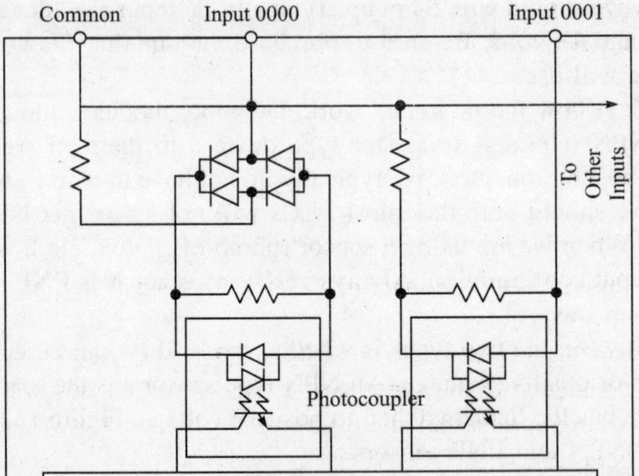

Figure 16.49 Sensor with Internal Circuit.

The only things accessible to the user are the terminals labelled COMMON, INPUT 0000, INPUT 0001, INPUTxxxx... The common terminal either gets connected to V+ or ground. Where it is connected depends upon the type of sensor used. When using an NPN sensor, this terminal is connected to V+. When using a PNP sensor, this terminal is connected to 0V (ground).

A common switch (i.e. limit switch, pushbutton, toggle, etc.) would be connected to the inputs in a similar fashion. One side of the switch would be connected directly to V+. The other end goes to the plc input terminal. This assumes the common terminal is connected to 0V (ground). If the common is connected to V+, then simply connect one end of the switch to 0V (ground) and the other end to the plc input terminal.

The photocouplers are used to isolate the PLC's internal circuit from the inputs. This eliminates the chance of any electrical noise entering the internal circuitry. They work by converting the electrical input signal to light and then by converting the light back to an electrical signal to be processed by the internal circuit.

AC inputs

Now that we understand how dc inputs work, let us take a close look at ac inputs. An ac voltage is non-polarized. Put simply, this means that there is no positive or negative to *worry about.* However, ac voltage can be quite dangerous to work with if we are careless. (*Remember when you stuck the knife in the toaster and got a shock? Be careful.*) Typically, ac input modules are available that will work with 24, 48, 110, and 220 volts. Be sure to purchase the one that fits your needs based upon the input devices (voltage) you will use.

AC input modules are less common these days than dc input modules. The reason being that today's sensors typically have transistor outputs. A transistor will not work with an ac voltage. Most commonly, the ac voltage is being switched through a limit switch or other switch type. If your application is using a sensor, it probably is operating on a dc voltage.

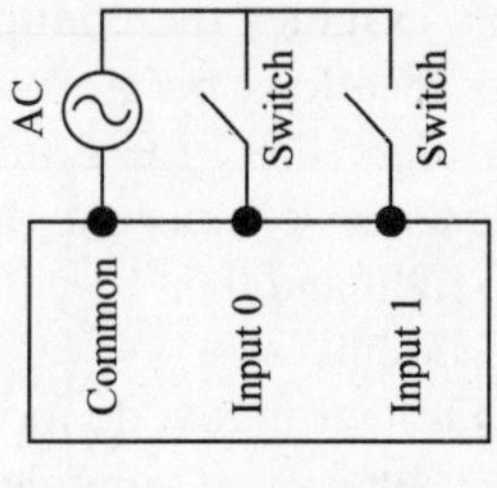

Figure 16.50 Connect an ac device to our input module.

We typically connect an AC device to our input module as shown in Figure 16.50. Commonly the ac "hot" wire is connected to the switch while the "neutral" goes to the PLC common. The ac ground (3rd wire where applicable) should be connected to the frame ground terminal of the PLC (not shown). As is true with DC, AC connections are typically colour coded so that the individual wiring the device knows which wire is which. This coding varies from country to country, but in the US it is commonly white (neutral), **black** (hot) and **green** (3rd wire ground when applicable). Outside the US, it is commonly coded as **brown** (hot), **blue** (neutral) and **green** with a **yellow** stripe (3rd wire ground where applicable).

The PLC's ac input module circuit typically looks like as shown in Figure 16.51.

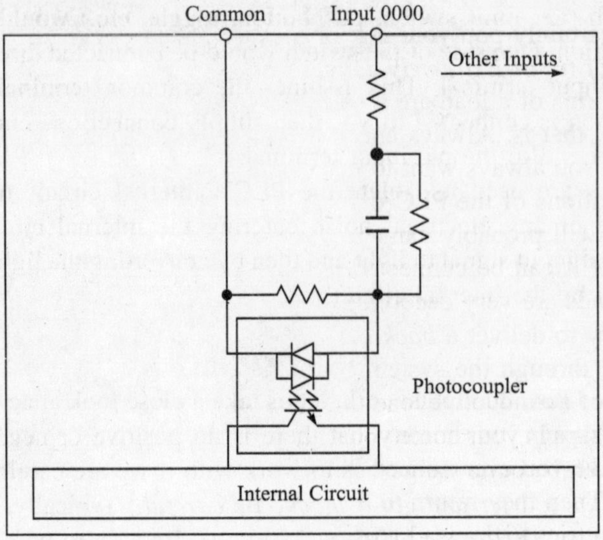

Figure 16.51 PLC's ac Input Module Circuit.

The only things accessible to the user are the terminals labelled COMMON, INPUT 0000, INPUTxxxx... The common terminal gets connected to the neutral wire.

A common switch (i.e. limit switch, pushbutton, toggle, etc.) would be connected to the input terminals directly. One side of the switch would be connected directly to INPUT XXX. The other end goes to the ac hot wire. This assumes the common terminal is connected to neutral. Always check the manufacturer's specifications before wiring, to be sure and safe.

The photocouplers are used to isolate the PLC's internal circuit from the inputs. This eliminates the chance of any electrical noise entering the internal circuitry. They work by converting the electrical input signal to light and then by converting the light back to an electrical signal to be processed by the internal circuit.

Finally, typically an ac input takes longer than a dc input for the PLC to see. In most cases, it does not matter to the programmer because an ac input device is typically a mechanical switch and mechanical devices are slow. It is quite common for a PLC to require that the input be on for 25 or more milliseconds before it is seen. This delay is required because of the filtering which is needed by the PLC internal circuit. Remember that the PLC internal circuit typically works with 5 or less volts dc.

Relay outputs

By now we should have a good understanding of how the inputs are used. Next up is the output circuits.

One of the most common types of outputs available is the relay output. A relay can be used with both ac and dc loads. A load is simply a fancy word for whatever is connected to our outputs. We call it a load because we are "loading the output" with something. If we connected no load to the output (i.e. just connect it directly to a power supply), we would certainly damage the outputs. This would be similar to replacing the light bulb in the lamp you are using to read this with a piece of wire. If you did this, the lamp would draw a tremendous amount of current from the outlet and certainly pop your circuit breaker or blow your fuse or your brains. (*Take our word. Please don't try it! Extremely dangerous!*)

Some common forms of a load are a solenoid, lamp, motor, etc. These "loads" come in all sizes. Electrical sizes, that is. Always check the specifications of your load before connecting it to the PLC output. You always want to make sure that the maximum current it will consume is within the specifications of the PLC output. If it is not within the specifications (i.e. draws too much current), it will probably damage the output. When in doubt, double check with the manufacturer to see if it can be connected without potential damage.

Some types of loads are very deceiving. These deceiving loads are called **inductive loads**. These have a tendency to deliver a *back current* when they turn on. This back current is like a voltage spike coming through the system.

A good example of an inductive load that most of us see about 6 months per year is an air-conditioning unit. Perhaps in your home you have an air-conditioner. (Unless you live in the arctic you probably do!) Have you ever noticed that when the air-conditioner *kicks on*, the lights dim for a second or two? Then they return to their normal brightness. This is because when the air-conditioner turns on, it tries to draw a lot of current through your wiring system. After this initial "kick", it requires less current and the lights go back to normal. This could be dangerous to your PLC's output relays. It can be estimated that this kick is about 30 times the rated current of the load. Typically a diode, varistor, or other "snubber" circuit should be used to help combat any damage to the relay. Enough said. Let us see how we can use these outputs in the "real plc world".

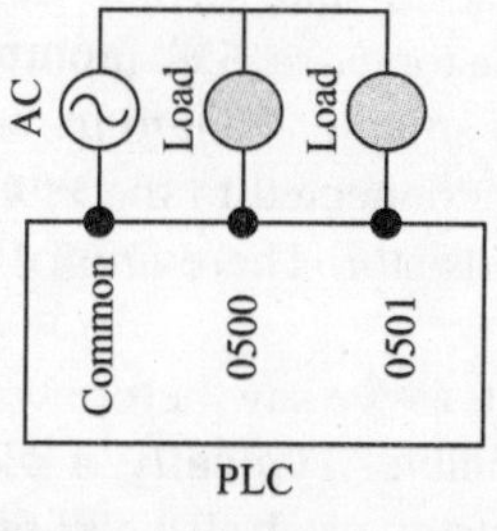

Figure 16.52 Connecting our Output to the PLC Relays.

Shown above (Figure 16.52) is a typical method of connecting our outputs to the PLC relays. Although our diagram shows the output connected to an ac supply, dc can be used as well. A relay is non-polarized and typically it can switch either ac or dc. Here the common is connected to the one end of our power supply and the other end of the supply is connected to the load. The other half of our load gets connected to the actual plc output you have designated within your ladder program.

The relay is internal to the plc. Its circuit diagram typically looks like as shown in Figure 16.53. When our ladder diagram tells the output to turn on, the PLC will internally apply a voltage to the relay coil. This voltage will allow the proper contact to close. When the contact closes, an external current is allowed to flow through our external circuit. When the ladder diagram tells the PLC to turn off the output, it will simply remove the voltage from the internal circuit thereby enabling the output contact to release. Our load will then have an open circuit and will therefore be off.

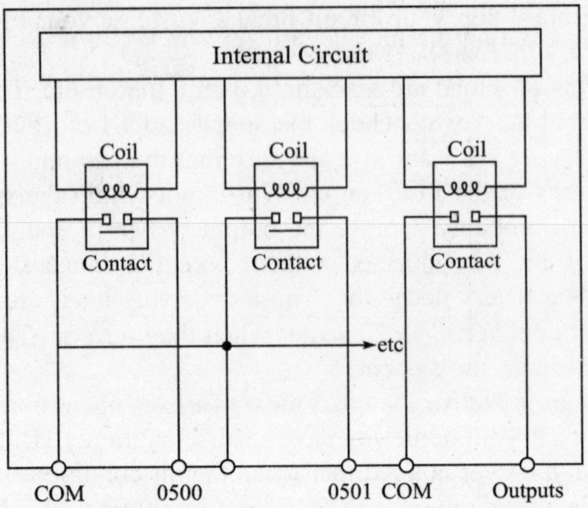

Figure 16.53 The Relay is Internal to the PLC.

Transistor outputs

The next type of output we should learn about is transistor type outputs. It is important to note that a transistor can only switch a dc current. For this reason, it cannot be used with an ac voltage.

We can think of a transistor as a solid-state switch. Or more simply put, an electrical switch. A small current applied to the transistor *base* (i.e. input) lets us switch a much larger current through its output. The PLC applies a small current to the transistor base and the transistor output *closes*. When it is closed, the device connected to the PLC output will be turned on. The above is a very simple explanation of a transistor. There are, of course, more details involved but we do not need to get too deep.

We should also keep in mind that, as we saw before with the input circuits, there is generally more than one type of transistor available. **Typically, a plc will have either NPN or PNP type outputs.** The *physical* type of transistor used also varies from manufacturer to manufacturer. Some of the common types available are BJT and MOS-FET. A BJT type (Bipolar Junction Transistor) often has less switching capacity (i.e. it can switch less current) than a MOS-FET (Metal Oxide Semiconductor- Field Effect Transistor) type. The BJT also has a slightly faster switching time. Once again, please check the output specifications of the particular PLC you are going to use. Never exceed the manufacturer's maximum switching current.

Shown alongside (Figure 16.54) is how we typically connect our output device to the transistor output. Please **note that this is an NPN type transistor**. If it were a PNP type, the common terminal would most likely be connected to V+, and V– would connect to one end of our load. Note that since this is a dc type output, we must always observe proper polarity for the output. One end of the load is connected directly to V+ as shown alongside.

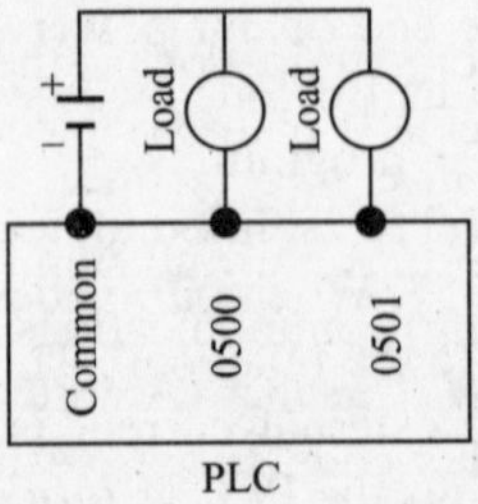

Figure 16.54 Connect our Output Device to the Transistor Output.

Let us wait a moment and see what happens inside the output circuit. Shown below (Figure 16.55) is a typical output circuit diagram for an NPN type output.

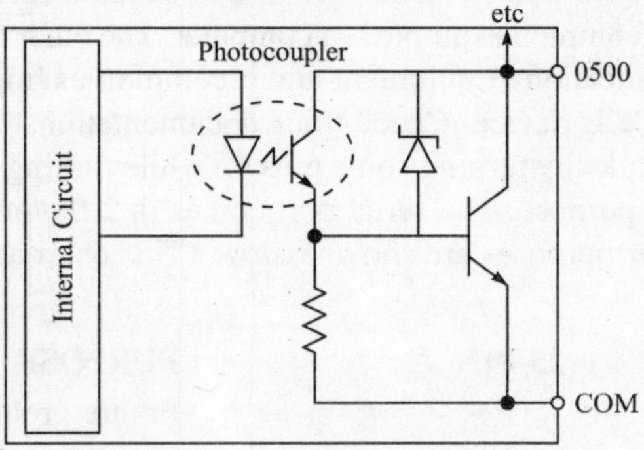

Figure 16.55 Output Circuit Diagram for an NPN Type Output.

Notice that as we saw with the transistor type inputs, there is a photocoupler isolating the *real-world* from the internal circuit. When the ladder diagram calls for it, the internal circuit turns on the photocoupler by applying a small voltage to the LED side of the photocoupler. This makes the LED emit light, and the receiving part of the photocoupler will see it and allow current to flow. This small current will turn on the base of the output transistor connected to output 0500. Therefore, whatever is connected between COM and 0500 will turn on. When the ladder tells 0500 to turn off, the LED will stop emitting light and hence the output transistor connected between 0500 and COM will turn off.

One other important thing to note is that a transistor typically cannot switch as large a load as a relay. Check the manufacturer's specifications to find the largest load it can safely switch. If the load current you need to switch exceeds the specification of the output, you can connect the plc output to an external relay. Then connect the relay to the large load. You may be thinking, "*why not just use a relay in the first place*?" The answer is because a relay is not always the correct choice for every output. A transistor gives you the opportunity to use external relays when and only when necessary.

To sum up, a transistor is fast, switches a small current, has a long lifetime, and works with dc only. Whereas a relay is slow, can switch a large current, has a shorter lifetime, and works with ac or dc. Select the appropriate one based upon your actual application needs.

RS-232 communications (hardware)

RS-232 communications is the most popular method of PLC to external device communications. Let us tackle it piece by piece to see how simple it can be when we understand it.

RS-232 is an asynchronous (a marching band must be "in sync" with each other so that when one steps, they all step. They are asynchronous in that they follow the band leader to keep their timing) communications method. We use a binary system (1's and 0's) to transmit our data in the ASCII format (American Standard Code for Information Interchange, pronounced ASS-KEY). This code translates human readable code (letters/numbers) into "computer readable" code (1's and 0's). Our PLC's serial port is used for transmission/reception of the data. It works by sending/receiving a voltage. **A positive voltage is called a MARK and a negative voltage**

is called a SPACE. Typically, the PLC works with +/– 15 volts. The voltage between +/– 3 volts is generally not used and is considered noise.

There are two types of RS-232 devices. The first is called a *DTE device*. This means Data Terminal Equipment and a common example is a computer. The other type is called a *DCE device*. DCE means Data Communications Equipment and a common example is a modem. Your PLC may be either a DTE or DCE device. Check your documentation.

The PLC serial port works by turning some pins on while turning others off. These pins each are dedicated to a specific purpose. The serial port comes in 2 flavours—a 25-pin type and a 9-pin type. The pins and their purposes are shown below. (This chart assumes your PLC is a DTE device.)

9-PIN	25-PIN	PURPOSE
1	1	Frame ground
2	3	Receive data (RD)
3	2	Transmit data (TD)
4	20	Data terminal ready (DTR)
5	7	Signal ground
6	6	Data set ready (DSR)
7	4	Request to send (RTS)
8	5	Clear to send (CTS)
9	22	Ring indicator (RI) *only for modems*

The purpose of each pin is discussed below.

- Frame ground—This pin should be internally connected to the chassis of the device.
- Receive data—This pin is where the data from the external device enters the PLC serial port.
- Transmit data—This pin is where the data from the PLC serial port leaves the PLC en route to the external device.
- Data terminal ready—This pin is a master control for the external device. When this pin is 1, the external device will not transmit or receive data.
- Signal ground—Since data is sent as + or – voltage, this pin is the ground that is referenced.
- Data set ready—Usually external devices have this pin as a permanent 0, and the PLC basically uses it to determine that the external device is powered up and ready.
- Request to send—This is part of hardware handshaking. When the PLC wants to send data to the external device, it sets this pin to a 0. In other words, it sets the pin to a 0 and basically says, "I want to send you data. Is it ok?" The external device says it is OK to send data by setting it clear to send pin to 0. The PLC then sends the data.
- Clear to send—This is the other half of hardware handshaking. As noted above, the external device sets this pin to 0 when it is ready to receive data from the PLC.
- Ring indicator—Only used when the PLC is connected to a modem.

What happens when your PLC and external device are both DTE (or both DCE) devices? They cannot talk to each other, that is what happens. Figure 16.56 shows why two same type devices cannot communicate with each other.

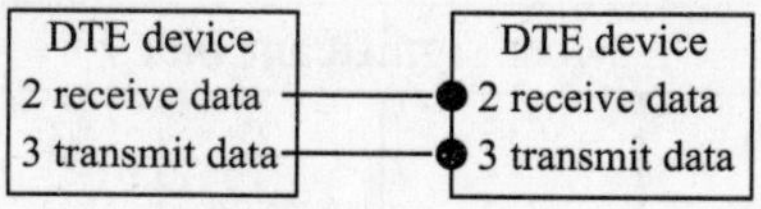

Figure 16.56 Communication of two same type device.

Notice that in the figure above, the receive data line (pin 2) of the first device is connected to the receive data line of the second device. And the transmit data line (pin 3) of the first device is connected to the transmit data of the second device. It is like talking through a phone with the wires reversed (i.e. your mouthpiece is connected directly to the other parties' mouthpiece and your earpiece is connected directly to the other parties' earpiece). Obviously, this will not work well!

The solution is to use a null-modem connection as shown in Figure 16.57. This is typically done by using a reverse (null-modem) cable to connect the devices.

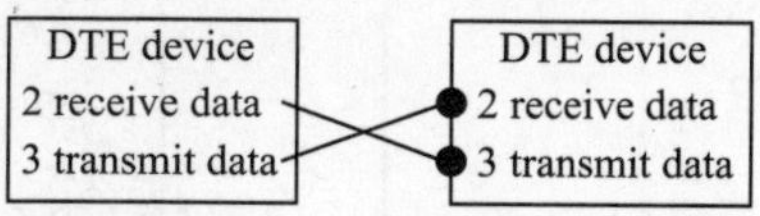

Figure 16.57 Null-modem connection.

To summarize everything, here is a typical communications session. Both devices are powered up. The PLC is DTE and the external device is DCE.

The external device turns on DSR which tells the PLC that is it is powered up and "there". The PLC turns on RTS which is like asking the external device, "are you ready to receive some data?" The external device responds by turning on its CTS which says it is ok for the PLC to send data. The PLC sends the data on its TD terminal and the external device receives it on its RD terminal. Some data is sent and received. After a while, the external device cannot process the data quick enough. So, it turns off its CTS terminal and the PLC pauses sending data. The external device catches up and then turns its CTS terminal back on. The PLC again starts sending data on its TD terminal and the external device receives it on its RD terminal. The PLC runs out of data to send and turns off its RTS terminal. The external device sits and waits for more data.

RS-232 communications (software)

Now that we understand the hardware part of the picture, let us go right into the software part. **ASCII** is a human-readable to computer-readable translation code (i.e. each letter/number is translated to 1's and 0's). It is a 7-bit (a bit is a 1 or a 0) code, so we can translate 128 characters (2^7 is 128). Character sets that use the 8th bit do exist, but they are not true ASCII. Given below is an ASCII chart showing its "human-readable" representation. We typically refer to the characters by using hexadecimal terminology. "0" is 30h, "5" is 35h, "E" is 45h, etc. (the "h" simply means hexadecimal).

Least significant bits		Most significant bits							
		0	1	2	3	4	5	6	7
	0			space	0	@	P	‘	p
	1		XON	!	1	A	Q	a	q
	2	STX		“	2	B	R	b	r
	3	ETX	XOFF	#	3	C	S	c	s
	4			$	4	D	T	d	t
	5		NAK	%	5	E	U	e	u
	6	ACK		&	6	F	V	f	v
	7			‘	7	G	W	g	w
	8			(	8	H	X	h	x
	9			)	9	I	Y	i	y
	A	LF		*	:	J	Z	j	z
	B			+	;	K	[	k	{
	C			,	<	L	\	l	|
	D	CR		–	=	M	]	m	}
	E			.	>	N	^	n	~
	F			/	?	O	_	o	

- **Start bit.** In RS-232 the first thing we send is called a *start bit*. This start bit (“invented” during World War I by Kleinschmidt) is a synchronizing bit added just before each character we are sending. This is considered a SPACE or negative voltage or a 0.
- **Stop bit.** The last thing we send is called a *stop bit*. This stop bit tells us that the last character was just sent. Think of it as an end-of-character bit. This is considered a MARK or positive voltage or a 1. The start and stop bits are commonly called *framing bits* because they surround the character we are sending.
- **Parity bit.** Since most PLCs/external equipment are byte-oriented (8 bits = 1 byte), it seems natural to handle data as a byte. Although ASCII is a 7-bit code, it is rarely transmitted that way. Typically, the 8th bit is used as a parity bit for error checking. This method of error checking gets its name from the math idea of parity (Remember the odd-even property of integers? I did not think so). In simple terms, parity means all characters will either have an odd number of 1’s or an even number of 1’s.

Common forms of parity are None, Even, and Odd. (Mark and Space are not very common, so I will not discuss them.) Consider these examples: send “E” (45h or 1000101b(binary)). In parity of None, the parity bit is always 0, so we send 10001010. In parity of even we must have an Even number of 1’s in our total character, so the original character currently has 3 1’s (1000101). Therefore, our parity bit we will add must be a 1 (1000101**1**). Now we have an even number of 1’s. In Odd parity we need an odd number of 1’s. Since our original character already has an odd number of 1’s (3 is an odd number, right?), our parity bit will be a 0 (1000101**0**).

During transmission, the sender calculates the parity bit and sends it. The receiver calculates parity for the 7-bit character and compares the result to the parity bit received. If the calculated and real parity bits do not match, an error occurred and we have to act appropriately.

It is strange that this parity method is so popular. The reason is because it is only effective half the time. That is, parity checking can only find errors that affect an odd number of bits. If the error affected 2 or 4 or 6 bits, the method is useless. Typically, errors are caused by noise which comes in bursts and rarely affects 1 bit. Block redundancy checks are used in other communication methods to prevent this.

- **Baud rate**-I will perpetuate the incorrect meaning since it is most commonly used incorrectly. Think of baud rate as referring to the number of bits per second that are being transmitted. So 1200 means 1200 bits per second are being sent and 9600 means 9600 bits are being transmitted every second. Common values (speeds) are 1200, 2400, 4800, 9600, 19200, and 38400.
- **RS232 data format**-(baud rate-data bits-parity-stop bits) This is the way the data format is typically specified. For example, 9600-8-N-1 means a baud rate of 9600, 8 data bits, parity of none, and 1 stop bit.

Figure 16.58 shows how data leaves the serial port for the character "E" (45h 100 0101b) and even parity.

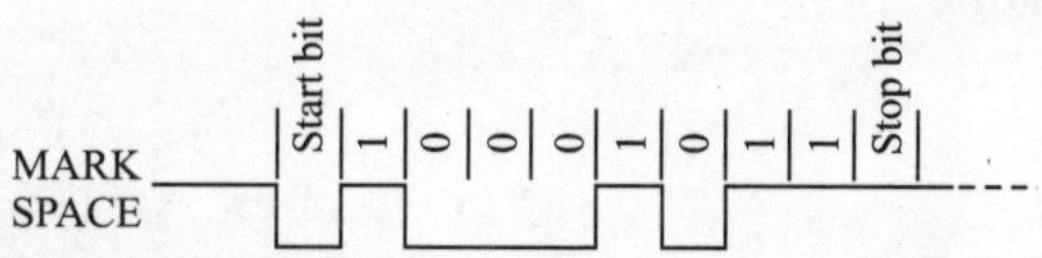

Figure 16.58 Shows how Data Leaves the Serial Port for the Character "E" (45h 100 0101b) and Even Parity.

Another important thing that is sometimes used is called *software handshaking* (flow control). Like the hardware handshaking, software handshaking is used to make sure both devices are ready to send/receive data. The most popular "character flow control" is called XON/XOFF. It is very simple to understand. Simply put, the receiver sends the XOFF character when it wants the transmitter to pause sending data. When it is ready to receive data again, it sends the transmitter the XON character. XOFF is sometimes referred to as the hold off character and XON as the release character.

The last thing we should know about is delimiters. A delimiter is simply added to the end of a message to tell the receiver to process the data it has received. The most common is the CR or the CR and LF pair. The CR (carriage return) is like the old typewriters (Remember them??). When you reached the end of a line while typing, a bell would sound. You would then grab the handle and move the carriage back to the start. In other words, you returned the carriage to the beginning. (This is the same as what a CR delimiter will do if you view it on a computer screen.) The plc/external device receives this and knows to take the data from its buffer. (Where the data is stored temporarily before being processed.) An LF (line feed) is also sometimes sent with the CR character. If viewed on a computer screen, this would look like what happens on the typewriter when the carriage is returned and the page moves down a line, so you do not type over what you just typed?

Sometimes an STX and ETX pair is used for transmission/reception as well. STX is "start of text" and ETX is "end of text". The STX is sent before the data and tells the external device that data is coming. After all the data has been sent, an ETX character is sent.

Finally, we might also come across an ACK/NAK pair. This is rarely used but it should be noted as well. Essentially, the transmitter sends its data. If the receiver gets it without error it sends back an ACK character. If there was an error, the receiver sends back a NAK character and the transmitter resends the data.

RS-232 has a lot of information to absorb, so feel free to reread it. It is not so difficult if you take your time and think about each step. Other communication methods get easier!

REVIEW QUESTIONS

1. What is PLC?
2. List out the programming involved in PLC.
3. What are the various applications of PLC?
4. Describe the major components of PLC system.
5. Briefly describe the operation of PLC.
6. Draw the ladder logic diagram in PLC.
7. What is concatenation?

Chapter 17

CONCURRENT ENGINEERING

17.1 INTRODUCTION

Concurrent Engineering is delivering better, cheaper and faster products to the market, by a lean way of working, using multi-disciplinary teams, right first time methods and parallel processing activities to continuously consider all constraints.

DR. STEPHEN EVANS; The Cranfield CIM Institute

Concurrent Engineering is a systematic approach to the integrated, concurrent design of products and their related processes, including manufacture and support. This approach is intended to cause the developers, from the outset, to consider all (relevant) elements of the product life cycle from conception through disposal, including quality, costs, schedule, and user requirements.

IDA Report R-338 (1988)
Institute for Defense Analyses
Alexandria, VA, USA

Concurrent Engineering is a systematic approach to creating a product design that considers all elements of the product life cycle from conception of the design to disposal of the product, and doing so defines the product, its manufacturing processes, and design processes, and all other processes required during the life cycle, such as logistics support. This takes place in a current or simultaneous manner as the product design evolves.

JOAN M. SMITH; An Introduction to CALS

In practice, the application of Concurrent Engineering methods means:
"*Team building*"—breaking down the cultural and organizational barriers between disciplines in the design, manufacture and support processes; a typical aim of this team building is to enable potential manufacturing problems to be addressed and solved at the design stage.
Integration of the systems used by the disciplines involved in these teams.

JULIAN FOWLER, CADDETC
University of Leeds, UK

Concurrent Engineering is a systematic communication between team members to enable consideration of all important product and process information in a timely manner. This implies structured information management. This in turn implies understanding of the design process and the development of product information.

DAVID G. ULLMAN
Oregon State University, USA

17.2 WHAT IS CONCURRENT ENGINEERING?

- A philosophy of product development: Integrating multiple design issues
- A method of product design: Integration of multi-disciplinary folks into the design team
- A method of managing people: Design issues are represented in the people
- What is not concurrent engineering?
  (a) It is not "over the wall"
  (b) Nor off the wall (it is being used and is here to stay)

17.3 IMPORTANCE OF CONCURRENT ENGINEERING

The goal of **Concurrent Engineering** is the interactive work of different disciplines that affect a product to make it better.

- *Minimize the product life cycle—Eliminate the redesign procedure.*
- *Decrease production cost—Results from the minimization of the product life cycle.*
- *Maximize product quality*—By spending more time and money initially in the **design** cycle and ensuring that the concept selection is optimized, the company can increase the prospect of delivering a quality product to the customer.
- *Team work*—Human resources are working together for a common product.

Basics of concurrent engineering

For a new-product work

Critical Steps	Activity
1. Create concept	Fundamental **Engineering**
2. Will it work?	Fundamental **Engineering**
3. Can it be made?	Basic **concurrent engineering**
4. Satisfy customer needs?	EQFD
5. Consistently?	Robust **design**
6. Profitable?	Strategy

17.4 NEED FOR CONCURRENT ENGINEERING

In today's business world, corporations must be able to react to the changing market needs rapidly, effectively, and responsively. They must be able to reduce their time to market and adapt to the changing environments. Decisions must be made quickly and they must be done right when first time out. Corporations can no longer waste time repeating tasks, thereby prolonging the time it takes to bring new products to market. Therefore, there is a need for concurrent engineering to bring rapid solutions to product design and development process.

Concurrent engineering is indisputably the wave of the future for new product development for all companies regardless of their size, sophistication, or product portfolio. In order to be competitive, corporations must alter their product and process development cycle to be able to complete diverse tasks concurrently. This new process will benefit the company, although it will require a large amount of refinement in its implementation. This is because concurrent engineering is a process that must be reviewed and adjusted for continuous improvements of engineering and business operation.

17.4.1 The Concurrent Engineering Approach

Concurrent engineering is a **business strategy** which replaces the traditional product development process with one in which tasks are done in parallel, and there is an early consideration for every aspect of a product's development process. This strategy focuses on the optimization and distribution of a firm's resources in the design and development process to ensure an effective and efficient product development process. It mandates major changes within the organizations and firms that use it, due to the people and process integration requirements. Collaboration is a must for individuals, groups, departments, and separate organizations within the firm. Therefore, it cannot be applied at leisure. A firm must be dedicated to the long-term implementation, appraisal, and continuous revision of a concurrent engineering process.

17.4.2 Strategic Plan of Concurrent Engineering

Concurrent engineering is recognized as a strategic weapon that businesses must use for effective and efficient product development. It is not a trivial task, but a complex strategic plan that demands full corporate commitment and therefore strong leadership and teamwork go hand in hand with successful concurrent engineering programs.

17.4.3 Commitment, Planning and Leadership

Concurrent engineering is not a trivial process to apply. If firms are going to commit to concurrent engineering, then they must first devise a plan. This plan must create organizational change throughout the entire company or firm. There must be a strong commitment from the firm's leadership in order to mandate the required organizational changes from the top down. Concurrent engineering without leadership will have no clear direction or goal. On the other hand, concurrent engineering with leadership, management support, and proper planning will bring success in today's challenging market place.

17.4.4 Continuous Improvement Process

Concurrent engineering is not a one-size-fits-all solution to a firm's development processes. There are many different aspects of concurrent engineering which may or may not fit in a corporation's development process. Concurrent engineering is only a set of process objectives and goals that have a variety of implementation strategies. Therefore, concurrent engineering is an evolving process that requires continuous improvement and refinement. This continuous improvement cycle consists of planning, implementing, reviewing, and revising. The process must be updated and revised on a regular basis to optimize the effectiveness and benefits of the concurrent engineering development process.

17.4.5 Communication and Collaboration

The implementation of concurrent engineering begins by creating a corporate environment that facilitates communication and collaboration not just between individuals, but also between separate organizations and departments within the firm. This may major structural changes, re-education of the existing workforce, and/or restructuring of the development process.

- Get a strong commitment from the senior management.
- Establish unified project goals and a clear business mission.
- Develop a detailed plan early in the process.
- Continually review your progress and revise your plan.
- Develop project leaders who have an overall vision of the project and goals.
- Analyze your market and know your customers.
- Suppress individualism and foster a team concept.
- Establish and cultivate cross-functional integration and collaboration.
- Transfer technology between individuals and departments.
- Break the project into its natural phases.
- Develop metrics.
- Set milestones throughout the development process.
- Collectively work on all parts of the project.
- Reduce costs and time to market.
- Complete tasks in parallel

17.5 WHEN CONCURRENT ENGINEERING IS USED?

Most of a product's costs are committed very early in the design and development process. Therefore, companies must apply concurrent engineering at the onset of a project. This makes concurrent engineering a powerful development tool that can be implemented early in the conceptual design phase where the majority of a product's costs are committed. There are several applications in which concurrent engineering may be used. Some primary applications include product research, design, development, re-engineering, manufacturing, and redesigning of existing and new products. In these applications, concurrent engineering is applied throughout the design and development process to enable the firm to reap the full benefits of this process.

What are some factors of good product development?

The first step in achieving a simple **design** is to develop a systematized product structure which standardizes the relationship between product function, form and fabrication. *What are important factors in* ***Concurrent Engineering*** *practice?*

17.5.1 Organizational Factor

(a) **Cross-functional teams**—Project team members gained a better understanding of the project priorities, process discipline, making risks and compromises visible for better control. The **design** team is composed of experts from **engineering**, production, marketing, and any other functional area which has a vested interest in the development project. The team is formed to work on a specific project, and stays together throughout the development of the product.

(b) **Liaison personnel**—Liaison personnel are not members of any functional piece of the organization, but rather people who are capable and prepared to address issues which span functional organizational boundaries. Liaison personnel have as their full-time job, the coordination of the disparate functions. Under this approach, they become the primary modes of accomplishing information transfer between functional areas.

(c) **Job rotation**—Job rotation means rotating personnel between functional categories. These personnel are assigned temporarily or permanently outside of their accustomed functional speciality, that is, a manufacturing engineer will work with **design** engineers or vice versa. Thus, it is possible to integrate the various knowledge bases without making significant structural changes to the organization. Job rotation does seem to have useful integration benefits.

17.5.2 Some Helpful Rule Based Methods

Product Design Methods

Design For Manufacturing (DFM)—DFM seeks to minimize manufacturing information content of a product **design** to the fullest extent possible within constraints imposed by functionality and performance.

- Minimize the total number of parts.
- Simplify the **design** to ensure that the remaining parts are easy to fabricate, assemble, handle and service.
- Standardize possible to facilitate desirable productibility characteristics such as interchangeability, interoperability, simplified interfaces, effective consolidation of parts and function, availability of components, and so forth.

Design for quality—It can be implemented in the system **design** step by intentionally designing the product and process to be tolerant of variation.

Design for cost—An effective analysis of the product or project costs and the ability to implement cost control management includes management of the product or project cost, and this requires a knowledge and understanding of the cost elements and their sensitivity to various control parameters. Cost analysis forms the basis for cost control, and without accurate and timely cost data, effective cost control is impossible.

Design for assembly (DFA)—It seeks to minimize the cost of assembly within constraints imposed by other **design** requirements. DFA has been the starting point for the development of a corporate DFM philosophy and the culture change that accompanies it.

Design for safety—The designer must develop the habit of constantly evaluating the design for safety, considering not only the design itself, but the personnel involved in fabricating the product, using the procedure, and in maintaining and repairing the product or system, as well as the end user or purchaser.

Design for reliability—Reliability is defined as the probability that a system, device, or component will successfully perform for:

- A given range of operating conditions,
- A specific environmental condition, and
- A prescribed economic survival time.

Design for X—It helps to ensure that parts and products to be produced are correctly designed, using a particular production process or method such as plastic injection molding or sheet metal stamping.

- Integrated computer analysis
- Produce Design method it is based on the recognition that steps in the development of a manufactured product are interrelated and can be accomplished effectively and effectively by using computers.
- DFM relationship comes about not only from the characteristics of the part being fabricated but also from the processes, specifications, instructions, and data that define and direct each step in the manufacturing process.

17.6 EVOLUTION OF CONCURRENT ENGINEERING

What things are at the core of **Concurrent Engineering**?

- Increased role of manufacturing process issues on product **design** decisions
- Formation of cross-functional teams
- Focus on meeting customer products requirements
- Lead time as competitive advantage
- **Concurrent engineering** practices are not warranted when the product is simple, or when the incremental change is marginal.
- **Concurrent engineering** is difficult, that is, it involves some administrative and communication overhead, as well as effort to change company culture.

17.6.1 Concurrent Engineering of Fundamentals

Alternative definitions of **concurrent engineering** are available. They are the increased role of manufacturing process **design** on product **design** decisions, the formation of cross-functional teams to accomplish the development process, a focus on the customer during the development process, and the use of lead time as a source of competitive advantage.

- All products have a need to incorporate constraints imposed by the manufacturing process on details of the product **design**.
- **Design**-marketing integration removing the **design**-manufacturing barrier, the designer can become more responsive to customer desires and thereby create a more successful product.
- Lead-time has proved to be a significant facet of modern competition. A lessened lead time creates a market advantage for those firms who are able to produce products rapidly.

17.6.2 Need for Integrating Design with other Functions

Product designs have existed for as long as mass production has existed. Early on, there arose a division of intellectual labour whereby the designer was responsible for producing the design and the manufacturer was responsible for making the actual product. A design which is thrown over the proverbial wall is generally difficult and costly to produce, and does not necessarily conform to the desires of the market. This functional separation and its resulting adverse effect on the resulting product design may be repeated with other functions (such as marketing, maintenance, or others). The remedy for this situation is to make the designer aware of others concerns within and the need to reduce development lead time. All of these justifications for pushing **concurrent engineering** ideas have deep historical antecedents. These justifications are discussed below.

17.6.3 Increasing Competition

One justification given for the need for increased cooperation in the product development process is an increased level of competition. There have been claims that the level of competition has increased 'recently' at times, which we no longer consider recent.

For example, the claim is made that the level of competition has increased since the beginning of the nineteenth century, and modern firms cannot afford to ignore **design**-manufacturing interaction issues. Similarly, the claim is made that the high level of competition in the 1950s required that **design** and manufacturing personnel cooperate on new product development. Economic competition is now, and has always been, fierce. This is not a new effect.

17.6.4 New Production Methods

As new production methods come into service, it becomes important for knowledge about the new production processes to affect the resulting product **design** to take advantage of it and respond to the limitations of the new processes. Knowledge about these processes must be made available to the product designer. This knowledge is often resident in the production engineer.

The situation where new production processes are used will often be an important area for ensuring that **design** engineers work closely with production engineers. Among new manufacturing processes, the development of automatic assembly techniques has been frequently cited as requiring a higher level of integration between design and manufacturing.

17.6.5 Lead Time

The time that the length of the development cycle is an important competitive advantage, and that addressing all aspects of the **design** problem simultaneously might lead to a shortened development cycle is a long-standing concept.

Implementing concurrent traditionally, design and manufacturing have been separate engineering departments. The trend today has been to combine these functions and form design teams. This systematic approach to product design has led to reduced lead times and reduced problems in manufacturing.

Some of the benefits that can be gained from concurrent engineering are:

- Customers are consulted during and throughout the design process; their expectations will more likely be met.
- Improved design quality, leading to fewer changes after manufacturing starts.
- Reduced product development and design times.
- Reduced production costs due to fewer design changes and faster product development.
- Reduced delays that often lead to loss of market share.
- Quicker product turnover.
- Increased reliability and customer turnover.

17.7 THE FOUR PILLARS OF CONCURRENT ENGINEERING

Obviously, the key element of concurrent design is the coordinated activities between all the stakeholders in the process. Beyond this, four areas have been identified upon which the success of any concurrent design rests.

17.7.1 Organization

People must be motivated to change their traditional work habits to share ideas, offer ideas, and accept the ideas of others. There must be an unhindered exchange of ideas and information. This challenges the way companies have been organised in the past, i.e. into discipline-based departments (mechanical, electrical, software, hardware, manufacturing, quality assurance, etc.)

The management must be flexible enough to build teams with representation from all departments. The organization of a company that implements concurrent practices is such that conventional departments are subservient to *project-based* teams.

17.7.2 Communication

The different groups must be able to understand each other (e.g. marketing and engineering), but the infrastructure, such as computer hardware and software to support them, must also be compatible. There must be an easy transfer of information relating to the product's development. For example, designers must be able to access marketing data, and CAD and CAM formats must be compatible. Workflows within departments must account for workflows in other departments to ensure that individual workers can participate in the *project*, not in the department's activities.

17.7.2.1 Product design specification

This is the "formal" description of the problem to be solved. Most importantly, it represents an agreement between the product developers and the clients/users on what exactly is needed of the product. It provides a goal towards which the whole team must work. Although in real life requirements can and do change, by having everyone working towards a "goal post", the number of design changes and iterations before production is reduced.

17.7.3 Product Development

Product development is the process used to get from the PDS to the total design of the product ready for production. The basis of concurrent engineering is the formation of a multi-disciplinary team. The best team members are not narrow specialists, but people with a good combination of breadth of knowledge and also a depth of some pertinent specialization. In addition to a constant core membership, individual specialists may be brought into the team from time to time to treat specific issues.

However, as the technologies being used and the products being developed became more complex, the need for technical specialists grew. This resulted in segmented companies where many cloistered groups of experts worked in isolation of each other. In *this* setting, the over the wall approach does not work any more because one cannot expect one's colleagues to have a reasonable understanding of anything except their own speciality. It is in order to address this problem that the concepts of concurrent engineering and team-based product development came into being.

In general, a member of a design team must:

- Understand the product requirements.
- Have close and constant communications with other team members.
- Be able to make decisions independently.

Depending on the size and complexity of the design project, "team" can refer to team. For example, in a new car design, there might be a team responsible for each major system (engine, suspension, interior, etc.) and a single core team composed of one representative of each sub team charged with coordinating all the work. Yet another example, the Boeing 777 employed some 5,000 engineers, arranged in four tiers of teams. Although for complex products there may be hundreds or thousands of team members, it is rare to have more than 20 workers in a single team; otherwise, one must resort to hierarchies of teams. Typically, teams have 5 or 6 members.

17.7.3.1 Group think

Unfortunately, teams can go wrong, with disastrous results. Team members can develop a strong desire to please each other, to demonstrate their "loyalty" to the team, at the expense of the product. This *group think* people with doubts to keep silent, to give in to the emerging group consensus, and to focus on the popular option only.

Overcoming these problems is not easy. It helps if members are aware of these hurdles, but often a good team leader can only resolve these problems.

Some principles of successful teams include:

1. Team members must try to like and respect other team members.
2. Teams should have diverse areas of expertise.
3. Teams must have a common vision.
4. All team members must understand and should accept the solution.
5. Teams should promote open-minded thinking, and should avoid premature consensus.
6. Teams should maintain a balance between individual and group work.
7. Teams must be systematic in their work.

8. Team members must communicate.
9. Some members should be comfortable working with conceptual ideas, others with specific details.
10. A good team leader empowers the team, he does not stifle or micro-manage it.

17.7.3.2 Team leaders

Oddly, strong leadership is a key to successful teams. Leadership is not about management, but rather about guidance and balance.

A good team leader:

- Has a broad scope of knowledge and responsibilities.
- Assumes responsibility for ensuring that the concept is accurately translated into technical detail.
- Communicates well with both team members and clients and users.
- Actively seeks to manage conflicts as they arise.
- Circulates among the team members informally.

The changing roles of team members

	CONCEPT DEVELOPMENT	DETAILED DESIGN	TESTING	PRODUCTION
MARKETING	Client surveys	Focus groups	Advertising and promotion	Follow-up surveys
DESIGN	Conceptual design	Detail and systems design	Design revisions	Assist Manufacturing
MANUFACTURING	Rough costing	Parts suppliers, machinery	Conduct tests	Manufacturing at production rates

17.8 THE RISKS OF CONCURRENT ENGINEERING

- Teams are more difficult to manage than individuals.
- Not all engineers are team players.
- Cost of bringing team members together.
- Cost of removing team members from "home" departments.
- Resistance to change.

Example: Xerox

The Xerox Document Centre 265 was the first product to emerge from the company's new product development process based on concurrent engineering principles. It was developed in response to a perceived need for an extensible, modular, digital platform that would serve as the basis for an entire family of new products. Though motivated essentially by business and corporate needs, this new system was developed as a technical rather than a business system, for achieving goals of market superiority. As such, the product development process had to be based on engineering needs—what the engineers needed to achieve the corporate goals—and not just marketing or business plans. The design facility was developed to have no doors and lots of windows. This encouraged the sense of 'team' that is needed for successful concurrent engineering projects. Also, the engineers in the teams, rather than having the goals dictated by the management, established product goals. This give the engineers the *control* they need to execute concurrent processes well. They moved away from metrics that measured defects (i.e. negative measures), to metrics that measured functionality (i.e. positive measures). This encouraged engineers to exceed design expectations rather than just minimize the things that can go wrong. Benchmarking of competitors' equipment was also used to identify opportunities to beat competing products. "Know your enemy." All corporate measures indicated a substantive success of this approach, in some cases by a factor of 2.

17.9 OVERVIEW OF CONCURRENT ENGINEERING

For the past two decades, throughout the world all engineering companies have faced similar challenges. These are ever more demanding customers, rapid technological change, environmental

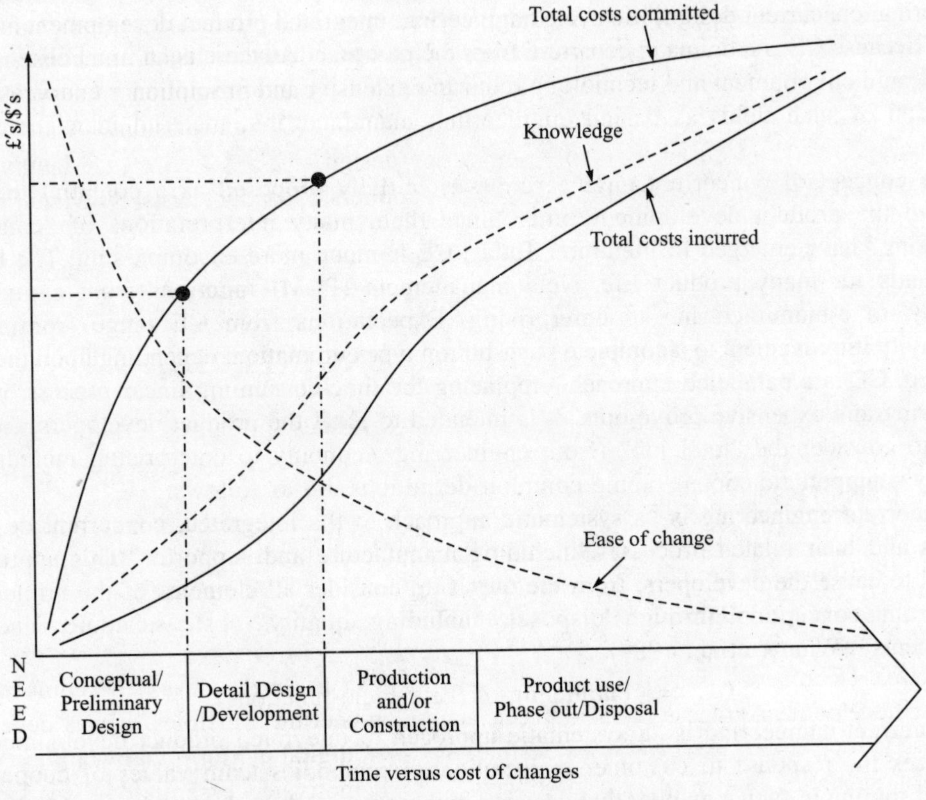

Figure 17.1 Concurrent Engineering.

issues, and competitive pressures on quality and cost, and shorter time to market with additional new product features. This is all happening with the majority of the Western world's common economic background of slow growth, excess capacity, increasing legislation compliance, demographic changes, market complexity and increasing globalization of industries.

In many cases the Japanese, who progressively made inroads into North America and Europe and in some cases dominated chosen markets, set the pace. The list of these chosen markets became longer year by year. Western companies were slow to recognize the basis of Japanese success, but eventually responded with a whole string of actions including CAD/CAE/CIM, robotics, automation, value analysis, quality programs, information technology, and so on. They sought to offset a perceived weakness—their workforces—by building on their apparent strength (i.e. technology, particularly computer-based technology).

This expensive technology was largely ineffective, because the new tools were used with existing structures, practices and attitudes. Products continued to arrive in the market place at unsatisfactory quality levels, and often too late to achieve sales and profit objectives. Their efforts were also undermined by short-term successes during market booms. There were also brief respites as a result of the appreciating yen, but the Japanese overcame this by aggressive cost reduction programmes and by investments in production capacity outside Japan—sometimes in countries with lower labour costs, sometimes in the key sales territories.

Concurrent engineering is also known by many other names, including simultaneous engineering, concurrent design, life cycle engineering, integrated product development and team design. Because CE is a dramatic departure from the past practice, there is an immediate need for a new design environment and technology requiring extensive interdisciplinary cooperation and integration of such fields as design engineering, manufacturing, material management and marketing.

The concept of concurrent engineering was initially proposed as a potential means to minimize the product development time. Since then, many interpretations of "concurrent engineering" have emerged in literature. Today, CE is much more encompassing. The letter E now stands for many product life cycle management (PLM) functions from engineering, economy, to e-manufacturing to enterprising. Expectations from CE range from modest productivity improvement to a complete push-button type automation, depending upon the views expressed. CE is a paralleled approach—replacing the time-consuming linear process of serial engineering and expensive prove-outs. It is intended to elicit the product developers, from the outset, to consider the "total job" (from engineering, economy to enterprising including the company's support functions). Some common definitions are as follows:

Concurrent engineering is "a systematic approach to the integrated, concurrent design of products and their related processes, including manufacture and support." "This approach is intended to cause the developers, from the outset, to consider all elements of the product life-cycle from conception through disposal, including quality, cost, schedule, and user requirements" [Winner et al., 1988].

CE <=> Paralleling-of (Life cycle-functions)

Concurrent engineering is "a systematic approach to integrated product development that emphasizes the response to customer expectations. It embodies team values of cooperation, trust, and sharing in such a manner that decision making proceeds with large intervals of parallel working by all life cycle perspectives early in the process, synchronized by comparatively brief exchanges to produce consensus." [Cleetus (CERC), 1992]

CE <=> Paralleling-of (Life cycle-functions) + Consensus + Cooperation

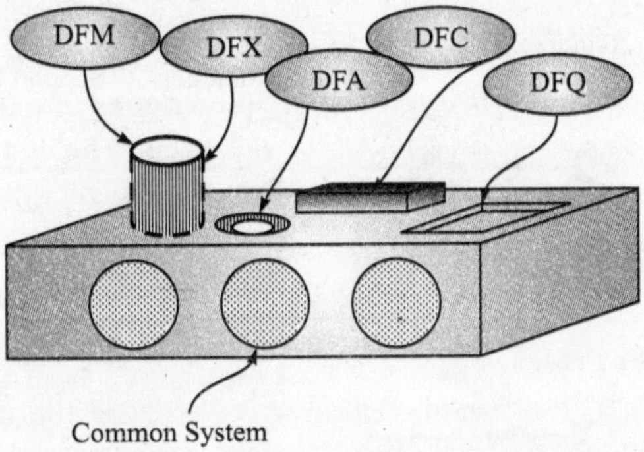

Figure 17.2 Common System.

The narrow view of concurrent engineering is "integrating product and process design." The wide view is "integrating over the product life," and the wider view is "integrating over the enterprise" [Finger, 1993].

CE <=> Integrating product and process design over the enterprise

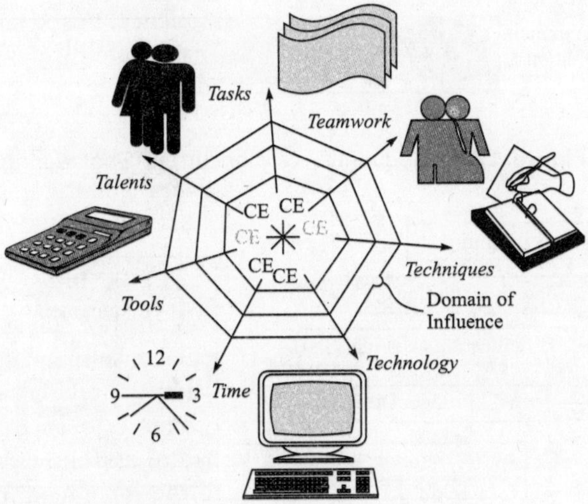

Figure 17.3 Influencing Agent-CE.

In general, there are greater affinity and dependence between pairs of activities, which are adjacent to each other. Farther away the activities are positioned from each other, the lesser would be the degree of affinity or the need for information transfer among them. For example, mission definition would be more closely related to concept definition but would have very little bearing with activities such as manufacturing. Similarly, manufacturing would be closely related to production operations and control but would be quite less sensitive to tracks such as engineering and analysis. The downward-slant arrows represent the resultant of enrichment (a horizontal component) and needs (a down-vertical component). The upward-slant arrows

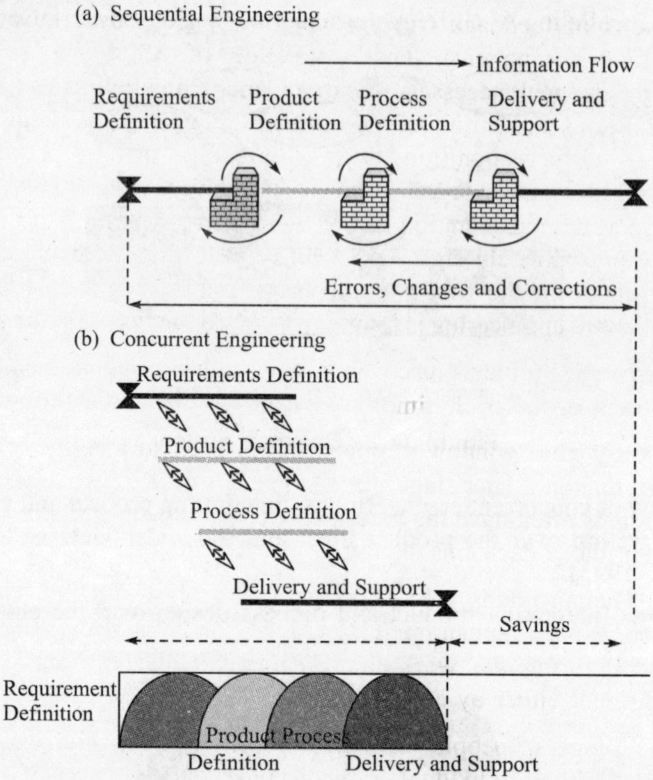

Figure 17.4 Sequential vs Concurrent Engineering.

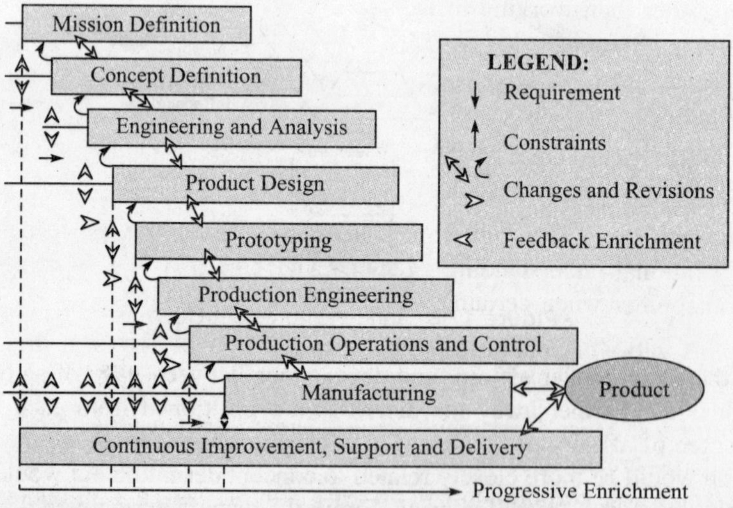

Figure 17.5 Concurrency Stages during Product Development.

indicate the net results of feedback (reverse horizontal component) and consequences and constraints (up-vertical component). The paralleling of life cycle activities in process restructuring is being deemed necessary by more and more industries to adapt quickly to changing market conditions and shrinking time-to-market targets. Additional savings may come from cost-avoidance—early recognition of possible faults, thereby preventing unnecessary issuance of the so-called "change orders" after the product is released.

Timing is an important consideration in CE. A lot rides on timing of decision-making and problem discovery. Approximately eighty per cent of product's life cycle development cost is driven by decisions made in the first twenty percent of the program effort [DARPA 1987; DARPA 1988]. Concurrent engineering is founded on the following eight fundamental principles:

Early problem discovery: Problems discovered early in the design process (particularly during the first twenty per cent of the cycle-time) are easier to solve than those discovered later.

Early decision-making: The "window of opportunity" to affect a design is much wider during an early design stage than at a later stage when some of the decisions are frozen and when the design is matured. Teams often have the natural tendency of making quick and novel decisions, which is good, except those decisions should be lasting as well.

Work structuring: Human minds practically cannot work on multiple tasks simultaneously, parallel computers do. What a human mind is good at is systematically structuring the work, or more importantly structuring the work-environment, so that each task can be performed independent of each other either by a human being, a machine or by a computer.

Teamwork affinity: Teams of people working in separate groups are likely to create designs, which may be optimal in their individual domains but will seldom remain optimal in a domain, which is a union-sum of those individual domains. Also, teams will have a better affinity if they trust each other. Trusting members, if they agree to accept responsibility for a task, do prefer to work together rather than working in isolation.

Knowledge leveraging: The domain of product design is often very large. It may be impossible to create a general-purpose "automated" or knowledge-based system, which will use appropriate tools and knowledge-driven rules (mostly computerized) to guide decision-making. Interlinking decision support tools with spurts of "human knowledge-base" will continue to be the most valuable tool for solving complex problems.

Common understanding: Teams will work better if they know what other members are doing. This includes operational understanding of all relevant interplay, e.g. what constraints a team member would encounter when certain parameters will be changed.

Ownership: Teams will work enthusiastically to make a good product if they are empowered to make decisions in shaping the design and are given "ownership" of what they produce.

Constancy of purpose: Most departments have a natural tendency to make their departments look good to others—create false profits, even though it may be detrimental to the overall corporate goals. The whole corporation will do even better if everyone works towards a common set of consistent goals, irrespective of departments they have allegiance to. This requires a change in thinking beyond the goals of one individual department or teams to those of the SBU's or the company's goals. The obligation of any supporting unit is not to suboptimize its own goals (such as unit's profit potential or sales) without a clear and direct relationship to the company's overall goals. It must contribute its best towards the system goals. Aiming towards constancy of purpose results in everyone contributing his or her best—working towards a common set of consistent goals.

17.10 WHAT IS INTEGRATED PRODUCT DEVELOPMENT (IPD)?

Concurrent engineering is structured around multifunctional teams that bring specialized knowledge necessary for the program. The multi-disciplinary set-up—called product development team (PDT)—is composed of several distinct technical subunits specializing in a variety of disciplines:

Product planners (T_{pp})
Product concept engineers (T_{ce})
Engineering and analysts (T_{ea})
Product designers (T_{pd})
Prototype engineers (T_{pe})
Production engineering planners (T_{ep})
Management and control (T_{mc})
Computer Integrated Manufacturing (CIM) and assemblers (T_{ma})
Delivery and support (T_{ds}) teams:

$$\text{PDTs} = \text{Ú}[T_{pp}, T_{ce}, T_{ea}, T_{pd}, T_{pe}, T_{ep}, T_{mc}, T_{ma}, \text{ and } T_{ds}]$$

where Ú indicates Union of and T stands for Talents. Each track is responsible for developing and integrating its own aspect to the product's life cycle as the program requires. However, there could be as many tracks/tasks and teams as needs arise. For example, experts from the volume production area must be involved in prototype production to identify as early as possible opportunities to improve process reliability. By using this multifunctional team approach to merge design and manufacturing, GE Aircraft engine division reduced design and fabrication lead time for some GE engine components from 22 to 3 weeks.

Although components of IPD have not been formally documented, it is purported that IPD has the origin in CE environment—spanned by its subunits—that yield a significant competitive advantage in product cost, quality, investments, and reduction in cycle time. Components distinguish whether a CE environment would be successful or not. The rewards and satisfaction in CE come from the success of its components. For instance, success of CE depends much on the part of those who participate in it. In CE, every participant is expected to contribute in the making of the final product: from planners and industrial designers, analysts and detailers, to production engineers and support personnel. In one direction, PDTs are supported by the customer on the top, and the infrastructure (organization) at the bottom. In a perpendicular direction, PDTs are sandwiched between product and process on one side, and tools and technology on the other.

Integrated Product Development (IPD) is a philosophy that systematically employs a teaming of functional disciplines to integrate and concurrently apply all necessary processes to produce

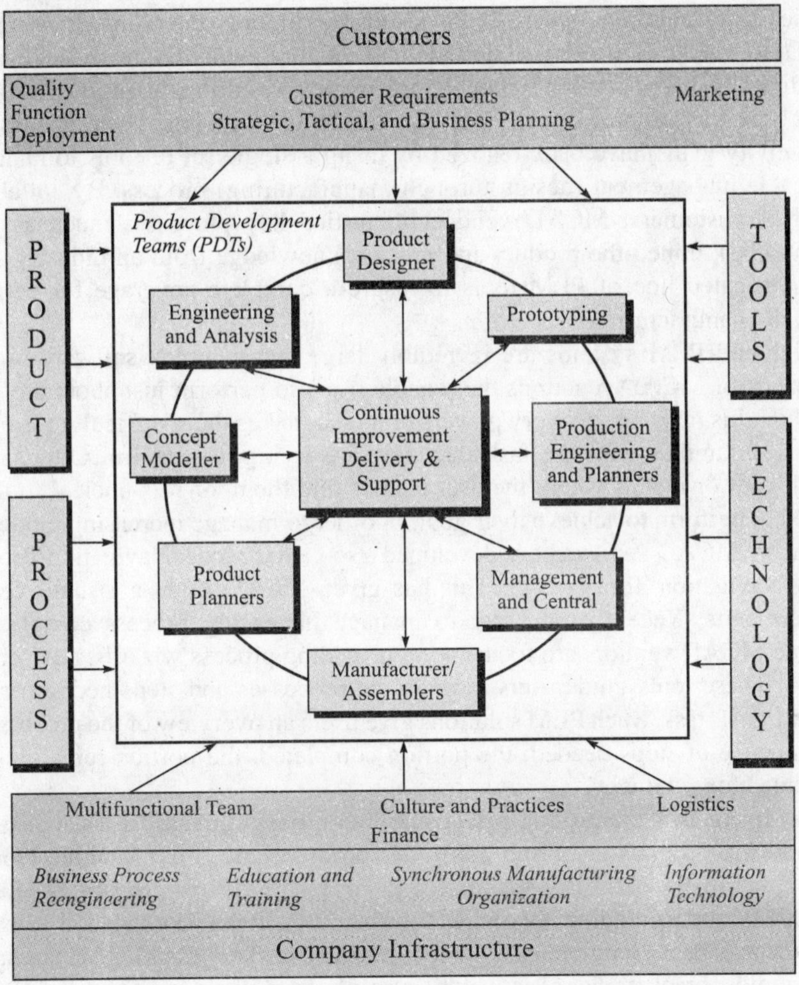

Figure 17.6 Typical Product Development Team.

an effective and efficient product that satisfies the customer's needs. There is no checklist for implementing IPD because there is no one solution...each application will be unique. [As defined in the USAFMC Guide on IPD, 1993]. Benefits of CE and IPD include 30% to 70% less development time, 65% to 90% fewer engineering changes, 20% to 90% less time to market, 200% to 600% higher quality, and 20% to 110% higher white collar productivity. [As reported by the National Institute of Standards & Technology, Thomas Group Inc., and Institute for Defense Analyses in *Business Week* April 30, 1990]

17.11 WHAT IS PRODUCT LIFE-CYCLE MANAGEMENT (PLM)

Over the past 30 years, OEMs and product manufacturers have been improving their design (product) process by leveraging Mechanical CAD (MCAD) tools and technology. In the 70s and

80s, companies were able to improve their design by digitally encoding drawing information (using MCAD). In the 80s and 90s, solid modelling, feature modelling and visualization enabled the user to digitally encode more product information resulting in productivity gains from processes such as virtual prototyping and concurrent engineering. Over the last five years, further productivity gains have been realized by using a single set of tools to handle the entire product life cycle management (design through manufacturing) process. By collaborating with OEMs and their customers, MCAD vendors identified key processes (such as Stamping or Injection Moulding), mined the product and process knowledge from an industry segment, and delivered sophisticated line of PLM tools to insure a complete coverage for entire life cycle (design through manufacturing) spectrum.

Today, high-end PLM systems are incredibly large and complex software products. They contain thousands of MCAD functions that enable users to perform just about any CAD/CAM/CAE task. While this makes users very powerful, it also makes them difficult to use, particularly when a user is unfamiliar with the task that they are trying to perform. Often, functions are hidden in dialogues or menus where the user cannot find them, or it is unclear to the user what steps they need to perform to achieve their goal. In order to manage increasing model complexity in problems, many MCAD vendors have developed a so-called Model Navigation Tool (MNT) and the Assembly Navigation Tool (ANT). This has given MCAD users a visual overview of the objects in their parts. Recently, in order to manage increasing process complexity in PLM problems, some MCAD vendors are focusing on providing process wizards, advisers, assistants, and navigators. These tools guide users through the processes and steps necessary to complete any CAD/CAM/CAE task. Such PLM solutions give users an overview of the processes available, the logical sequence of steps needed, the portion completed, the portion remaining, as well as options available along the way.

PLM vendors are now identifying and developing process threads to assist customers in the automotive, aerospace, machinery, and consumer product industries. Ranging from simple to extremely complex, these processes help users to cope with distributed large-scale design and production. OEMs are leveraging this expertise to bring process knowledge to the forefront. Process wizards provide a common, easy-to-use interface to CAD/CAM/CAE tasks. By creating items such as molds, brackets, joggles, welds, even photo-realistic images are simplified due to the addition of process wizards. Digital knowledge capabilities and process navigator tools in emerging PLM tools (and CAD/CAM systems) are providing a visual view of every task, empowering the user with the knowledge of what to do, how to do it, and how long it will take. Individual tasks are combined into larger processes that customers can immediately take advantage of, regardless of their area of expertise.

Eventually, MCAD customers are able to incorporate their own custom processes into the PLM system. Rather than just use processes predefined by a CAD/CAM tool, OEM customers are able to create, modify, and adapt processes to suit their own needs. Using an existing model, a previously defined template, or just starting from scratch, customers are able to create proprietary processes for everything from automobile tire tread layout to turbine design. Such process templates are reusable—can be used over and over again to speed up production and reduce training time.

Benefits

Having a well-conceived digital knowledge that guides the designers towards the product's life cycle goals is becoming a key differentiator for concurrent product developers and design companies of the future. It makes the product more deployable because its captured digital knowledge is adaptive and becomes process-specific when needed. Process wizards and navigators (for instance) capture the expertise of many design individuals and make them available to every member of the concurrent work groups involved in the designing of that product. These wizards, in the form of design assistants, provide a clear, visual picture of the product/problem objectives and the steps required to get there. The work group members learn by example, and such digital knowledge (in the form of wizards) provides a process template—to walk them through a set of unfamiliar tasks/territories.

17.12 THE TEAM APPROACH

Working in teams is not a usual practice for most of the Western world's companies. In order to achieve effective team working, a major structural and organizational development process needs to take place. All the team members must work as part of a group with the same objectives; hence the selection process undertaken by the management must be precise. The team created

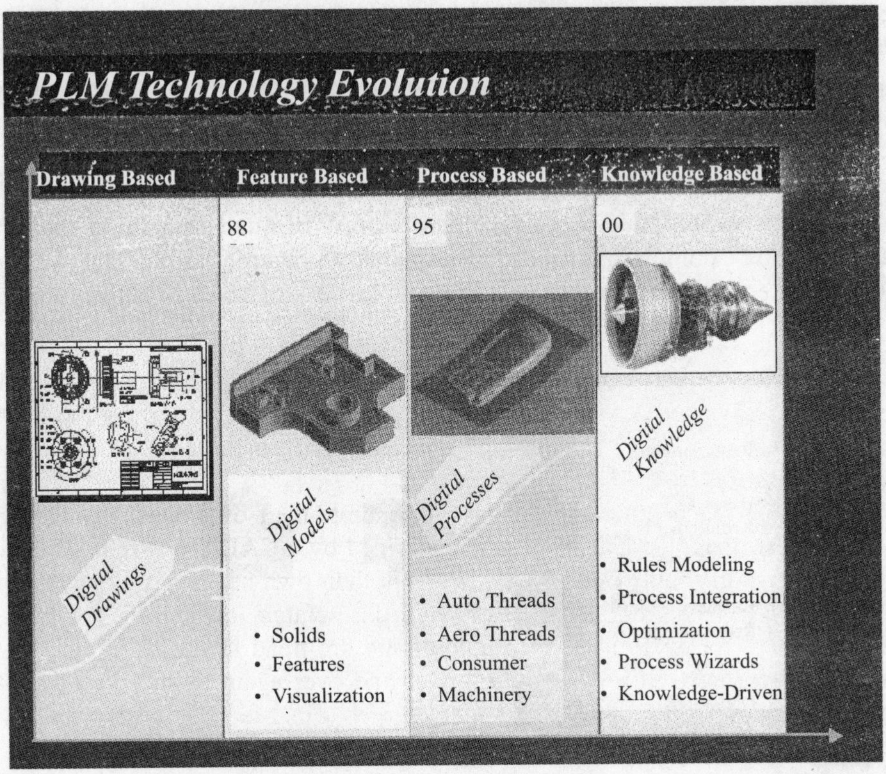

Figure 17.7 PLM Evolution.

should appreciate that their role is not of a committee, but actually to do what is required to successfully complete the project in hand. The CE practice demands multi-disciplinary team approach.

Design for manufacture/design for assembly (DFM/DFA) is also a formal methodology that can assist in building team spirit. The quantitative evaluation approaches used in many methodologies now available necessitate collectively agreeing on scores that need close cooperation of the team members from different disciplines.

17.13 CE DESIGN METHODOLOGIES

While discussing design methodologies, it is important to understand that what is being described is a set of conceptual approaches that are used to develop the product realization process. In light of the coupled decision areas in composites and the need for early design decisions, the design process itself should be thought of as one that stretches from conceptual design to actual process design and fabrication. The methodologies outlined below, with the exception of the Total Quality Design (TQD) process, were, however, developed for generic use, but have been found to be universally applicable. There are two major schools of thought regarding product development methodologies: (i) the scientific, as represented by Dixon (1988), and (ii) the engineering method, as represented by Koen (1985). In the former, prescriptive methodologies are only developed following the development of accurate descriptive methodologies that lead to testable theories of product development. The latter prescribes that the prescriptive methodology be put forth based on the best available information, and then modified as necessary. Again, there is perhaps more evidence of their use in the area of composites in Japan, than there is in the US.

QFD in concurrent engineering

The approach which we promote was developed because of our corporate background of 25 years spent promoting Computer-Aided Engineering and Concurrent Engineering. Basically, the QFD in concurrent engineering process (see Figure 17.8) is structured to convey the following:

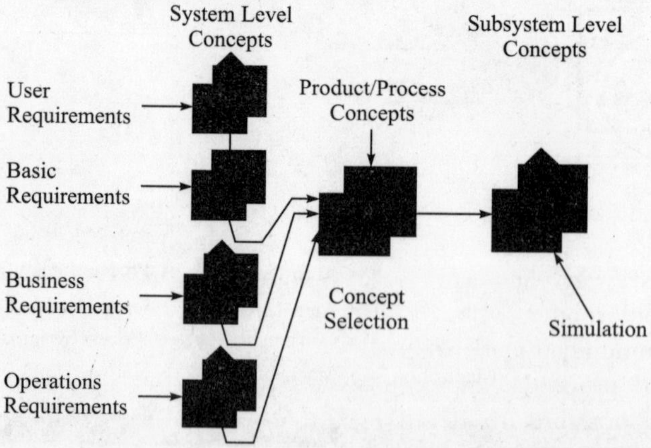

Figure 17.8 QFD in Concurrent Engineering.

Both internal and external customers have needs that must be addressed. These needs cannot be easily mixed, so a separate matrix is required for each different kind of customer. Only after all requirements have been converted to measurable product characteristics, should the team attempt to select a product/process concept.

The product/process concept should be selected using a rigorous selection technique and not gut feel. QFD should only be performed to the level required. Do not think that you have to apply QFD throughout the process of refining the product design into parts, detailed processes, etc. You may not get the high payback you expect.

Conceptually, this strikes a strong chord with those who are familiar with product development. They recognize that the process will get all disciplines involved in a project early in the process so that they all buy into the decisions, which are made during the development process. They also recognize that this is a good approach for tying the strategic planning process into product development. The weakness of the approach is that not all potential customers are equally important and some mathematical magic is required before all of the product characteristics can be compared to each other.

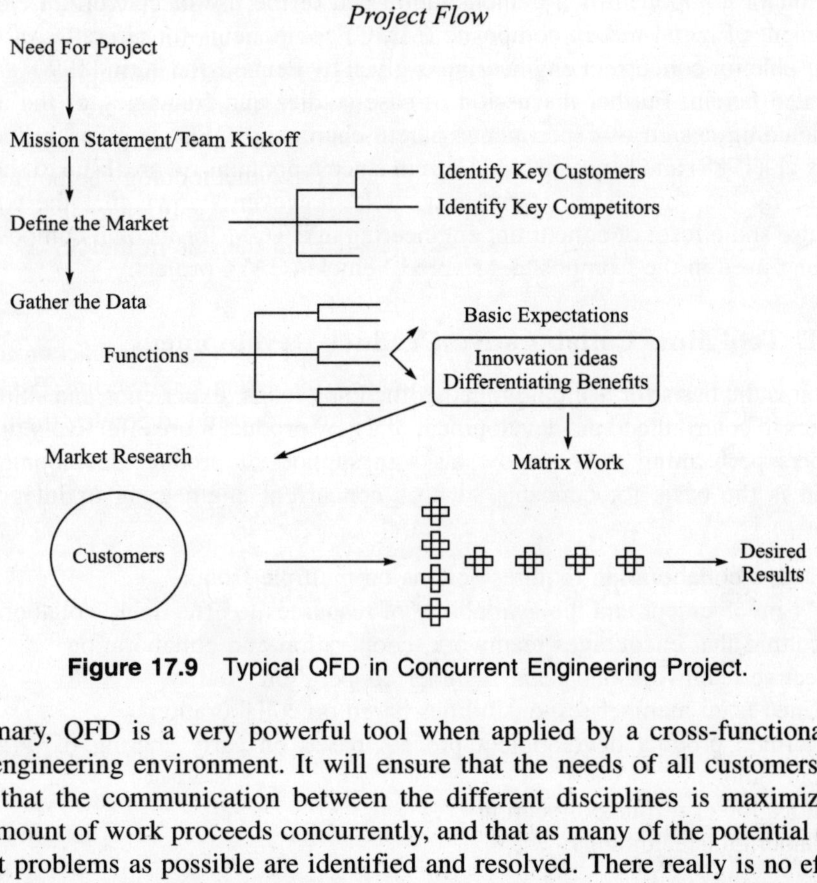

Figure 17.9 Typical QFD in Concurrent Engineering Project.

In summary, QFD is a very powerful tool when applied by a cross-functional team in a concurrent engineering environment. It will ensure that the needs of all customers are at least considered, that the communication between the different disciplines is maximized, that the maximum amount of work proceeds concurrently, and that as many of the potential downstream development problems as possible are identified and resolved. There really is no effective way to implement concurrent engineering principles without this key enabling technology.

17.14 COLLABORATIVE PRODUCT DEVELOPMENT

A systematic approach to the integrated, concurrent design of products and their related processes, including manufacture and support. This approach is intended to cause the developers, from the outset, to consider all elements of the product life cycle from conception through disposal, including quality, cost, schedule, and user requirements (Winner 1988).

In the context of this chapter, the term concurrent engineering is taken to represent the approach of collaborative product development with input from each functional group involved in the product development process. The approach taken herein is to include as many tools as necessary, which may include the methodologies mentioned before. The need for the use of a number of methodologies is due to the fact that no single methodology addresses all the concerns for successful product development. A snapshot of this diversity is provided through a comparison made by Henshaw (1989) of the methodologies based on a number of key factors as needed in composites product development.

A test bed for the use of the concurrent engineering approach through software integration has been established at University of West Virginia at the Concurrent Engineering Research Center (CERC). A major component is a demonstration bed of the use of concurrent engineering for the development of metal-matrix composite (MMC) components for aircraft. An extensive list of 26 tools usable for concurrent engineering is given by Pennell and Akin (1990) and hence will not be repeated herein. Further discussion of case studies and frequency of the use of formal methods (including design of experiments, pareto charts and QFD) is given in the IDA report by Winner et al. (1988) and is not included herein since it contains no break-up to show the actual extent of use by the composites community. It is, however, known that the CPD routinely attempts to use some form of concurrent engineering and TQM for its own composites projects. It is also being used in the Composite Armored Vehicle (CAV) project.

17.4.1 CE Tool Box Collaborative Product Development

Collaboration is the basis for bringing together the knowledge, experience and skills of multiple team members to contribute to the development of a new product more effectively than individual team members performing their narrow tasks in support of product development. As such, collaboration is the basis for concepts such as concurrent engineering or integrated product development.

- Effective collaboration requires actions on multiple fronts
- Early involvement and the availability of resources to effectively collaborate
- A culture that encourages teamwork, cooperation and collaboration
- Effective teamwork and team member cooperation
- Defined team member responsibilities based on collaboration
- A defined product development process based on early sharing of information and collaboration
- Collocation or virtual collocation
- Collaboration technology

In an increasingly competitive and global environment, it is not uncommon for corporations to pursue joint partnerships, and strategic alliances in order to rapidly introduce products and services into the market place. These partnerships can and do occur between corporations that are located in countries with significant cultural and societal differences. Inherent in these partnerships is the continuing need to design and develop products and processes for rapid introduction into the market place. **Concurrent** engineering and **collaborative** engineering are approaches that have proven effective for bringing high quality products into the market place very quickly. Because the **concurrent** engineering process relies extensively on communications and various interactions among diverse team members, understanding how cultural and societal differences impact this process may prove critical to its successful implementation where strategic alliances or partnerships are concerned. What happens when team members are not culturally and socially identical? What are the potential barriers to successful use of **concurrent** engineering in this scenario? We will seek to identify some of those barriers in this chapter. This discussion is based in part on case study research conducted in Japan and the United States specifically for packaging design. Background material on **concurrent** engineering and packaging/packaging systems is also provided.

Collaboration requires effective team work. Team members must trust and respect one another. There must be open communication and a willingness to accept input from others. There are often conflicting goals in product development. Therefore, decision-making must be based on a collaborative approach. This is shown in Fig. 17.10.

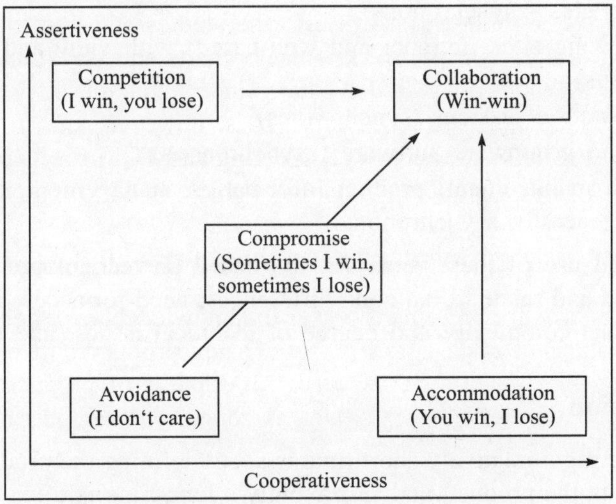

Figure 17.10 Co-Op vs Assertiveness.

Figure 17.10 represents two axes: cooperativeness and assertiveness. Different approaches to dealing with an issue are mapped against these two axes. A low degree of assertiveness and cooperativeness represents avoidance of an issue or the approach of "I don't care". A high degree of cooperativeness and a low degree of assertion represents accommodation or the approach of "You win, I lose". A high degree of assertiveness and a low degree of cooperativeness represents competition or the approach of "I win, you lose".

Many people believe a compromise is the ideal. Compromise represents a moderate degree of both assertiveness and cooperativeness. It represents the approach of "Sometimes I win and sometimes I lose". This, however, is not the ideal. A good team includes people that have strong beliefs and are professionally committed and what the team members believe is important to the company and the development program (but not necessarily to just the narrow functional perspective of their discipline). We also want a high degree of cooperation. This represents the basis for a collaborative approach or the "win-win" approach. The key to the win-win approach is to creatively search for solutions that can mutually satisfy the needs of the team rather than focusing on just two competing solutions that involve trade-offs or are mutually exclusive.

17.14.2 Collaboration Technology

There are a variety of tools and technologies to facilitate communication and collaboration. Collaboration can happen synchronously where all participants view information and/or meet at the same time or asynchronously where participants view information and provide feedback at different points in time. These tools and technologies include:

- Email exchange of drawings, models and project information (asynchronous)
- Teleconferencing and videoconferencing (synchronous)
- Web-hosted meetings (synchronous)
- Project hosting tools to create one pool of all released project documentation, with email alerts for updates (asynchronous)
- Drawing viewing sites (intranet and web-based) with view and mark-up capabilities (asynchronous)
- CAD collaboration sessions (synchronous)
- Workflow and groupware software (asynchronous)
- Product data management, product information management, collaborative product commerce (generally asynchronous)

The selection and use of these tools will be based on technology availability and cost; geographic dispersion and related time zone differences; need for access by partners, suppliers and customers; product complexity and degree of technical issues; and other factors.

17.14.3 Conclusion

In traditional engineering a relatively short time is spent defining the product. A relatively long time is spent designing the product, and surprisingly a long time is often spent redesigning the product. The key to shortening the overall design time is to better define the product and better document the design process.

Traditionally, the development of a product had been seen as a cycle of plan...do...check... act...(adjust). Concurrent engineering is a process in which appropriate disciplines are committed to work interactively to conceive, approve, develop, and implement product programs that meet predetermined objectives.

Concurrent engineering is relatively a recent term which is applied to the engineering design philosophy of cross-functional cooperation in order to create products which are better, cheaper, and more quickly brought to market. This new trend reunites technical and non-technical disciplines such as engineering, marketing and accounting. Always focusing on satisfying the customer, these representatives work together in defining the product to be manufactured.

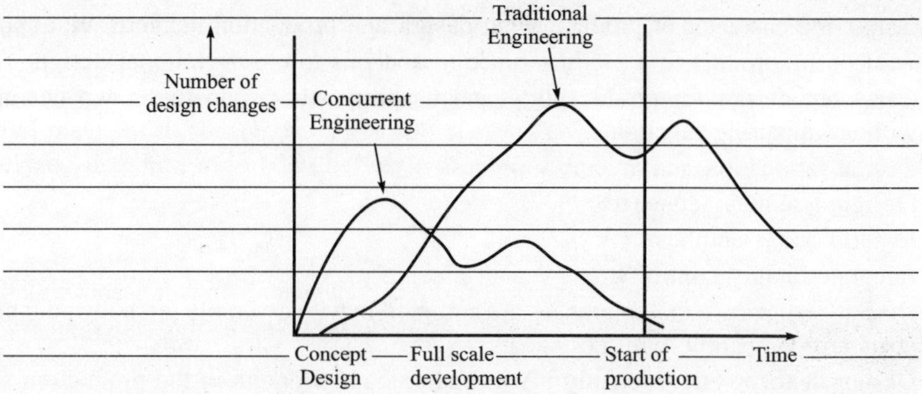

Figure 17.11 Example of Design Changes as a Function of Time for an American and Japanese Automobile.

Good ideas that are novel have a unique motivating quality. People get excited about them and eventually there will be competition and disagreement about their origin. Nonetheless, most ideas—good or bad—are never acted upon either by individuals, groups, and, especially, enterprises.

To practice CE:

1. From the start, include all domains of expertise as active participants in the design effort.
2. Resist making irreversible decision before they must be made.
3. Perform continuous optimization of product and process.
4. Identify product concepts that are inherently easy to manufacture.
5. Focus on component design for manufacture and assembly.
6. Integrate the manufacturing process design and product design that best match needs and requirements.
7. Convert concept to manufacturability, salable, usable design by stating all constraints.
8. Anticipate fabrication and assembly methods and problems.
9. Reduce the number of parts.
10. Increase interchangeability between models.
11. Define subassemblies to allow models to differ by the subassemblies.
12. Standardize fastener types and sizes; use low-cost, irreversible fasteners only where a skilled serviceperson would work.
13. Improve robustness of the product and process.
14. Identify difficult process steps for which costs and process times cannot be predicted.
15. Use existing processes and facilities so that product yield is high.
16. Break down products and processes into self-contained modules and assembly lines.
17. Adjust tolerances to eliminate failures during assembly.
18. Identify testable areas.
19. Make assembly easier by minimizing set-ups and reorientations.

20. Design parts for feeding and insertion.
21. Determine character of product; what design and production methods are appropriate.
22. Subject the product to a product function analysis to ensure rational design.
23. Carry out design for producability and usability study; can these two be improved without impairing function ?
24. Design fabrication and assembly process.
25. Design assembly sequence.
26. Identify subassemblies.
27. Integrate quality control strategy with assembly.
28. Design each part so that tolerances are compatible with assembly method and fabrication costs are compatible with cost goals.
29. Design factory system to fully involve production workers in the production strategy, operate on minimum inventory, and integrate with vendor capabilities.

17.15 FUTURE TOOLS FOR CONCURRENT ENGINEERING

- More computer tools are becoming available for breaking down the barriers between the 'Islands' of automation.
- Some tasks have been defined for evolving software to support concurrent engineering.
  - Interoperable tools and tasks—Develop standards for file transfer, so that software may pass information freely between various engineering systems.
  - Interoperable computing environment—The differences between different computers, from different vendors should be eliminated, or at least made to be insignificant. This will overcome the need for separated systems, which occurs now. This also extends to other aspects of computing technology, such as user interfaces, database interfaces, files management systems.
  - Data management—Basically integrate all information into a single framework, so that transferring information is no longer an issue. An example of this is a single database. This would serve all software in a facility, so that there is no need for transferring designs, and other information.
  - Process management—At present human judgment is still required to initiate any, and all processes in an automated environment. If the computers could be made to automatically review designs, or forward information to the interested parties, then the computer would create an automated framework for concurrent engineering.
  - Decision support—While most of the developments are concerned with the design information, we eventually would require tools that examine the work done by the group, and determine which aspects of the design still need to be performed, or are insufficient. In effect, the computer takes on some of the management tasks.
- As the five stages above develop, the focus changes from data to knowledge, and from simple storage of files, to shared collaboration of data.

REVIEW QUESTIONS

1. What is concurrent engineering?
2. What is the importance of concurrent engineering?
3. What are the steps involved in concurrent engineering?
4. Write short notes on:
   - (i) Commitment
   - (ii) Planning
   - (iii) Leadership
5. When concurrent engineering is used?
6. What is organizational factor?
7. Briefly explain the concept of QFM and QFD.
8. What is design for assembly?
9. What are the core things of concurrent engineering?
10. What are the important roles of four pillars of concurrent engineering?
11. Discuss typical example in concurrent engineering.
12. Compare sequential vs concurrent engineering.
13. What is Integrated Product Development (IPD)?
14. What is Product Life-cycle Management (PLM)?
15. What is CE methodology?
16. What do you mean by collaborative product development?

Chapter 18

LOCAL AREA NETWORK COMMUNICATIONS

18.1 INTRODUCTION

A communication network is a collection of equipment and physical media that interconnects two or more computers.

Local Area Networks (LANs) are generally used to interconnect computers within the same building or organization and to transmit data at a speed up to 100 Mbps. In LAN computers are networked for small area. Most LANs are confined to a single building or group of buildings. However, one LAN can be connected to other LANs over any distance via telephone lines and radio waves. A system of LANs connected in this way is called a **Wide Area Network (WAN).**

Most LANs connect workstations and personal computers. Each node (individual computer) in a LAN has its own CPU with which it executes programs, but it is also able to access data and devices anywhere on the LAN. This means that many users can share expensive devices, such as laser printers, as well as data. Users can also use the LAN to communicate with each other, by sending e-mail or engaging in chat sessions.

There are many different types of LANs, ethernets being the most common for PCs. Most Apple Macintosh networks are based on Apple's AppleTalk network system, which is built into Macintosh computers.

The following characteristics differentiate one LAN from another.

18.2 TOPOLOGY

Topology means the geometric arrangement of devices on the network. For example, devices can be arranged in a ring or in a straight line.

There are four principal topologies used in LANs.

- **Bus topology:** All devices are connected to a central cable called the bus or backbone. Bus networks are relatively inexpensive and easy to install for small networks. Ethernet systems use a bus topology.

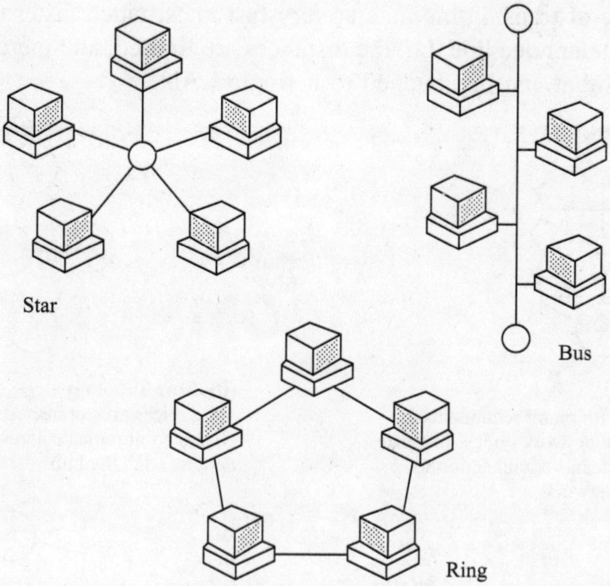

Figure 18.1 The Shape of a Local Area Network (LAN) or other Communications System.

- **Ring topology:** All devices are connected to one another in the shape of a closed loop, so that each device is connected directly to two other devices, one on either side of it. Ring topologies are relatively expensive and difficult to install, but they offer high bandwidth and can span large distances.
- **Star topology:** All devices are connected to a central *hub*. Star networks are relatively easy to install and manage, but bottlenecks can occur because all data must pass through the hub.
- **Tree topology:** A tree topology combines the characteristics of linear bus and star topologies. It consists of groups of star-configured workstations connected to a linear bus backbone cable.

These topologies can also be mixed. For example, a bus-star network consists of a high-bandwidth bus, called the *backbone,* which connects a collection of slower-bandwidth star segments.

Topology refers to the shape of a network, or the network's layout. How different nodes in a network are connected to each other and how they communicate are determined by the network's topology. Topologies are either *physical* or *logical.* Figure 18.2 shows the five most common network topologies. The important terms used in networking are:

- **Protocols:** The rules and encoding specifications for sending data. The protocols also determine whether the network uses a peer-to-peer or client/server architecture.
- **Media:** Devices can be connected by twisted-pair wire, coaxial cables, or fiber optic cables. Some networks do without connecting media altogether, communicating instead via radio waves.

LANs are capable of transmitting data at very fast rates, much faster than the data that can be transmitted over a telephone line; but the distances are limited, and there is also a limit on the number of computers that can be attached to a single LAN.

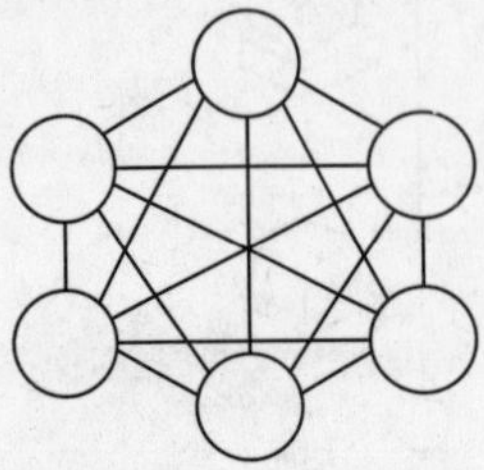

(a) **Mesh Topology**
Devices are connected with many redundant interconnections between network nodes. In a true mesh topology every node has a connection to every other node in the network.

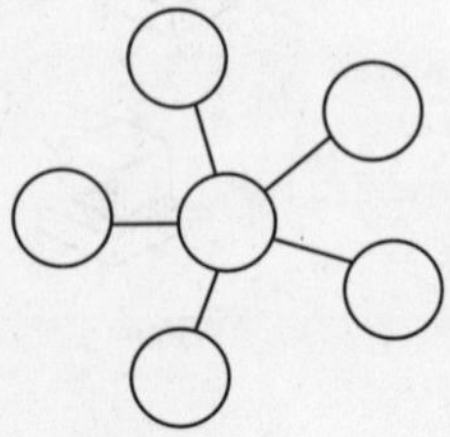

(b) Star Topology
All devices are connected to a central hub. Nodes communicate across the network by passing data through the hub.

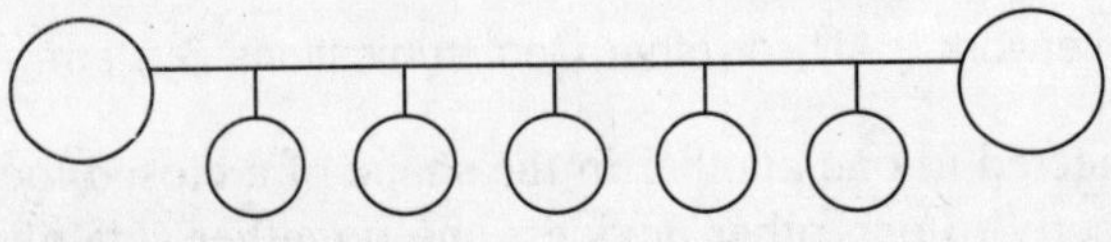

(d) **Bus Topology**
All devices are connected to a central cable, called the bus or backbone.

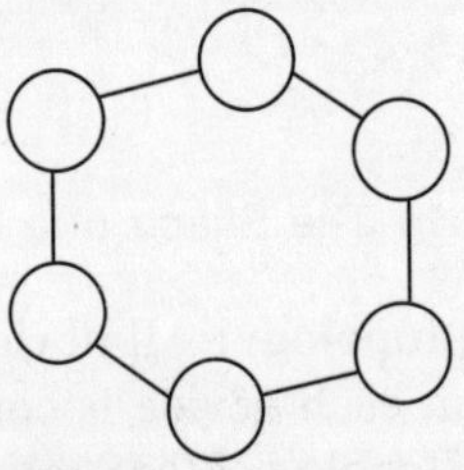

(c) **Ring Topology**
All devices are connected to one another in the shape of a closed loop, so that each device is connected directly to two other devices, one on either side of it.

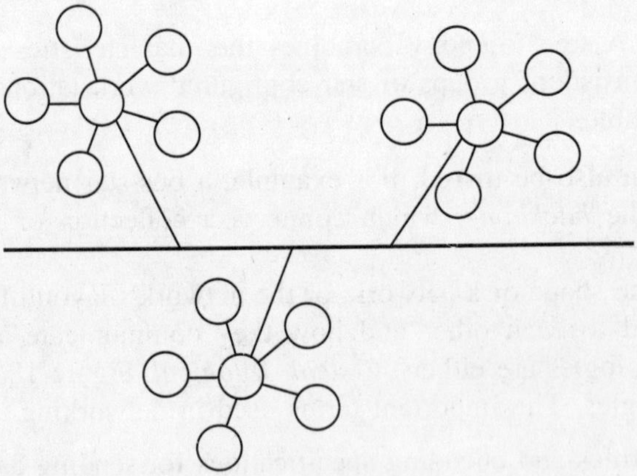

(e) **Tree Topology**
A hybrid topology. Groups of star-configured networks are connected to a linear bus backbone

Figure 18.2 Network Topology.

18.3 DATA STRUCTURES

Type of data in which each element is attached to one or more elements *directly beneath* it. The connections between elements are called *branches*. Trees are often called *inverted trees* because they are normally drawn with the *root* at the top.

The elements at the very bottom of an inverted tree (that is, those that have no elements below them) are called *leaves*. Inverted trees are the data structures used to represent hierarchical file structures. In this case, the leaves are files and the other elements above the leaves are directories.

A binary *tree* is a special type of inverted tree in which each element has only two branches below it.

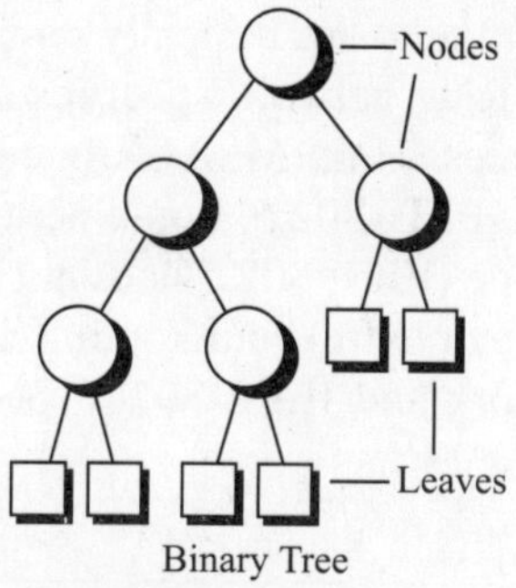

Figure 18.3 Binary Tree.

18.4 TOKEN BUS NETWORK

A token bus network is similar to a token ring network in that a station must have possession of a token before it can transmit on the network (see "Token and Token-Passing Access Methods"). However, the topology and token-passing method are different. The IEEE 802.4 Committee has defined token bus standards as broadband networks, as opposed to Ethernet's baseband transmission technique. The Manufacturing Automation Protocol (MAP) developed by General Motors for manufacturing floor networks uses IEEE 802.4. ARCNET (Attached Resource Computing NETwork) is a token bus network, although it does not conform to the IEEE 802.4 standards.

The topology of the network can include groups of workstations connected by long trunk cables. These workstations branch from hubs in a star configuration, so the network has both a bus and star topology. Token bus topology is well suited to groups of users that are separated by some distance. IEEE 802.4 token bus networks are constructed with 75-ohm coaxial cable using a bus topology. The broadband characteristics of the 802.4 standard support transmission over several different channels simultaneously.

The token and frames of data are passed from one station to another following the numeric sequence of the station addresses. Thus, the token follows a logical ring rather than a physical ring. The last station in numeric order passes the token back to the first station. The token does not follow the physical ordering of workstation attachment to the cable. Station 1 might be at one end of the cable and station 2 might be at the other, with station 3 in the middle.

While the token bus is used in some manufacturing environments, Ethernet and token ring standards have become more prominent in the office environment.

18.5 TOKEN RING/IEEE 802.5

The Token Ring network was originally developed by IBM in the 1970s. It is still IBM's primary Local Area Network (LAN) technology. The related IEEE 802.5 specification is almost identical to and completely compatible with IBM's Token Ring network. In fact, the IEEE 802.5 specification was modeled after IBM Token Ring, and it continues to shadow IBM's Token Ring development. The term *Token Ring* is generally used to refer to both IBM's Token Ring network and IEEE 802.5 networks. Here, we discuss both Token Ring and IEEE 802.5.

Token Ring and IEEE 802.5 networks are basically compatible, although the specifications differ in minor ways. IBM's Token Ring network specifies a star, with all end stations attached to a device called a Multi Station Access Unit (MSAU). In contrast, IEEE 802.5 does not specify a topology, although virtually all IEEE 802.5 implementations are based on a star. Other differences exist, including media type (IEEE 802.5 does not specify a media type, although IBM Token Ring networks use twisted-pair wire) and routing information field size. Figure 18.4 summarizes IBM Token Ring network and IEEE 802.5 specifications.

	IBM Token Ring network	IEEE 802.5
Data rates	4.16 Mbps	4.16 Mbps
Stations/segment	280 (shielded) twisted pair 72 (unshielded twisted pair)	250
Topology	Star	Not specified
Media	Twisted pair	Not specified
Signaling	Baseband	Baseband
Access method	Token passing	Token passing
Encoding	Differential manchester	Differential manchester

Figure 18.4 Although Dissimilar in Some Respects, IBM's Token Ring Network and IEEE 802.5 Are Generally Compatible.

Physical connections

IBM Token Ring network stations are directly connected to MSAUs, which can be wired together to form one large ring (see Figure 18.5). Patch cables connect MSAUs to adjacent MSAUs, while lobe cables connect MSAUs to stations. MSAUs include bypass relays for removing stations from the ring.

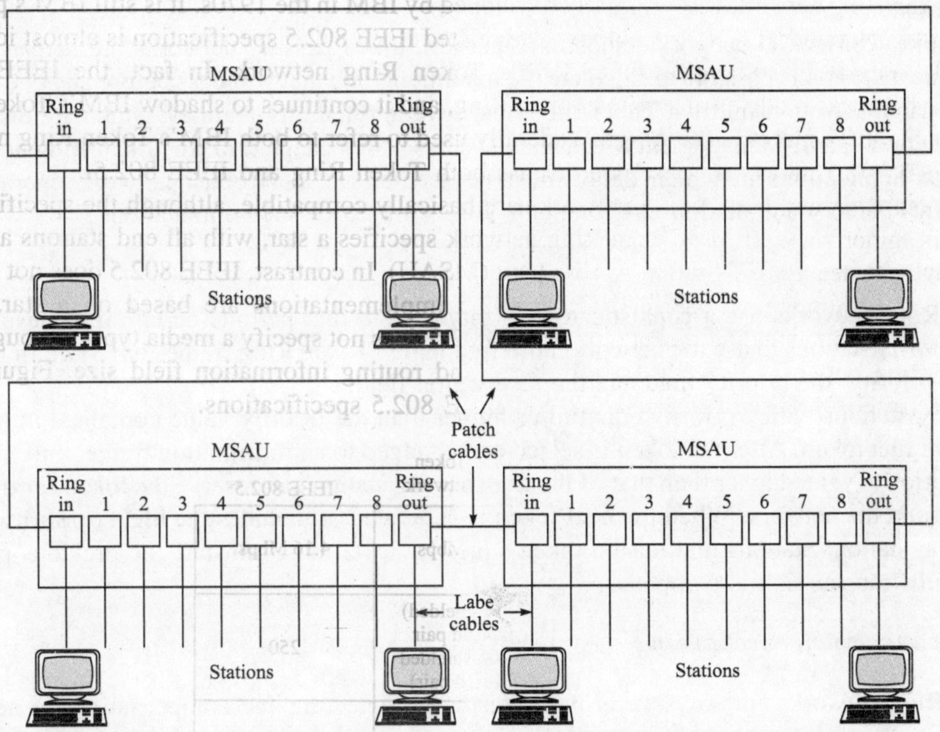

Figure 18.5 MSAUs Can Be Wired Together to Form One Large Ring in an IBM.

18.6 TOKEN RING NETWORK

Token ring operation

Token Ring and IEEE 802.5 are two principal examples of token-passing networks (FDDI is the other). *Token-passing networks* move a small frame, called a token, around the network. Possession of the token grants the right to transmit. If a node receiving the token has no information to send, it passes the token to the next end station. Each station can hold the token for a maximum period of time.

If a station possessing the token does have information to transmit, it seizes the token, alters 1 bit of the token (which turns the token into a start-of-frame sequence), appends the information that it wants to transmit, and sends this information to the next station on the ring. While the information frame is circling the ring, no token is on the network (unless the ring supports early token release), which means that other stations wanting to transmit must wait. Therefore, collisions cannot occur in Token Ring networks. If early token release is supported, a new token can be released when frame transmission is complete.

The information frame circulates the ring until it reaches the intended destination station, which copies the information for further processing. The information frame continues to circle the ring and is finally removed when it reaches the sending station. The sending station can check the returning frame to see whether the frame was seen and subsequently copied by the destination.

Unlike CSMA/CD networks (such as Ethernet), token-passing networks are *deterministic*, which means that it is possible to calculate the maximum time that will pass before any end station will be capable of transmitting. This feature and several reliability features, which are discussed in the section "Fault-Management Mechanisms" later in this chapter, make Token Ring networks ideal for applications in which delay must be predictable and robust network operation is important. Factory automation environments are examples of such applications.

Priority system

Token Ring networks use a sophisticated priority system that permits certain user-designated, high-priority stations to use the network more frequently. Token Ring frames have two fields that control priority: the priority field and the reservation field.

Only stations with a priority equal to or higher than the priority value contained in a token can seize that token. After the token is seized and changed to an information frame, only stations with a priority value higher than that of the transmitting station can reserve the token for the next pass around the network. When the next token is generated, it includes the higher priority of the reserving station. Stations that raise a token's priority level must reinstate the previous priority after their transmission is complete.

Fault-management mechanisms

Token Ring networks employ several mechanisms for detecting and compensating for network faults. For example, one station in the Token Ring network is selected to be the *active monitor*. This station, which potentially can be any station on the network, acts as a centralized source of timing information for other ring stations and performs a variety of ring-maintenance functions. One of these functions is the removal of continuously circulating frames from the ring. When a sending device fails, its frame may continue to circle the ring. This can prevent other stations from transmitting their own frames and essentially can lock up the network. The active monitor can detect such frames, remove them from the ring, and generate a new token.

The IBM Token Ring network's star topology also contributes to overall network reliability. Because all information in a Token Ring network is seen by active MSAUs, these devices can be programmed to check for problems and selectively remove stations from the ring, if necessary.

A Token Ring algorithm called *beaconing* detects and tries to repair certain network faults. Whenever a station detects a serious problem with the network (such as a cable break), it sends a beacon frame, which defines a failure domain. This domain includes the station reporting the failure, its Nearest Active Upstream Neighbour (NAUN), and everything in between. Beaconing initiates a process called *autoreconfiguration,* in which nodes within the failure domain automatically perform diagnostics in an attempt to reconfigure the network around the failed areas. Physically, the MSAU can accomplish this through electrical reconfiguration.

18.7 SIGNALLING METHODS

A pre-paid subscriber account system for use with wireless telephone systems is disclosed. The system, which monitors a subscriber's call, deducts the cost of the call from the subscriber's pre-paid account in real-time, warns the subscriber during a call when the account is nearing depletion, and terminates the call when the account is depleted. The system can also prevent the initiation of a new call when the account is depleted. Various options can also apply different charging rates, can charge only when specified criteria are met, or can charge only for calls to/from specified areas or telephone numbers. The system uses signalling techniques that will allow the metering or billing of the call, along with any authorization or restrictions, to be done remotely from the actual switching of the call. Call events or chargeable events are transmitted to the pre-paid control system while the communications path of the call is held at the switching system awaiting control information. The identification of the pre-paid service subscriber is established using existing devices that currently are used to authenticate the user. The wireless phone user does not have to enter any additional codes or identification to obtain access to the pre-paid service.

Simple network management protocol overview

Simple Network Management Protocol (SNMP) is a network management standard that defines a strategy for managing TCP/IP and, more recently, Internet Packet Exchange (IPX) networks.

SNMP uses a distributed architecture that includes:

Multiple managed nodes, each with an SNMP entity called an *agent* which provides remote access to management instrumentation.

At least one SNMP entity referred to as a *manager* which runs management applications to monitor and control managed elements. Managed elements are devices such as hosts, routers, and so on; they are monitored and controlled by accessing their management information. A management protocol, SNMP, is used to convey management information between the management stations and agents. *Management information* refers to a collection of managed objects that reside in a virtual information store called a Management Information Base (MIB).

SNMP messages

To communicate host information, management systems and agents use SNMP messages. These messages are sent using the User Datagram Protocol (UDP) and are routed between the management system and host by using the Internet Protocol (IP).

A Management Information Base (MIB) contains the information requested by the management system. The MIB for a networked computer may include information on the configuration and performance of the network interface card, the available hard drive space, the version of drivers and applications, and so on. Additional MIBs may be written and loaded, to expose the data that is specified for collection, as long as the system itself supports the collection of the requested information.

REVIEW QUESTIONS

1. Define network.
2. What is LAN?
3. What is binary tree?
4. Write short notes on
   (i) Token ring
   (ii) Token Bus
5. Discuss the application signalling in network.
6. Compare LAN and WAN.
7. Discuss (i) Tree topology, (ii) Bus topology, and (iii) Ring topology.
8. What is star network?
9. Define mesh topology.

Appendix A

A SIMULATION CASE STUDY OF AN FMS CELL

The case study presented in this section is based on the work of Musil and Akbay (1989). The flexible manufacturing cell used in their simulation study consists of an ASEA six-axis robot, Kearney and Trecker horizontal spindle mill, Mori Seiki lathe, gripper change station and rough part unloading area.

In this case, the cell is set up to produce three components: base, jaw and screw for the assembly of a vise. Each part is machined on one particular machine—the base on the Kearney and Trecker mill, the jaw on a Bridgeport mill and the screw on the Mori Seiki lathe. The components arrive at the cell as rough castings in the loading area. The robot follows a preprogrammed route to transport the parts to their respective machines, load them onto the machines, reposition the parts for additional operations on the same machine, and transport the finished parts to the unloading area. The gripper change station is necessary because the base and the jaw require that the robot use a straight gripper while the screw requires the use of a curved gripper.

Simulation model

The simulation model of the flexible manufacturing cell was developed in the Simulation Language for Alternative Modelling (SLAM II) and the Extended Simulation Support System (TESS). TESS adds a relational database management system and animation capabilities to the simulation.

Analysis of the current system

The model of the flexible manufacturing cell was run over a simulated time span of 4800 min, or ten 8-hour shifts. Key data values or statistics, with regard to the intended objectives, were:

1. Time in the system for each component part
2. Number of observations (product output)
3. Average waiting time for each resource
4. Percentage of time that each resource was captured.

The statistics for the simulation run of the current system as well as policies experimented with are summarized in Table A.1. The absolute minimum time required to produce a particular part can be determined by summing all the necessary activities for that part throughout the system. This would result in minimum times of 5.19, 6.56 and 2.90 min for the base, jaw and screw, respectively. These times could be achieved only if an entity never had to wait for a resource. With the help of this information and the statistics in Table A.1, we can obtain the average waiting time and percentage of time that each resource was captured.

As seen from Table A.1, the jaw spent the longest time in the system, 10.87 min per part. A key aim was to reduce this critical or system-dependent time, without increasing the processing time of the base or screw above the processing time of the jaw. After initial runs, it was checked that the output of the simulated model coincided with that of the actual system, validating the model. Experiments with the model could now be performed with the expectation of realistic results.

TABLE A.1 Results for the Case Study

	Current system	*Policy 1*	*Policy 2*	*Policy 3*
Time in system for base (min)	9.62	8.66	8.51	6.49
Time in system for jaw (min)	10.87	9.91	9.76	7.43
Time in system for screw (min)	5013	5.13	6.55	8.79
System output (cycle)	441.00	484.00	491.00	546.00
Percentage increase	–	9.75	11.33	23.81

The objective of this study was to increase the output of the cell by making only scheduling changes. The way to alter the scheduling order of the component parts in the model is to change the priorities associated with the resources. The only resource in the cell that requires prioritizing is the robot, because it is the only resource that is shared by more than one component part. In model, the priorities of the parts requesting the robot resource can be changed, making it possible to "experiment with the programming" of the robot without affecting the cell itself.

A comparison of the order in which the parts are processed by the robot in the current system and in the three policies to be considered is as follows:

Current policy	:	B-J-S-B-J-W-J-B-J
Policy 1	:	B-J-S-J-B-J-B-J
Policy 2	:	B-J-S-J-B-S-J-B-J
Policy 3	:	S-J-B-J-W-B-W-J-B-J-S

where **B** stands for base, **J** for jaw, **S** for screw and **W** for wait.

Policy 1 In this case the order of the base and the jaw was switched after processing the screw. This change caused the order of the remaining processing for two parts to be reversed and eliminated the time during which the robot was waiting for the second operation on the jaw to be finished. The results of this simulation run are summarized in Table A.1. The results demonstrate that the changes made in this experiment provided an increase in output by 9.75 per cent and that the time in the system for the jaw was reduced from 10.87 to 9.91 min per production cycle. This simple change in the processing order of the base and the jaw made a significant impact on the productivity of the cell.

Policy 2 The next question to be considered was: What if the screw processing was split up into multiple steps? The intent of Policy 2 was to answer this question. The results of this simulation run are summarized in Table A.1. The output was raised to 491 units, an increase of 50 units over the current system, which translates to an increase in productivity of 11.3 per cent. It is apparent that dividing the processing of the screw into two steps decreased the waiting time of both the base and the jaw while raising the processing time of the screw only slightly. This increased time, however, is still less than the critical time for the cell, which is the time for the jaw.

Policy 3 The results thus far are acceptable, but could the output be improved even further without going beyond scheduling changes? Considering that the screw has the shortest processing time of the three parts, it would be best if the screw did most of the waiting, thereby reducing the waiting time of the jaw. In this experiment, the second step in the processing of the screw was moved all the way to the end, after completion of both the base and the screw. In order to do this, the robot was made to wait for a short time at two points in the processing of the other parts. The simulation run produced the data shown in Table A.1. The critical time was no longer the time spent in the system by the jaw, because the time to process the screw had increased to a point above that of the jaw. However, this time was well below the critical time in the other cases, providing an output of 545 units, which is an increase in productivity of 23.8 per cent over the current system.

Study results

With simulation modelling, it is possible to create and test countless scenarios. In this study, among the scenarios considered, the processing order in Policy 3 provided the best scheduling to meet the intended objectives. Although Policy 3 may or may not be the best possible processing order, it provides an excellent improvement over the current system. With the help of this simulation model, it was shown that a 24 per cent increase in productivity would be realized by modifying the processing order as shown in the third policy.

Appendix **B**

BAR CODE TECHNOLOGY

Character recognition fonts are divided into two groups, Magnetic Ink Character Recognition (MICR) and Optical Character Recognition (OCR). However, this distinction is artificial; some applications use magnetic (actually, easily magnetizable) ink for OCR, while the MICR fonts can be recognized optically.

The history of these fonts can be traced to the financial world. With automatic processing of cheques at other than the native banks, automatic reading of information became important. Moreover, a bar code would obfuscate what was actually written, so in the fifties the first font (E-13B) was devised that would result in nearly unambiguous reading. In banking, generally magnetic reading is used because that is more reliable.

We describe the following fonts here: two MICR, E-13B and CMC-7 and two OCR, OCR-A and OCR-B. All four are used in banking. E-13B is most dominantly used.

E-13B

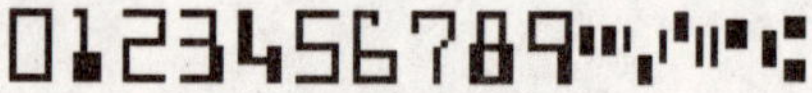

CMC-7

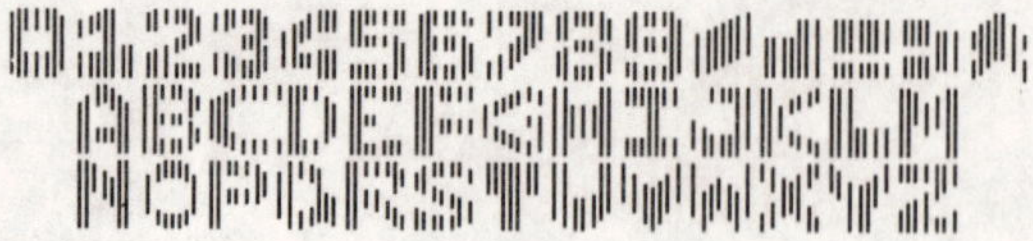

Optical character recognition, is a computer software designed to translate images of handwritten or typewritten text (usually captured by a scanner) into machine-editable text, or to translate pictures of characters into a standard encoding scheme representing them (e.g. ASCII or Unicode). OCR began as a field of research in pattern recognition, artificial intelligence and machine vision. Though academic research in the field continues, the focus on OCR has shifted to implementation of proven techniques.

Optical character recognition (using optical techniques such as mirrors and lenses) and digital character recognition (using scanners and computer algorithms) were originally considered separate fields. Because very few applications survive that use true optical techniques, the optical character recognition term has now been broadened to cover digital character recognition as well.

Early systems required "training" (essentially, the provision of known samples of each character) to read a specific font. Currently, though, "intelligent" systems that can recognize most fonts with a high degree of accuracy are now common. Some systems are even capable of reproducing formatted output that closely approximates the original scanned page including images, columns and other non-textual components.

E-13B

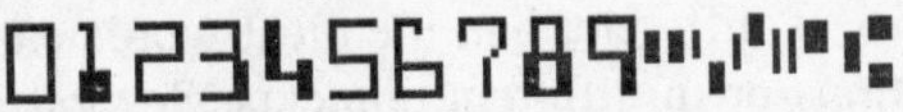

The first font [E-13B] is used in automated banking. It contains the digits and four special symbols used in banking. The four additional symbols are called (from left to right): "dash", "amount", "on*us", and "transit". The font was specially designed so that magnetic pulses would be read unambiguously. That is the reason for some of the heavy black features of some symbols. For this purpose, a grid of 7 by 11 squares was used that had to be either white or black. The filling was done so that a magnetic scan would give a pulse signal that was very distinctive. In computing, a technique is used that enables special characters printed in magnetic ink to be read and input rapidly to a computer. MICR is used extensively in banking because magnetic ink characters are difficult to forge and are therefore ideal for marking and identifying cheques.

CMC-7

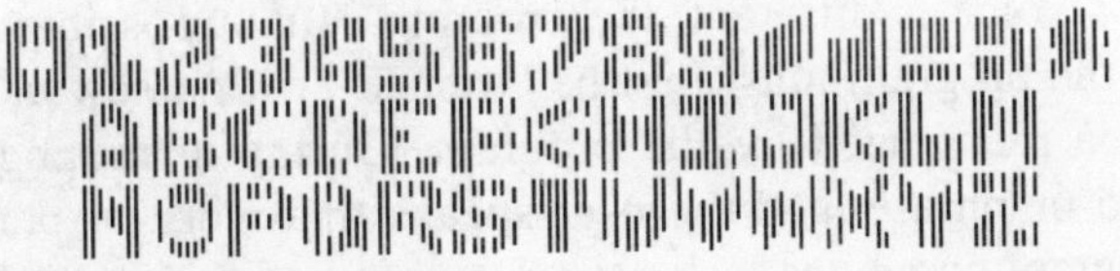

B.1 VOICE RECOGNITION

Voice recognition is the process of taking the spoken word as an input to a computer program. This process is important to virtual reality because it provides a fairly natural and intuitive way of controlling the simulation while allowing the user's hands to remain free. Here, we will delve into the uses of voice recognition in the field of virtual reality, examine how voice recognition is accomplished, and list the academic disciplines that are central to the understanding and advancement of voice recognition technology.

B.1.1 What is Voice Recognition, and Why is it Useful in a Virtual Environment?

Voice recognition is "the technology by which sounds, words or phrases spoken by humans are converted into electrical signals, and these signals are transformed into coding patterns to which meaning has been assigned" [ADA90]. While the concept could more generally be called "sound recognition", we focus here on the human voice because we most often and most naturally use our voices to communicate our ideas to others in our immediate surroundings. In the context of a virtual environment, the user would presumably gain the greatest feeling of immersion, or being part of the simulation, if they could use their most common form of communication, the voice. The difficulty in using voice as an input to a computer simulation lies in the fundamental differences between human speech and the more traditional forms of computer input. While computer programs are commonly designed to produce a precise and well-defined response upon receiving the proper (and equally precise) input, the human voice and spoken words are anything but precise. Each human voice is different, and identical words can have different meanings if spoken with different inflections or in different contexts. Several approaches have been tried, with varying degrees of success, to overcome these difficulties.

B.1.2 How is Voice Recognition Performed?

The most common approaches to voice recognition can be divided into two classes: "template matching" and "feature analysis". Template matching is the simplest technique and has the highest accuracy when used properly, but it also suffers from the most limitations. As with any approach to voice recognition, the first step is for the user to speak a word or phrase into a microphone. The electrical signal from the microphone is digitized by an "analog-to-digital (A/D) converter", and is stored in memory. To determine the "meaning" of this voice input, the computer attempts to match the input with a digitized voice sample, or template, that has a known meaning. This technique is a close analogy to the traditional command inputs from a keyboard. The program contains the input template, and attempts to match this template with the actual input using a simple conditional statement.

Since each person's voice is different, the program cannot possibly contain a template for each potential user, so the program must first be "trained" with a new user's voice input before that user's voice can be recognized by the program. During a training session, the program displays a printed word or phrase, and the user speaks that word or phrase several times into a microphone. The program computes a statistical average of the multiple samples of the same word and stores the averaged sample as a template in a program data structure. With this approach to voice recognition, the program has a "vocabulary" that is limited to the words or phrases used in the training session, and its user base is also limited to those users who have trained the program. This type of system is known as *speaker-dependent*. It can have vocabularies on the order of a few hundred words and short phrases, and recognition accuracy can be about 98 per cent.

A more general form of voice recognition is available through feature analysis, and this technique usually leads to "speaker-independent" voice recognition. Instead of trying to find an exact or near-exact match between the actual voice input and a previously stored voice template, this method first processes the voice input using "Fourier transforms" or "linear predictive coding (LPC)", then attempts to find characteristic similarities between the expected inputs and the actual digitized voice input. These similarities will be present for a wide range of speakers, and so the system need not be trained by each new user. The types of speech differences that the speaker-independent method can deal with, but which pattern matching would fail to handle, include accents, and varying speed of delivery, pitch, volume, and inflection. Speaker-independent speech recognition has proven to be very difficult, with some of the greatest hurdles being the variety of accents and inflections used by speakers of different nationalities. Recognition accuracy for speaker-independent systems is somewhat less than for speaker-dependent systems, usually between 90 and 95 per cent.

Another way to differentiate between voice recognition systems is by determining if they can handle only discrete words, connected words, or continuous speech. Most voice recognition systems are discrete word systems, and these are easiest to implement. For this type of system, the speaker must pause between words. This is fine for situations where the user is required to give only one word responses or commands, but is very unnatural for multiple word inputs. In a connected word voice recognition system, the user is allowed to speak in multiple word phrases, but he or she must still be careful to articulate each word and not slur the end of one word into the beginning of the next word. Totally natural, continuous speech includes a great deal of "coarticulation", where adjacent words run together without pauses or any other apparent division between words. A speech recognition system that handles continuous speech is the most difficult to implement.

B.1.3 What Disciplines are Involved in Voice Recognition?

The template matching method of voice recognition is founded on the general principles of digital electronics and basic computer programming. To fully understand the challenges of efficient speaker-independent voice recognition, the fields of phonetics, linguistics, and digital signal processing should also be explored.

Optical character recognition refers to the branch of computer science that involves reading text from paper and translating the images into a form that the computer can manipulate (for example, into ASCII codes). An OCR system enables you to take a book or a magazine article, feed it directly into an electronic computer file, and then edit the file using a word processor.

All OCR systems include an optical scanner for reading text, and sophisticated software for analyzing images. Most OCR systems use a combination of hardware (specialized circuit boards) and software to recognize characters, although some inexpensive systems do it entirely through software. Advanced OCR systems can read text in large variety of fonts, but they still have difficulty with handwritten text.

The potential of OCR systems is enormous because they enable users to harness the power of computers to access printed documents. OCR is already being used widely in the legal profession, where searches that once required hours or days can now be accomplished in a few seconds.

B.2 DATA ACQUISITION SYSTEMS

Data acquisition systems, as the name implies, are products and/or processes used to collect information to document or analyze some phenomenon. In the simplest form, a technician logging the temperature of an oven on a piece of paper is performing data acquisition. As technology has progressed, this type of process has been simplified and made more accurate, versatile, and reliable through electronic equipment. Equipment ranges from simple recorders to sophisticated computer systems. Data acquisition products serve as a focal point in a system, tying together a wide variety of products, such as sensors that indicate temperature, flow, level, or pressure. Some common data acquisition terms are shown below: ADC, DIO, DA **Differential Input General Purpose Interface Bus (GPIB) RS232 RS485.**

Bar Code

Appendix C

STEPPER MOTORS

A stepper, or stepping motor, converts electronic pulses into proportionate mechanical movement. Each revolution of the stepper motor's shaft is made up of a series of discrete individual steps. A step is defined as the angular rotation produced by the output shaft each time the motor receives a step pulse. These types of motors are very popular in digital control circuits, such as robotics, because they are ideally suited for receiving digital pulses for step control. Each step causes the shaft to rotate a certain number of degrees. A step angle represents the rotation of the output shaft caused by each step, measured in degrees. Figure C.1 below illustrates a simple application for a stepper motor. Each time the controller receives an input signal, the paper is driven a certain incremental distance. In addition to the paper drive mechanism in a printer, stepper motors are also popular in machine tools, process control systems, tape and disk drive systems, and programmable controllers.

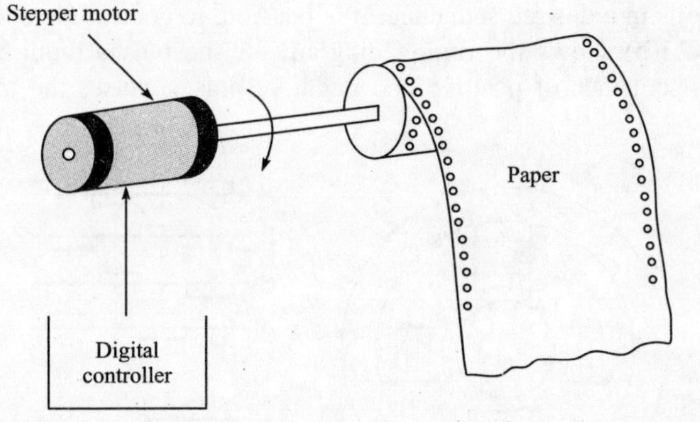

Figure C.1 Paper drive mechanism using stepper machine.

C.1 PERMANENT-MAGNET (PM) STEPPER MOTORS

The permanent-magnet stepper motor operates on the reaction between a permanent-magnet rotor and an electromagnetic field. Figure C.2 shows a basic two-pole PM stepper motor. The rotor shown in Figure C.2(a) has a permanent magnet mounted at each end. The stator is illustrated in Figure C.2(b). Both the stator and the rotor are shown as having teeth. The teeth on the rotor surface and the stator pole faces are offset so that there will be only a limited number of rotor teeth aligning themselves with an energized stator pole. The number of teeth on the rotor and stator determines the step angle that will occur each time the polarity of the winding is reversed. The greater the number of teeth, the smaller the step angle.

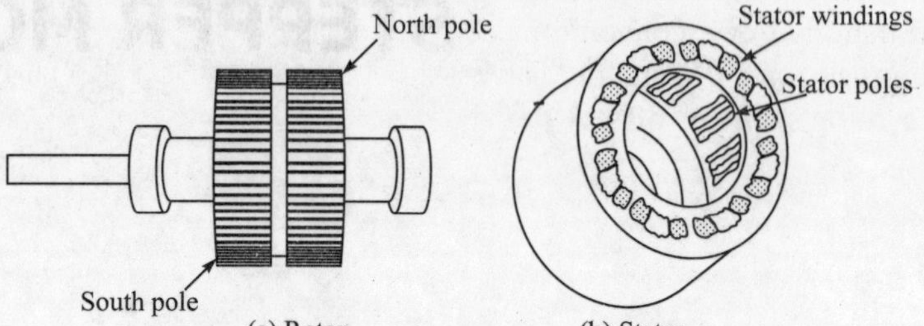

Figure C.2 Components of a PM stepper motor.

When a PM stepper motor has a steady DC signal applied to one stator winding, the rotor will overcome the residual torque and line up with that stator field. The holding torque is defined as the amount of torque required to move the rotor one full step with the stator energized. An important characteristic of the PM stepper motor is that it can maintain the holding torque indefinitely when the rotor is stopped. When no power is applied to the windings, a small magnetic force is developed between the permanent magnet and the stator. This magnetic force is called a *residual*, or *detent torque*. The detent torque can be noticed by turning a stepper motor by hand and is generally about one-tenth of the holding torque.

Figure C.3(a) shows a permanent magnet stepper motor with four stator windings. By pulsing the stator coils in a desired sequence, it is possible to control the speed and direction of the motor. Figure C.3(b) shows the timing diagram for the pulses required to rotate the PM stepper motor. This sequence of positive and negative pulses causes the motor shaft to rotate

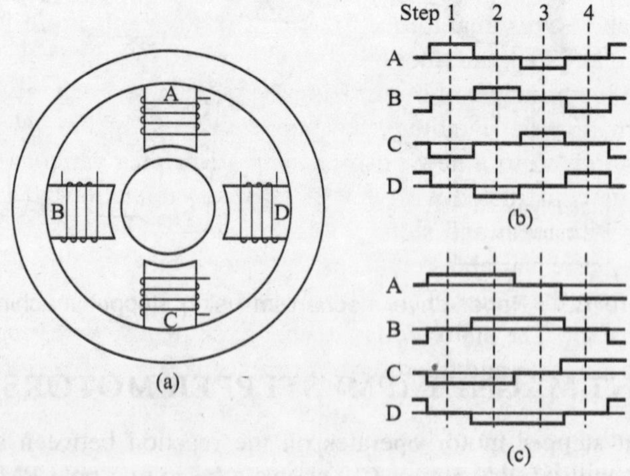

Figure C.3 (a) PM stepper motor; (b) 90° step; (c) 45° step.

counterclockwise in 90° steps. The waveforms of Figure C.3(c) illustrate how the pulses can be overlapped and the motor made to rotate counterclockwise at 45° intervals.

A more recent development in PM stepper motor technology is the thin-disk rotor. This type of stepper motor dissipates much less power in losses such as heat than the cylindrical rotor and as a result, it is considerably more efficient. Efficiency is a primary concern in industrial circuits such as robotics, because a highly efficient motor will run cooler and produce more torque or speed for its size. Thin-disk rotor PM stepper motors are also capable of producing almost double the steps per second than that of a conventional PM stepper motor. Figure C.4 shows the basic construction of a thin-disk rotor PM motor. The rotor is constructed of a special type of cobalt-steel, and one-half a rotor segment offsets the stator poles.

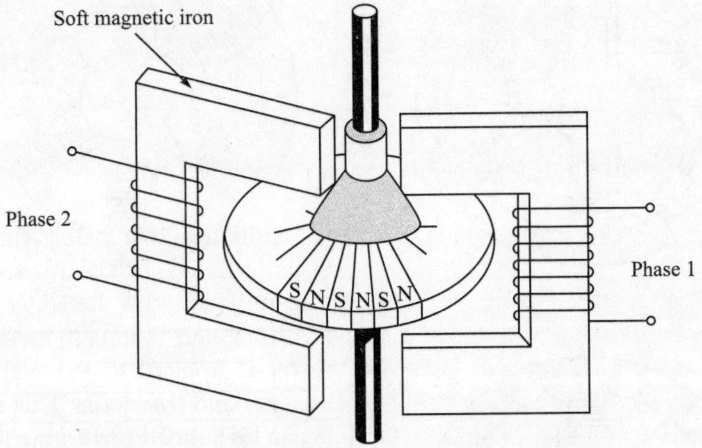

Figure C.4 Thin-disk rotor PM stepper motor.

C.2 VARIABLE-RELUCTANCE (VR) STEPPER MOTORS

The variable-reluctance (VR) stepper motor differs from the PM stepper in that it has no permanent-magnet rotor and no residual torque to hold the rotor at one position when turned off. When the stator coils are energized, the rotor teeth will align with the energized stator poles. This type of motor operates on the principle of minimizing the reluctance along the path of the applied magnetic field. By alternating the windings that are energized in the stator, the stator field changes, and the rotor is moved to a new position. The stator of a variable-reluctance stepper motor has a magnetic core constructed with a stack of steel laminations. The rotor is made of unmagnetized soft steel, with teeth and slots.

Figure C.5 shows a basic variable-reluctance stepper motor. In this circuit, the rotor is shown with fewer teeth than the stator. This ensures that only one set of stator and rotor teeth will align at any given instant. The stator coils are energized in groups referred to as *phases*. In Figure C.5, the stator has six teeth and the rotor has four teeth. The rotor will turn 30° each time a pulse is applied.

Figure C.5(a) shows the position of the rotor when phase A is energized. As long as phase A is energized, the rotor will be held stationary. When phase A is switched off and phase B is energized, the rotor will turn 30° until two poles of the rotor are aligned under the north and south poles established by phase B. The effect of turning off phase B and energizing phase C is shown in Figure C.5(c). In this circuit, the rotor has again moved 30° and is now aligned under the north and south poles created by phase C. After the rotor has been displaced by 60° from its starting point, the step sequence has completed one cycle. Figure C.5(d) shows the switching sequence to complete a full 360° of rotation for a variable-reluctance motor with six stator poles and four rotor poles.

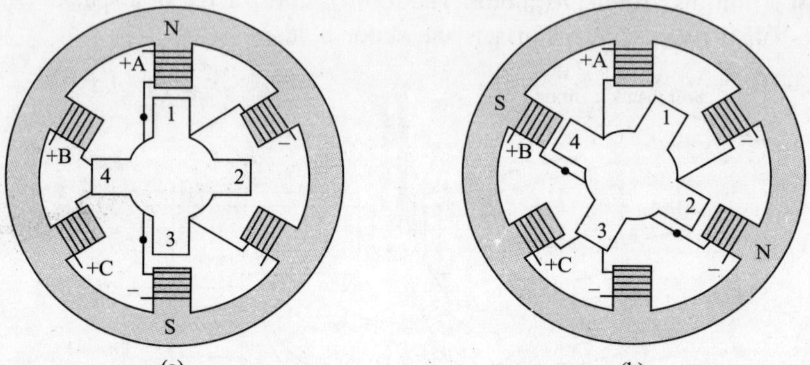

(a) (b)

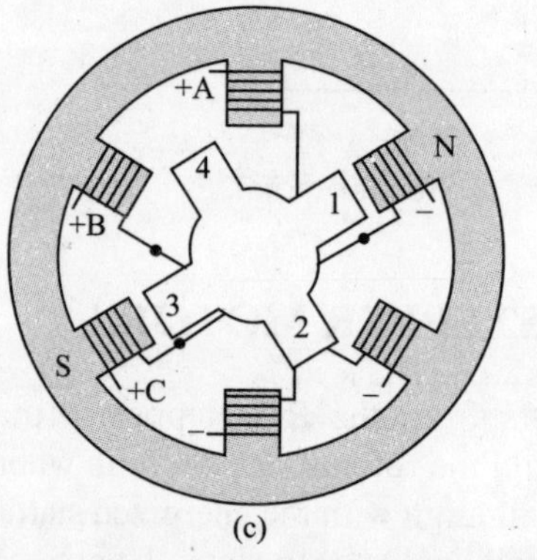

(c)

Cycle	Phase A	Phase B	Phase C	Position
1	ON	OFF	OFF	0°
	OFF	ON	OFF	30°
	OFF	OFF	ON	60°
2	ON	OFF	OFF	90°
	OFF	ON	OFF	120°
	OFF	OFF	ON	150°
3	ON	OFF	OFF	180°
	OFF	ON	OFF	210°
	OFF	OFF	ON	240°
4	ON	OFF	OFF	270°
	OFF	ON	OFF	300°
	OFF	OFF	ON	330°
5	ON	OFF	OFF	360°

(d)

Figure C.5 Variable-reluctance stepper motor and switching sequence.

The VR stepper motors mentioned up to this point are all single-stack motors. That is, all the phases are arranged in a single stack or plane. The disadvantage of this design for a stepper motor is that the steps are generally quite large (above 15°). Multistack stepper motors can produce smaller step sizes because the motor is divided along its axial length into magnetically isolated sections or stacks. Each of these sections is excited by a separate winding or phase. In this type of motor, each stack corresponds to a phase, and the stator and rotor have the same tooth pitch.

C.3 HYBRID STEPPER MOTORS

The hybrid step motor consists of two pieces of soft iron, as well as an axially magnetized, round permanent-magnet rotor. The term *hybrid* is derived from the fact that the motor is operated under the combined principles of the permanent magnet and variable-reluctance stepper motors. The stator core structure of a hybrid motor is essentially the same as its VR counterpart. The main difference is that in the VR motor, only one of the two coils of one phase is wound on one pole, while a typical hybrid motor will have coils of two different phases wound on the same pole.

The two coils at a pole are wound in a configuration known as a *bifilar connection*. Each pole of a hybrid motor is covered with uniformly spaced teeth made of soft steel. The teeth on the two sections of each pole are misaligned with each other by a half-tooth pitch. Torque is created in the hybrid motor by the interaction of the magnetic field of the permanent magnet and the magnetic field produced by the stator.

Stepper motors are rated in terms of the number of steps per second, the stepping angle, and load capacity in ounce-inches and the pound-inches of torque that the motor can overcome. The number of steps per second is also known as the *stepping rate*.

Appendix **D**

AC VARIABLE DRIVE

An AC motor is an electric motor that is driven by an alternating current. An AC motor consists of two basic parts.

- An outside stationary stator having coils supplied with AC current to produce a rotating magnetic field
- An inside rotor attached to the output shaft that is given a torque by the rotating field.

There are two types of AC motors, depending on the type of rotor used.

- The *synchronous motor*, which rotates exactly at the supply frequency or a submultiple of the supply frequency. The magnetic field on the rotor is either due to current transported with slip rings or a permanent magnet.
- The *induction motor*, which turns slightly slower than the supply frequency. The magnetic field on the rotor of this motor is created by an induced current.

Three-phase AC induction motors

Where a polyphase electrical supply is available, the three-phase (or polyphase) AC induction motor (Figure D.1) is commonly used, especially for higher-powered motors. The phase differences between the three phases of the polyphase electrical supply create a rotating electromagnetic field in the motor.

Through electromagnetic induction, the rotating magnetic field induces a current in the conductors in the rotor, which in turn sets up a counterbalancing magnetic field that causes the rotor to turn in the direction the field is rotating. The rotor must always rotate slower than the rotating magnetic field produced by the polyphase electrical supply; otherwise, no counterbalancing field will be produced in the rotor.

Induction motors are the workhorses of industry and motors up to about 500 kW (670 horsepower) in output are produced in highly standardized frame sizes, making them nearly completely interchangeable between manufacturers (although European and North American standard dimensions are different). Very large induction motors are capable of tens of thousands of kW in output, for pipeline compressors, wind-tunnel drives and overland conveyor systems.

There are two types of rotors used in induction motors: squirrel cage rotors and wound rotors.

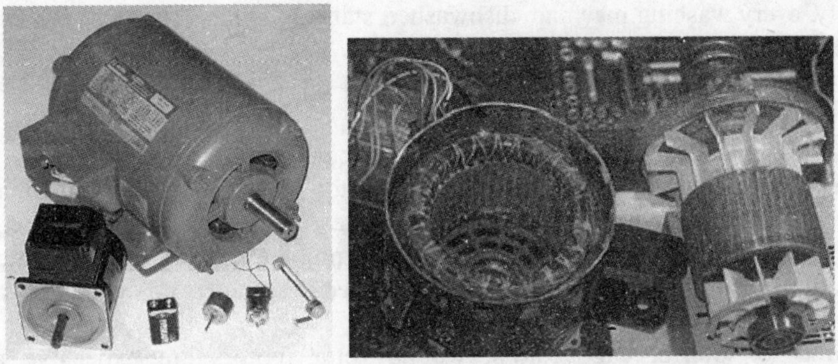

(a) (b)

Figure D.1 (a) Three-phase AC induction motors rated 1 Hp (746 W) and 25 W with small motors from CD player, toy and CD/DVD drive reader head traverse. (b) Disassembled 250 W motor from a washing machine. The 12 stator windings are in the housing on the left. Next to it is the "squirrel cage" rotor on its shaft.

Squirrel-cage rotors

Most common AC motors use the squirrel cage rotor, which will be found in virtually all domestic and light industrial alternating current motors. The squirrel cage takes its name from its shape—a ring at either end of the rotor, with bars connecting the rings running the length of the rotor. It is typically cast aluminum or copper poured between the iron laminates of the rotor, and usually only the end rings will be visible. The vast majority of the rotor currents will flow through the bars rather than the higher-resistance and usually varnished laminates. Very low voltages at very high currents are typical in the bars and end rings; high efficiency motors will often use cast copper in order to reduce the resistance in the rotor.

In operation, the squirrel cage motor may be viewed as a transformer with a rotating secondary—when the rotor is not rotating in sync with the magnetic field, large rotor currents are induced; the large rotor currents magnetize the rotor and interact with the stator's magnetic fields to bring the rotor into synchronization with the stator's field. An unloaded squirrel cage motor at synchronous speed will consume electrical power only to maintain rotor speed against friction and resistance losses; as the mechanical load increases, so will the electrical load—the electrical load is inherently related to the mechanical load. This is similar to a transformer, where the primary's electrical load is related to the secondary's electrical load.

This is why, as an example, a squirrel cage blower motor may cause the lights in a home to dim as it starts, but does not dim the lights when its fanbelt (and therefore mechanical load) is removed. Furthermore, a stalled squirrel cage motor (overloaded or with a jammed shaft) will consume current limited only by circuit resistance as it attempts to start. Unless something else limits the current (or cuts it off completely), overheating and destruction of the winding insulation is the likely outcome.

In order to prevent the currents induced in the squirrel cage from superimposing itself back onto the supply, the squirrel cage is generally constructed with a prime number of bars, or at least a small multiple of a prime number (rarely more than 2). There is an optimum number of bars in any design, and increasing the number of bars beyond that point merely serves to increase the losses of the motor particularly when starting.

Virtually, every washing machine, dishwasher, standalone fan, record player, etc. uses some variant of a squirrel cage motor.

Wound rotor

An alternate design, called the *wound rotor*, is used when variable speed is required. In this case, the rotor has the same number of poles as the stator and the windings are made of wire, connected to slip rings on the shaft. Carbon brushes connect the slip rings to an external controller such as a variable resistor that allows changing the motor's slip rate. In certain high-power variable speed wound-rotor drives, the slip-frequency energy is captured, rectified and returned to the power supply through an inverter.

Compared to squirrel cage rotors, wound rotor motors are expensive and require maintenance of the slip rings and brushes, but they were the standard form for variable speed control before the advent of compact power electronic devices. Transistorized inverters with variable-frequency drive can now be used for speed control, and wound rotor motors are becoming less common. (Transistorized inverter drives also allow the more-efficient three-phase motors to be used when only single-phase mains current is available, but this is never used in household appliances, because it can cause electrical interference, and because of high power requirements.)

Several methods of starting a polyphase motor are used. Where the large inrush current and high starting torque can be permitted, the motor can be started across the line, by applying full line voltage to the terminals (Direct-on-line, DOL). Where it is necessary to limit the starting inrush current (where the motor is large compared with the short-circuit capacity of the supply), reduced voltage starting using either series inductors, an autotransformer, thyristors, or other devices are used. A technique sometimes used is (Star-Delta, Y?) starting, where the motor coils are initially connected in wye (Y) for acceleration of the load, then switched to delta when the load is up to speed. This technique is more common in Europe than in North America. Transistorized drives can directly vary the applied voltage as required by the starting characteristics of the motor and load.

This type of motor is becoming more common in traction applications such as locomotives, where it is known as the *asynchronous traction motor.*

The speed of the AC motor is determined primarily by the frequency of the AC supply and the number of poles in the stator winding, according to the relation:

$$N_s = 120F/p$$

where

N_s = Synchronous speed, in revolutions per minute
F = AC power frequency
p = Number of poles per phase winding.

Actual RPM for an induction motor will be less than this calculated synchronous speed by an amount known as *slip*, that increases with the torque produced. With no load, the speed will be very close to synchronous. When loaded, standard motors have between 2–3% slip, special motors may have up to 7% slip, and a class of motors known as *torque motors* are rated to operate at 100% slip (0 RPM/full stall).

The slip of the AC motor is calculated by:

$$S = (N_s - N_r)/N_s$$

where

N_r = Rotational speed, in revolutions per minute
S = Normalised slip, 0 to 1.

As an example, a typical four-pole motor running on 60 Hz might have a nameplate rating of 1725 RPM at full load, while its calculated speed is 1800 RPM.

The speed in this type of motor has traditionally been altered by having additional sets of coils or poles in the motor that can be switched on and off to change the speed of magnetic field rotation. However, developments in power electronics mean that the frequency of the power supply can also now be varied to provide a smoother control of the motor speed.

GLOSSARY

Accuracy: The ability of a robot to reach a given point in space repeatedly and how far off it will be in the worst-case.

Actuator: An **actuator** is the mechanism by which an agent acts upon an environment. The agent can be either an artificial intelligence agent or any other autonomous being (human, other animal, etc). A mechanism that puts something into automatic action is called an actuator.

Adaptive control: A system that can modify its behaviour in response to changes in the dynamics of the process and the character of the disturbances.

Agile manufacturing: A means of thriving in an environment of continuous change, by managing complex inter and intra-firm relationships through innovations in technology, information, and communication, organizational redesign and new marketing strategies.

AGV: These trailers can be used to move raw materials into line to get them ready to be manufactured. The AGV can also store objects on a bed. The objects can be placed on a set of motorized treads and then pushed off by reversing them. Some AGVs use fork lifts to lift objects for storage. Transporting materials such as medicine in a hospital.

ARM: Robot arms come in all shapes and sizes. The arm is that part of the robot that positions the end-effector and sensors to do their pre-programmed business.

Artificial intelligence: The concept that computers can be programmed to assume capabilities such as learning, reasoning, adaptation and self-correction. Also known as **machine intelligence** and often abbreviated as **AI,** is intelligence exhibited by any manufactured system. The term is often applied to general-purpose computers and also in the field of scientific investigation into the theory and practical application of AI.

Automation: (1) The conversion to and implementation of procedures, processes, or equipment by automated means. (2) Industrial open- or closed-loop control systems in which the manual operation of controls is replaced by servo operation.

Autonomous system: A network that is administered by a single set of management rules that are controlled by one person, group or organization. Autonomous systems often use only one routing protocol, although multiple protocols can be used. The core of the Internet is made up of many autonomous systems.

Axis: Axis in robotics means an individual component which is free to move relative to the other components.

Ball screw: A device for transforming rotary motion to linear, or vice versa, incorporating a threaded rod portion and a nut consisting of a cage holding many ball-bearings.

Benchmark: A fixed point of reference or a standard for comparison and used to achieve excellence within the manufacturing firm. An outstanding example, appropriate for use as a model.

Bill of Material: A listing of all the subassemblies, parts, and raw materials that go into a parent assembly.

CAD: Computer Aided Design is the use of high-resolution graphics in a wide range of design activities, allowing quick evaluation and modification of intent. Computer-Aided Design is the use of computers in interactive engineering drawing and storage of designs.

CAM: Computer Aided Manufacturing is the use of computer technology to generate data to control part or all of a manufacturing process. Use of computers to program, direct and control production equipment in the fabrication of manufactured items.

Capacity Requirements Planning: CRP, used in manufacturing (MRP), is a technique for projecting resource requirements for critical work stations, generated from scheduled receipts and planned order releases.

Cartesian coordinates: A means for giving the location of a point in space by measuring its distance from a reference point along straight lines. The lines are at right angles to each other and all meet at the reference point. Named after its creator, Rene Descartes.

Cell control: A manufacturing unit consisting of two or more workstations or machines and the materials, transport mechanisms, and storage buffers that interconnect them.

Changeover time: The time required to modify a system or workstation, usually including both tear down time for the existing condition and set-up time for the new condition. Typically associated with a switch to a new product.

Chain drive: Any device that transmits power by use of a chain and sprocket.

CIM: Computer Integrated Manufacturing, known as CIM, is the phrase used to describe the complete automation of a manufacturing plant, with all processes functioning under computer control and digital information tying them together. The heart of computer integrated manufacturing is CAD/CAM.

CIM architecture: A graphical representation of organization and connectivity of elements that participate in a CIM system.

CIM-OSA: A European organization which is attempting to develop Open System Architecture (OSA) standards for CIM (Computer Integrated Manufacturing) environments.

CIM program proposal: A logical series of projects to achieve key business requirements.

CIM system: Refers to an implementation of the CIM architecture to integrate an enterprise. Proper selection of CIM products and standards as part of the system will require characteristics of the particular enterprise and attributes of its data requirements to be defined.

Closed-loop control: Robot control which uses a feedback loop to measure then compare actual system performance with desired performance, and then makes adjustments accordingly.

CNC: Computer Numerical Control allows the control of motion in an accurate and programmable manner through the use of a dedicated computer within a numerical control unit, with a capability of local data input such that machine tools are freed from the need for hard-wired controllers.

CNC control: The CNC control will interpret a CNC program and activate the series of commands in sequential order. As it reads the program, the CNC control will activate the appropriate machine functions, cause axis motion, and in general, follow the instructions given in the program.

Computer-aided engineering: Computer-aided engineering (often referred to as CAE) is a broad term describing the use of computer technology to aid in the design, manufacture, handling, or transport of goods.

Communication: The transfer of information and understanding from one point or person to another. The basic elements in the process of communication are an information source, encoding, transmission, reception and decoding.

Computer Graphics: CG is the field of visual computing, where one utilizes computers both to generate visual images synthetically and to integrate or alter visual and spatial information sampled from the real world.

Computer Numerical Control (CNC): The use of the dedicated computer within a numerical control unit that provides data input for the machine.

Computer simulation: Computer simulation exercises are performed to identify system bottlenecks, the utilization of fixed resources such as machine tools, and variable resources such as operators, tools and material transporters.

Contact sensor: A device that detects the presence of an object or measures the amount of force or torque applied by the object through physical contact with it.

Controller: Every robot is connected to a computer, which keeps the pieces of the arm working together. This computer is known as the controller. The controller functions as the "brain" of the robot. The controller also allows the robot to be networked to other systems, so that it may work together with other machines, processes, or robots.

Controls IDEF: Inputs to the top surface of the IDEF block representing those items which exercise controls or constraints over the transformation represented by the block.

Concurrent Engineering: Concurrent engineering is a business strategy which replaces the traditional product development process with one in which tasks are done in parallel, and there is an early consideration for every aspect of a product's development process. This strategy focuses on the optimization and distribution of a firm's resources in the design and development process to ensure effective and efficient product development process.

Continuous-flow production: Lotless production where products flow continuously rather than being proportioned into lots.

Continuous improvement: A philosophy of frequent and small changes to production processes developed in Japan. The cumulative results lead to high levels of quality and efficiency.

Continuous process control: The use of transducers (sensors) to monitor a process and make automatic changes in operations through the design of appropriate feedback control loops. Such devices have been historically mechanical or electromechanical; there is now widespread utilization of computers and centralized control.

Control system: The deliberate guidance or manipulation of the elements in a system in order to achieve a prescribed value or performance of a system to complete a defined process.

CSMA/CD: Carrier Sense Multiple Access with Collision Detection, a method of assigning mastership in a communications network.

CSU/DSU: Communications Service Unit/Data Service Unit. A device which serves to interface a LAN with an external communications service such as an ISDN data line.

Data acquisition system: Any instrument or computer that acquires data from sensors via amplifiers, multiplexers, and any necessary analog to digital converters. Typically associated with process industries.

Database: A collection of structured data, independent of any application.

DCS: Distributed Control Systems evolved from centralized process control computers common in the 1960s. The systems were developed for continuous-flow processes that required loop, analog, and limited discrete control. A DCS is a real-time, fault-tolerant system for continuous and complex batch-process applications.

Degrees of freedom: The number of independent ways in which the end-effector can move, defined by the number of rotational or translational axes through which motion can be achieved.

Direct Numerical Control (DNC): The use of a computer for providing data inputs to several remote numerically controlled machine tools.

Decision-support tools: PC, client or application server-based systems that use memory-based processing to perform rapid simulations using data drawn from business-transaction processing systems such as enterprise resources planning.

Dexterity: A measure of the robot's ability to follow complex paths.

Dependent demand: In manufacturing there is demand for an item that results from the production decisions for its parent items. An order for a parent part can only be completed if all of its components are available.

Discrete-event simulation: Material-flow or discrete-event simulators have most often been used by engineers in the design and modification of production systems. The models output statistical estimates of performance, using graphic animation to help create a greater understanding of system dynamics.

Distribution management: Determines optimal quantities of each product to be made at each plant and to be distributed to each warehouse, such that manufacturing and distribution costs are minimized and customer demand is met.

DNC system: A DNC system is nothing more than a computer that is networked with one or more CNC machines. Until only recently, rather crude serial communications protocol (RS-232c) had to be used for transferring programs.

Document management system: Allows users to store, search and manipulate documents electronically, and to maintain a library of text and images in a compact space. Most systems also provide a means for passing documents across a network.

Drive: The drive is the "engine" that drives the links (the sections between the joints) into their desired position. Without a drive, a robot would just sit there, which is not often helpful. Most drives are powered by air, water pressure, or electricity.

EDI: Electronic Data Interchange is a standard for automated exchange of business documents. Using EDI, purchasers and suppliers can exchange digital paperwork including purchase orders, invoices, and other business documents, and perform electronic funds transfers.

Encoder: An electronic device used to measure small units of motion, often using a light source and detector.

End-effector: The end-effector is the "hand" connected to the robot's arm. It is often different from a human hand—it could be a tool such as a gripper, a vacuum pump, tweezers, scalpel, blowtorch, just about anything that helps it do its job. Some robots can change end-effectors, and be reprogrammed for a different set of tasks.

Enterprise management: Enterprise management is the planned, coordinated automation of IT functions across all the computing resources within an enterprise.

EPROM: Electronically Programmed Read Only Memory.

Equality constraint: A restriction that requires the displacement or motion of the robot to equal a specified value. Equality constraints specify the position and orientation of the robot's end-effector.

Error function: The error function assigns a single value that represents the difference between the desired and actual values of one or several dependent variables.

ERP: Enterprise Resource Planning is an extension of the manufacturing resources planning concept standardized by APICS (American Production and Inventory Control Society).

External sensor: A feedback device for detecting locations, orientations, forces, or shapes of objects outside the robot's immediate surroundings.

Expert system: A computer program that uses knowledge and reasoning techniques to solve problems normally requiring the abilities of human experts.

Fault tolerance: The ability of a system to execute tasks regardless of strategic components failure.

Feedback: It is a process whereby some proportion, or in general, function of the output signal of a system is passed (fed back) to the input. Often this is done intentionally, in order to control the dynamic behaviour of the system. Feedback is observed or used in various areas dealing with complex systems, such as engineering, economics and biology.

A signal from the robot equipment about conditions as they really exist, rather than as the computer has directed them to exist.

Feedback loop: A signal orientating in the computer that is changed by the robot and given back as feedback.

Finishing: In many ways similar to semi-finishing, but different in terms of what it starts from. After semi-finishing, the excess material is left at a constant thickness over the final surface, so during this process there is little change in metal removal rate, allowing the results to be accurate. (Like wall plastering, where the final layer is done against an almost smooth base.) Common strategies, in addition to those in semi-finishing, are pencil milling, rest machining.

Finite Element Analysis: A CAD method that determines the structural integrity of a mechanical part or physical construction under design by mathematical simulation of the part and its loading condition.

Flexible automation: Short set-up times and the ability to switch quickly from one product to another.

Flexible machine centre: Usually an automated system, consisting of CNC machines, with robots loading and unloading parts conveyed into and through the system.

Flexible Manufacturing System: A manufacturing process designed so that the production line may be rebalanced often, rapidly matching output to changes in demand.

FMS: Flexible machining systems (FMSs) are extensions of group technology and cellular manufacturing concepts. Using integrated CAD/CAM, parts can be designed and programmed in half the time it would normally take to do the engineering. The part programs can be downloaded to a CNC machining centre under the control of an FMS host computer. The FMS host can schedule the CNC and the parts needed to perform the work.

Force sensor: A device that measures and detects the magnitude of the force exerted by an object upon contacting.

Fully constrained robot: A robot with as many independent joints as there are equality constraints on the placement of the end-effector.

Fuzzy logic: A method used to model linguistic expressions that have non-binary truth values. It has been used with PID algorithms in process control, especially where process relationships are non-linear.

G-Code: G-code is also the name of any word in a CNC program that begins with the letter G, and generally is a code telling the machine tool what type of action to perform, such as: rapid move, controlled feed move in straight line or arc and series of controlled feed moves that would result in a hole being drilled or bored.

Generative CAPP: A process planning system, including a database and decision logic, that will automatically generate a process plan from graphical and textual information on the part.

Globalization: The process whereby the competition has expanded geographically through the expansion in freer trade arrangements and improvements in transportation and communications.

Group technology: Group technology was based on the principle of using the common properties of similar parts to rectify shared design and manufacturing problems. This involved the creation of a coding system to classify parts into groups given their shape, overall dimension, required accuracy, surface finish, manufacturing requirements, and material composition.

Hard automation: Automated machinery that is fixed, or dedicated, to one particular manufacturing task throughout its life.

Hybrid: A robot that is part picks and place and part servo controlled, or has the same abilities.

ICAM: Abbreviated as Integrated Computer-Aided Manufacturing, it was an initiative managed by the United States Air Force out of Wright Patterson AFB, Materials Laboratory and was part of their Technology Modernization efforts, specifically the Computers in Manufacturing (CIM) initiative.

Interface: A boundary between the robot and machines, transfer lines, or parts outside immediate environment. The root must communicate with these items through input/output signals provided by sensors.

Interpolation: The process of automatically selecting a path in space based on the positions of the end points of the path; can be used for circular or linear paths.

Integrated manufacturing systems: Linking together a unique technology, such as laser, with machine controls, sensors and drive systems.

Inverse kinematics: The inverse kinematics problem is to find the robot's joint displacements given position and orientation constraints on the robot's end-effector.

Island of automation: Stand alone pocket of automation (robots, CAD/CAM system, CNC machines) that are not connected into a cohesive system.

JIT: Just-in-Time is an approach to manufacturing that stresses the benefits inherent in a system, where material is brought to the work site only when it is needed. To achieve this goal, each operation must be synchronized with those subsequent to it.

Joint space: A coordinate system used to describe the state of the robot in terms of its joint states. Inverse kinematics may also be thought of as a mapping from end-effector space to joint space.

Kanban: The Japanese methodology for achieving JIT, often involving the use of kanbans, or cards, to indicate parts status.

Kinematics influence coefficients: These coefficients describe the total influence the N input joints have on the motion of the robot and allow a direct statement of the complex and coupled non-linear differential equations controlling the response of the system.

Knowledge-based system: Software that uses artificial intelligence techniques and a base of information about a specialized activity to control systems or operations.

LAN: Local Area Networks span a limited geographical area to connect computers and terminals, usually at moderate-to-high data rates.

Lagrange multipliers: A mathematical technique for transforming equality constraints into performance criteria, thus expressing a constrained problem as an unconstrained problem.

Lean production: The system of organization production developed in Japan that stresses quick tool changeover times, minimum parts and work-in-progress inventories, high levels of quality and continuous improvement.

Linearly dependent: A correspondence between quantities or functions that can be described by simply adding, subtracting, or multiplying a scalar.

Load cell: A transducer for the measurement of force or weight. Action is based on strain gauges mounted within the cell on a force beam.

Machine Vision: A computer perception of a visually-based sensory output used to produce a concise description of an image. Devices used for optical non-contact sensing to automatically receive and interpret an image of a real scene in order to obtain information or to control a process.

Maintenance management systems: Automated software systems for handling maintenance work orders, as well as associated inventory, purchasing, accounting, and human resources functions.

Manual programming: A means of teaching a robot by physically moving the aim by teach pendant.

Manufacturing: Software requirements include master production scheduling, distribution requirements planning, material requirements planning and capacity requirements planning (plan); purchasing and receiving, shop floor control and inventory control (execute); and performance measurement (measure).

Manufacturing Automation Protocol: MAP is a specification for a suite of communication standards for use in manufacturing automation developed under the auspices of General Motors Corporation.

Mass customization: A system of production stressing the production of small lots of customized goods as opposed to large volumes of standardized products.

Master production schedule: An expression of the operational plan of production by period, indicating the timing and size of end item quantities (used in MRP).

Micro-Electro-Mechanical Systems (MEMS): Micro-electrical-mechanical systems allow sensors, electronics, and actuators to be batch fabricated on a single chip. MEMS can perform extremely delicate operations such as drilling holes with bits finer than the human eye can see unaided.

MES: Manufacturing Execution Systems use network computing to automate production control and process automation. They do so by downloading recipes and work schedules, and uploading production results, thereby bridging the gap between business and plant-floor or process control systems.

MRP: Materials Requirements Planning was the first phase in the development of MRP II. An MRP software module uses the bill of material, inventory data, and the master production schedule (MPS) to calculate requirements for materials. It makes recommendations to release replenishment orders (POs) for materials.

MRP II: Manufacturing Resources Planning is a computerized method for planning the use of a company's resources, including scheduling raw materials, vendors, production equipment, and processes. System includes financial, manufacturing, and distribution management. It is a method for the effective planning of all resources of a manufacturing company.

Multiplexer: The portion of a robot that allows communication with a large number of outside lines, using only a few lines.

Nanotechnology: Nanotechnology is molecular manufacturing, or building things one atom at a time. A nanometre is one billionth of a metre, and approximately ten atoms fit inside one nanometre. Utilizing the physical properties of atoms and molecules, nanotechnology proposes the construction of nanosize devices possessing extraordinary properties.

NC: Numerical Control is a technique of operating machine tools or similar equipment in which motion is generated in response to numerically-ordered commands generated by CAD systems, punched tapes, or other communication media.

Network: (1) Any system of computers and peripherals. (2) In an electrical or hydraulic circuit, any combination of circuit elements.

Networked organization: A company organized in the form of a group of peers rather than a hierarchy or a system of departments.

Neural network: A processing architecture derived from models of neuron interconnections of the brain. Unlike typical computers, neural networks are supposed to incorporate learning, rather than programming and parallel, rather than sequential, processing.

Next generation manufacturing systems: Systems used to link customers, suppliers, and the many functions within a company with other companies in the enterprise. Next Generation Manufacturing Systems assume that next generation manufacturing enterprises are organized into distributed work units, that the work units have a high degree of autonomy, and that they behave intelligently.

Numerical Control (NC): A means of providing prerecorded information that gives complete instructions for the operation of a machine.

Numerical methods: Iterative methods of solving problems on a computer. Numerical methods may have an analytical basis or they may involve heuristics.

Object-oriented database: A database used to store objects that form the bases of object-oriented computing, in which data as well as references to the procedures used to perform operations on that data are combined.

Object-oriented software: Results from a kind of modular programming. Each object is a software package containing a collection of related procedures and data that can be reused, shortening application development time. Objects also make it easier to customize software systems to mirror actual business processes, and to do so without negatively impacting the ability to migrate to later software releases.

On-line programming: A means of programming a robot on a computer that directly controls the robot. The programming is performed in real time.

Optimization: Calculating the independent variables in a function so as to generate the best function value for a given set of conditions. Optimization usually involves maximizing or minimizing a function.

Optimized production technology: OPT is a technique for production scheduling and inventory control that recognizes capacity constraints (which MRP does not). The goal of OPT is to simultaneously increase throughput and decrease inventory and operating expenses.

Parallel development: Design and production engineers at a company and at its suppliers work jointly from beginning to end, not in a sequential manner.

Pallet: A flat bed, usually made from wood, used for transportation and storage of goods.

Pick and place robot: A simple category of robot used to pick parts and place them down somewhere else.

PID control: Proportional plus integral plus derivative control is used in processes where the controlled variable is affected by long downtimes.

Planning and scheduling systems: A new generation of planning and scheduling tools which, unlike MRP II, includes constraint models that deal with both materials and capacity. These technologies can be applied along a continuum extending from short-term plant-floor scheduling to strategic planning of supply chains.

PLC: A Programmable Logic Controller controls large numbers of discrete elements using very fast I/O (Input/Output) scan times. Today, even the smallest PLCs may be equipped with serial communication and analog control capabilities, and perform arithmetic functions.

Point-to-point motion: A type of robot motion in which a limited number of points along a path of motion is specified by the controller, and the robot moves from point to point rather than in a continuous, smooth path.

Precision engineering: Sensing, control, and fabrication at scales comparable to or smaller than optical wavelengths.

Programming language: Any of the computer instruction codes that simplify programming for people.

Process: A natural phenomenon marked by gradual changes that lead towards a particular result. A series of actions or operations leading to an end. A continuous operation or treatment in manufacturing. Continuous and regular production executed in a definite, uninterrupted manner.

Process automation: Includes objectives of control and also those of enterprise management. This requires an integrated approach to plant operations and enables a variety of applications such as production scheduling, inventory control, performance monitoring, statistical process control, maintenance management and environmental audit.

Process planning: Process planning translates design information into the process steps and instructions to efficiently and effectively manufacture products. As the design process is supported by many computer-aided tools, computer-aided process planning (CAPP) has evolved to simplify and improve process planning and achieve more effective use of manufacturing resources.

Process control: Automatic monitoring and control of a process by an instrument or computer programmed to respond appropriately to feedback from the process.

Process simulation: Use of a mathematical model by a computer program to envision process design scenarios with real-time visual and numerical feedback. Process optimization and the ability to forecast potential problems are the results.

Productivity: A measurement of output for a given amount of input(s). Increases in productivity are considered critical to raising living standards.

Program: (1) A complete structured sequence of program statements directing a computer to implement an algorithm. (2) All software programming necessary to solve a problem.

Protocol: A standard set of procedures to allow data to be transferred among systems.

Pseudo inverse: A simple method of inverting a matrix that is not square. As commonly applied to redundant robots, the pseudo inverse minimizes the two-norm of the joint speeds.

Quality circle: A small group of people who normally work as a unit and meet frequently for the purpose of uncovering and solving problems with the quality of the items produced, process capability, or process control.

Rapid prototyping: A series of approaches to obtaining an initial or prototype of a part or assembly quickly. See, for example, stereolithography. Rapid prototyping is also known as **solid freeform fabrication** is the automatic construction of physical objects in 3D by, Stereolithography machines or selective laser sintering systems. Rapid prototyping is a type of computer-aided manufacturing (CAM) and is one of the components of rapid manufacturing.

Real time: Refers to tasks or functions executed so rapidly that the feedback is used to guide task completion. The performance of computations concerning a physical process such that their results are used in controlling the process.

Redundancy: More independent variables than constraints.

Re-engineering: The redefinition of a company's internal processes.

Repeatability: The variability of the end-effector's position and orientation as the robot makes the same moves under the same conditions (load, temp, etc.).

Resolver: An electrical device used to measure the position of a shaft as it rotates.

Resolved-rate: An extremely simple inverse kinematics method at the velocity level. Scale—changing magnitude by linear operation, i.e. multiplying by a scalar. Self-motion—the robot's ability to move its intermediate links while holding the placement of the end-effector constant.

Robot: A reprogrammable multifunctional manipulator designed to move material, parts, tools, or specialized devices through variable programmed motion for performance of a variety of tasks.

Robotics: The study of the design and use of robots, particularly for their use in manufacturing and related processes.

Roll-forming: The process of bending a continuous strip of metal through a series of shaped rolls. The process gradually forms metal into a predetermined shape. The nature of forming metal by this process allows for very tight tolerances to be met and maintained.

Roughing: This process begins with a cuboids block of stock or a billet, and cuts it very roughly to the shape of the final model. Often, the result gives the appearance of terraces, because the strategy has taken advantage of the ability to cut the model horizontally. Common strategies are zig zag clearing, offset clearing, plunge roughing, rest-roughing.

Rule-based system: A functional system in which knowledge is stored in the form of simple if-then or condition-action rules.

Serial robot: A serial robot is a single chain of joints connected by links.

Semi-finishing: This process begins with a rough part that unevenly approximates the model and cuts to within a fixed offset distance from the model. Common strategies are raster passes, waterline passes, constant step-over passes.

Sensor: A transducer whose input is a physical phenomenon and whose output is a quantitative measure of the phenomenon. Most robots of today are nearly deaf and blind. Sensors can provide some limited feedback to the robot, so it can do its job. Compared to the senses and abilities of even the simplest living things, robots have a very long way to go. The sensor sends information, in the form of electronic signals back to the controller. Sensor also give the robot controller information about its surroundings and lets it know the exact position of the arm, or the state of the world around it.

Sequence control: The control of a series of machine movements, with the completion of one movement initiating the next. The extent of movements is typically not specified by numerical input data.

Servo mechanism: An automatic device for controlling large amounts of power by means of small amounts of power.

Servo-motor: A power-driven mechanism that supplements a primary control operated by a comparatively feeble force (as in a servo-mechanism).

Servo valve: A valve used to control the speed or position of another component by changing its flow.

Shoulder: The manipulator arm link joint that is attached to the base.

Simulation: (1) A device, system, or computer program that represents certain features of the behaviour of a physical or abstract system. Vendors of planning and scheduling, forecasting and demand management, and other types of decision-support systems make growing use of simulation to compare the consequences of alternative courses of action. See discrete-event simulation. (2) A CAD/CAM program that simulates the effect of structural, thermal, or kinematic conditions on a part under design. (3) Within MRP II, utilizing a mathematical model to test various design configurations.

Singularity: A position in the robot's workspace where one or more joints no longer represent independent controlling variables. Commonly used to indicate a position where a particular mathematical formulation fails.

SMART: Self Monitoring Analysis and Reporting Technology is a drive technology that reports its own degradation enabling the operating system to warn the user of a potential failure.

Software: The entire set of programs, procedures, and related documentation associated with a computer.

Solenoid: An electrical coil, with an iron section inside, that will pull or push when current goes through the coil.

Solid modelling: Three-dimensional modelling in which solid characteristics of an object are built into the database so that complex internal structures can be realistically represented.

SPC: Statistical Process Control is a quality control method focusing on continuous monitoring of the process rather than the inspection of finished products, with the intent to achieve control of the process and eliminate defective product.

SQC: Statistical Quality Control applies the laws of probability and statistical techniques to the observed characteristics of a product or process.

Standardization: To promote conformity by means of a standard. Establishing criteria for uniform practices.

Stereolithography: A form of rapid prototyping involving the building of solid plastic objects from a CAD file by projecting a laser-generated beam of ultraviolet radiation onto the surface of a vat of photosensitive resin.

Supervisory control: The use of microcomputers and workstations to accomplish operator interface, data acquisition, process monitoring, and some degree of production control.

Supply-chain management: The use of information technology to endow automated intelligence to an ever-growing network of cash registers, delivery vehicles, distribution centres, factories, and raw material suppliers.

Supply-chain optimization: Software that allows the management to look at the relationships among the various elements of the supply-chain (manufacturing plants, distribution centres, points of sale, as well as raw materials, relationships among product families, and other factors) to synchronize supply chains at a strategic level.

Systems integration: The ability of computers, instrumentation, and equipment to share data or applications with other components in the same or other functional areas.

TCP: An artificial point in space relative to the robot arm, where the centre of end-of-arm tooling will be also known as **Tool Centre Point.**

Tracking system: A set of rails on which a robot can travel from one place to another.

Transducer: A device that converts signals from one physical form to another.

Transportation planning: Optimizes assignments from plants to distribution centres, and from distribution centres to stores. The systems combine moves to ensure that the most economical means are employed on a scale that no human planner could match.

Touch sensor: A sensor that detects the presence of an object by coming into contact with it.

Value engineering: It is a systematic method to improve the "value" of goods and services by using an examination of FUNCTION. Value, as defined, is the ratio of function to cost.

Variant or retrieval CAPP: A process planning system that creates new plans by retrieving and modifying a standard process plan for a given part family. Typically requires input from a human planner.

Walkthrough programming: A method of programming a robot by physically moving the manipulator arm through a complete operating cycle. This is typically used for continuous path robots.

Warehouse management systems: Software that integrates mechanical and human performed activities with an information system to effectively manage warehouse business processes and direct warehouse activities.

Work cell: A group of machines all working together on a common part and physically located together.

Work envelope: The area in which a robot can work, the limits to its motions.

Work-in-progress: An accounting term used to express the value of material taken up continuously by the work process.

Workstation: A stationary position where work is performed, the work piece moving to the station.

Workspace: The maximum reach space refers to all of the points the robot can possibly reach. The dexterous workspace is all of the possible points the robot can reach with an arbitrary orientation. The dexterous workspace is usually a subspace of the maximum reach space.

Wrist: The manipulator arm joint to which an end-effector is attached.

XYZ coordinates: A reference to the most common names given to the lines forming a Cartesian solid.

REFERENCES

Anderson, M.M. and L. Hein, *Integrated Product Development*, Springer-Verlag, 1987.

Chang, T.C., *An Expert Process Planning System*, Prentice-Hall of India, New Delhi, 1985.

Chang, Tien-Chien and Richard A. Wysk, *An Introduction to Automated Process Planning Systems*, Prentice-Hall of India, New Delhi, 1985.

Chung, P.W.H. and G. Lovegrove, *Industrial and Engineering Application of Artificial Intelligence and Expert Systems*, Gardon Breach Science Publication, New Delhi, 1993.

Cleetus, J., *Design for Concurrent Engineering*, Concurrent Engg. Research Centre, Morgantown, WV, New Delhi, 1992.

Craig, J.J., *Introduction to Robotics Mechanics and Control*, Addison-Wesley, New Delhi, 1999.

Date, C.J., *An Introduction to Database Systems*, Narosa Publishing House, New Delhi, 1997.

Deb, S.R., *Robotics Technology and Flexible Automation*, Tata McGraw-Hill, New Delhi, 1994.

Fu, K.S., R.C. Gonzalez and C.S.G. Lee, *Robotics Control, Sensing, Vision, and Intelligence*, McGraw-Hill, New Delhi, 1987.

Gideon, Halevi and Roland D. Weill, *Principles of Process Planning: A logical approach*, Chapman & Hall, New Delhi, 1995.

Grahens, T. Smith, *CNC Machining Technology*, Springer-Verlag, 1993.

Groover, M.P., *Automation, Production Systems and Computer Integrated Manufacturing*, 3rd ed., Prentice-Hall of India, New Delhi, 1996.

Groover, M.P., *Industrial Robotics Technology, Programming and Applications*, McGraw-Hill New Delhi, 1995.

Ibrahim, Zeid, *CAD/CAM: Theory and Practice*, Tata McGraw-Hill, New Delhi, 1998.

Kalpakjian, Serope, *Manufacturing Engineering and Technology*, Addison-Wesley, New Delhi, 1995.

Kerr, R., *Knowledge-Based Manufacturing Management*, Addison-Wesley, 1991.

Klafter, R.D., T.A. Chmielewski and M. Negin, *Robot Engineering: An Intergrated Approach*, Prentice-Hall of India, New Delhi, 1994.

Koran, Yoram, *CNC Machines*, Khanna Publishers, New Delhi, 1992.

Koren, Yorem, *Computer Integrated Manufacturing Systems*, McGraw-Hill, 1983.

Kusaik, Andrew, *Concurrent Engineering: Automation Tools and Technology*, John Wiley & Sons Inc., 1992.

Luca, G. Sartori, *Manufacturing Information Systems*, Addison-Wesley, 1988.

Newman, William M. and Robert F. Sproull, *Principles of Interactive Computer Graphics*, McGraw-Hill International, New Delhi, 1984.

Orlicky, G., *Material Requirements Planning*, McGraw-Hill, New Delhi, 1975.

Prasad, B., *Concurrent Engineering Fundamentals: Integrated Product Development*, Prentice-Hall of India, New Delhi, 1996.

Pressman, I. R.S. and J.E. Willaim, *Numerical Contol and CAD*, John Wiley & Sons Inc., NewYork, 1997.

Radhakrishnan, P., *Computer Numerical Control Machines*, New Central Book Agency, New Delhi, 1992.

Radhakrishnan, R., *Computer Integrated Manufacturing*, New Central Book Agency, New Delhi, 2000,

Rao, P.N., *Computer Aided Mnufacturing*, McGraw Hill, New Delhi, 2000.

Singh, Naua, *Systems Approach to Computer Intergrated Design and Manufacturing*, John Wiley & Sons, New Delhi, 1996.

Sinha, G. Sammy, *Successful Implementation of Concurrent Product and Process*, John Wiley & Sons Inc., 1998.

INDEX